Christopher Columbus

JOURNAL OF
THE FIRST VOYAGE
(Diario del primer viaje)

1492

Edited and Translated with an Introduction and Notes by
B. W. Ife

together with an essay on Columbus's language by
R. J. Penny

Aris & Phillips Ltd – Warminster – England

© B. W. Ife 1990. All rights reserved. No part of this publication may be reproduced or stored in a retrieval system or transmitted by any means or in any form including photocopying without the prior permission of the publishers in writing.

ISBNs 085668 350 7 (cloth)
       085668 351 5 (limp)

The publishers gratefully acknowledge the financial assistance of the Dirección General del Libro y Bibliotecas of the Ministerio de Cultura de España with this translation.

Printed and published in England by Aris & Phillips Ltd., Teddington House, Warminster, Wiltshire BA12 8PQ.

CONTENTS

Introduction v

    Text history v
    The role of Bartolomé de las Casas vi
    Las Casas's working methods viii
    Verbatim transcription and summary x
    The aim of the Journal xiii
    The objectives of the 1492 voyage xvi
    The preparations for the 1492 voyage xviii
    The landfall and its aftermath xix
    Landscape xx
    Native inhabitants xxii

Editorial Note xxvi

The Language of Christopher Columbus xxvii

    Columbus's linguistic background xxvii
    Columbus's written Spanish xxviii
    The language of the 1492 Journal xxviii
    Influence of Genoese and Portuguese xxix
    Influence of Genoese alone xxx
    Influence of Portuguese alone xxxi
    Evidence of Columbus's imperfect Spanish xxxiii
    Columbus and late 15th-century practice xxxv
    Amerindian words borrowed by Columbus xxxix
    Idiosyncrasies of Columbus's language xxxix
    Conclusion xl

Bibliography xli
Maps xliii

Text and translation 1
Notes to the Journal 242

PREFACE

There was a time when the inclusion of a historical document such as Columbus's Journal in a series dedicated to Spanish Golden-Age prose fiction and drama might have required some comment. To put Columbus alongside Cervantes, Quevedo and Calderón might have been taken to imply that the contents of the Journal were just so much fiction or, conversely, that the editors were taking an essentially documentary view of the other works included in the series. Nowadays we have a much less compartmentalised approach to the notion of 'text' - one which is more in tune with the expectations of Renaissance writers and readers -, and much has been gained by bringing the techniques of literary textual analysis and criticism to bear on a wide variety of texts, whether written, spoken or non-verbal forms of cultural expression.

The purpose of this new edition of the Spanish text of Columbus's Journal of the 1492 voyage, published together with a new translation, is to make available to the general reader as well as the specialist historiographer one of the most important texts ever written in Spanish. Columbus's 1492 Journal, even in the truncated and partially summarised form in which it has survived, gives an unrivalled insight into the events of the voyage, Columbus's first impressions of a people and a culture which failed in so many ways to live up to his expectations, and the creation of many of the myths surrounding the New World which have coloured its view of itself down to the present day.

Columbus's Spanish is not that of a native-speaker. Even after several transcriptions at the hands of Spanish-speaking copyists, it retains many features which have an important bearing on our understanding of Columbus's cultural and linguistic formation, and on such issues as the reliability of the Journal in the form in which we have it. I am grateful to my colleague Ralph Penny for agreeing to contribute a short study of the most important features of Columbus's language. Some of the material of the Introduction derives from my Inaugural Lecture, *Writing and Conquest*, given at King's College London on 1 May 1990.

This edition and translation is dedicated to Henry Maxwell.

BWI

INTRODUCTION

by B.W. Ife

Text history

When Columbus set sail for the Far East in August 1492 he decided, in view of the significance of what he was about to attempt, to make a documentary record of the voyage in the form of charts and a log book:

> ... *I decided to write down the whole of this voyage in detail, day by day, everything that I should do and see and undergo, as will be seen in due course.* (Prologue)[1]

Keeping such a Journal was by no means routine at the time and did not become a legal requirement for captains of vessels flying the Spanish flag until 1575. The importance which Columbus attached to the accurate day-to-day recording of the events of the first voyage cannot be underestimated. By setting the voyage down in writing he ensured a place for himself in history which others have disputed but from which no one has succeeded in displacing him. The written record has become the touchstone of his achievement.

On returning to Spain in the spring of 1493 Columbus presented his record of the voyage to Queen Isabel. She had it copied, retained the original, and gave the copy to Columbus before he set out on the second voyage in the autumn of 1493. The original has not been seen since 1504, the year in which the Queen died.

In 1506, on the Admiral's death, the copy passed to Columbus's eldest son Diego, and then in 1526 to Diego's son, Luis, the Third Admiral of the Indies. Luis was granted permission to publish the Journal in 1554, though it did not in fact appear. This is thought to indicate that he sold the manuscript, as he did that of his uncle Ferdinand's biography of the Admiral, in order to subsidise his

---

[1] Throughout this Introduction and the text of the Journal itself, Columbus's verbatim text is printed in *italic*.

legendary debauchery. Whatever the explanation, it is clear that both the original Journal, and the only copy known to have been made of it, have both disappeared.

The role of Bartolomé de las Casas

We should have very little knowledge indeed about the conduct and events of the 1492 voyage had it not been for the intervention of the historian Bartolomé de las Casas. Las Casas, whose father and uncle had accompanied Columbus on the second voyage in 1493, began collecting material for a history of the Indies as early as 1502. After his conversion in 1514 he dedicated himself to exposing in writing and by personal advocacy the oppression of the Indians and the illegitimacy of the Spanish presence in the New World. In 1527 he began his great *Historia de las Indias*. Chapters 35 to 75 of the *Historia* rely heavily on the evidence of Columbus's Journal. It is not clear when Las Casas consulted it,[1] though from remarks made in the *Historia* about scribal errors and confusions, we may be sure that what he consulted was a copy, possibly Columbus's own copy, and not the original. The access which Las Casas had to the Journal was evidently restricted. However he came by it, he was evidently not able to take it away with him or to keep it over a period of time. He therefore made an extensive digest for his own use, summarising the majority of the text, but copying out word-for-word those parts of the original which he thought were particularly interesting or worthy of quotation in full. Failing the discovery of the full text, Las Casas's summary, preserved in the National Library in Madrid, is the closest we are likely to get to Columbus's original.

The major textual and historiographical problem surrounding the Journal is therefore easily stated: how much of what we have is Columbus and how much Las Casas? On the face of it, the evidence is not encouraging. At best, the manuscript is at two removes from the original: a digest of a copy of the original, which may itself have been a fair copy rather than the actual log-book which Columbus wrote up from day to day on board ship. We can only assume that the copy from which Las Casas worked was reasonably faithful, although he was

---

[1] This must, however, have been between 1513, the date of the discovery of Florida, alluded to in the entry for 21 November, and 1527, the year Las Casas began the *Historia*.

Introduction

himself aware of inaccuracies and mistranscriptions. In the entry for 13 January, concerning Columbus's astrological observations, Las Casas writes in the margin:

> ... here it seems that the Admiral knew something about astrology, although these planets do not seem to be in their proper positions, due to bad transcription by the copyist ... (13.1)[1]

Other remarks made both in the text and in the margin suggest that Las Casas was less than confident in the accuracy of what he was reading:

> He steered WSW and they made about 11 and a half or 12 leagues during the day and night and it seems that at times during the night they were making 15 miles an hour, if the text is to believed. (8.10)

The major doubts, however, must concern Las Casas's own working methods. Las Casas was a tendentious historian and the *Historia de las Indias* is a work of extreme political and moral commitment. Cecil Jane, for one, has accused Las Casas of 'deliberate misstatement of fact' and reliance on 'a memory which was either curiously defective or singularly convenient'.[2] Can an avowed champion of the Indians' cause be relied upon to summarize accurately, without distortion and editorialising, the work of a pioneer colonist like Columbus?

Since virtually everything we know about the 1492 voyage has come down to us from Columbus via Las Casas's digest, it is perhaps surprising that a serious answer to such a fundamental question appears not so far to have been sought. Historians have not always shown a proper circumspection in their treatment of the text, and, until recently, successive generations of editors have failed to improve significantly on the text first published by Martín Fernández de Navarrete in 1825.

A more serious failing among scholars, however, has been the lack of any systematic attempt to evaluate the role of Las Casas as intermediary or to use the physical and linguistic evidence of the manuscript to establish how much of

---

[1] References to the Journal are made in the form of day.month, i.e. '15.10' is 15 October 1492, and '15.2' is 15 February 1493.
[2] *The Voyages of Christopher Columbus*, London, 1930, p. 63.

Columbus's original has survived the process of being copied and then summarized. Such a study is beyond the scope of this Introduction, but it is worthwhile to give some indication of the issues involved because they help to illuminate the nature of the Journal itself as well as the textual and interpretative problems which it poses. Broadly speaking, there are two main areas of interest: the evidence of Las Casas's working methods derived from the manuscript itself; and comparative analysis of linguistic and descriptive evidence in the summary and verbatim sections of the Journal.

Las Casas's working methods

One of the most impressive features of Las Casas's digest is its length. The manuscript consists of 67 folios (133 pages) with a total text length of nearly 54,000 words. It is abstracted on a day-to-day basis and covers the period 3 August 1492 to 15 March 1493, that is, the full extent of the outward voyage, including the preparations, the progress through the Bahamas, to the north coast of Cuba and Hispaniola and the return voyage. There is an entry in the digest for the majority of the days covered by the period of the voyage. The main omissions are the period 9 August to 6 September while the fleet was fitting out and provisioning in the Canaries, but the intervening period is summarized. There is another omission for the period 6-12 November when Columbus was unable to sail through bad weather. 17 February also has no entry in the digest. Otherwise, there are only a couple of small lacunae in the text, probably attributable to damage to or the illegibility of the original. The day-to-day structure of the Journal imposed a similar constraint on the digest and seems to have prevented significant loss of coverage. This perhaps is an encouraging sign.

Also encouraging is the fact that the manuscript we have is clearly not a fair copy of a ready-made digest; Las Casas was making the summary as he wrote. There are many corrections in the text, and in the margins. Sometimes errors were detected immediately, sometimes later, when they had to be squeezed in between the lines or put in the margin. In all, there are just over 1,000 corrected errors in the manuscript, most of them quite legible, and a full analysis of them gives a vivid picture of Las Casas struggling to capture the essence of the original text as fully and as succinctly as possible, going back and correcting often quite trivial details where he senses that he has misrepresented the emphasis of the original text. Occasionally, however, as in the case of the correction of 'dezía' to

Introduction

'fingía' on 25 September,[1] Las Casas betrays some misunderstanding or misinterpretation of what he is reading.

Las Casas is also careful, as far as is possible, to separate fact from opinion. Overt comment is restricted to the margins of the text, and takes various forms:

- notes or short summaries to assist in locating the more important events, such as the marginal note marking the first landfall on 12 October.

- clarifications or explanations made with the benefit of hindsight. Las Casas had lived in the Caribbean for several years before he began abstracting the text of the Journal and is often able to correct Columbus's first impressions. When, on 17 October, Columbus describes the straw crowns on the roofs of the native houses as chimneys, Las Casas records the mistake in the text and notes the correct explanation in the margin (see Note 56, p. 247).

- criticism of the Admiral's actions and praise of the Indians. When Columbus says that 1,000 Indians live together in fifty huts, Las Casas comments in the margin that this is a sign that they are amicable (6.11), and when Columbus records that an Indian who had been released from captivity on the understanding that he would return the next day had failed to come back, Las Casas observes in the margin 'What a fool!' (6.11)

- the word 'no[ta]', used to indicate a point of interest or one which will require explanation at some later date. Many of these instances are precisely those which Las Casas later expanded when writing up the digest into the finished version of the *Historia de las Indias*.

Las Casas's use of the margin of the manuscript as he proceeds seems, then, to indicate in general a feeling for the distinction between fact and comment and a willingness to keep the two apart as far as is possible.

---

[1] See Note 23 on page 245.

Verbatim transcription and summary

Las Casas began the digest by assuming that he would make a summary of the entire Journal. He writes at the top of the first page:

> ... This is the first voyage with the courses and route which the Admiral don Christopher Columbus took when he discovered the Indies, set down in an abbreviated form, except for the prologue to the Monarchs which is given in full and begins ...

That is, everything will be summarized, except the prologue, which will be given verbatim. Las Casas promptly forgot this distinction. The first entry immediately following the prologue, 3 August, is also written in the first person, and thereafter a substantial portion of the Journal is transcribed verbatim, or at least, in the first person. Usually this is indicated by words which introduce direct speech ('he says', 'says the Admiral at this point') or which refer back ('those are his own words'). Very often small stretches of verbatim text are not introduced as such and are detectable only by changes in point of view and in the person of the verb. There are also many cases where the text is a mixture of direct and indirect speech:

> Here the Admiral says that those indications came from the west, *where I hope that Almighty God, in whose hands all victories are found, will soon grant us land.* (17.9)

On arrival in the New World, whole entries are written in the first person. All the entries from 11-24 October are in what are ostensibly the Admiral's own words, as are the entries for 6, 12, 27 November, and several of the December entries, when Columbus was in Hispaniola, contain extensive verbatim sections. In all, about 20% of the digest is in the first person and appears to record the Admiral's own words.

The two parts of the Journal, first-person verbatim text and third-person summary, therefore provide a means of contrasting Columbus's contribution with that of Las Casas, and of judging how much of Columbus's original input is still detectable in the summary. Here the linguistic evidence, summarised by Ralph Penny at the end of this Introduction, is very important. There are many indications both in the summary and verbatim sections of non-standard usage in lexis, morphology and syntax which have survived at least two stages of transcription.

Introduction

As we might expect, the errors are those commonly committed by foreign learners of Spanish: pronouns, relatives, subjunctives. One of Columbus's most endearing errors is his mangling of the phrase 'desnudos como sus madres los parió' ('naked as their mothers bore them') which he consistently uses with a singular verb, and which Las Casas respects in the digest but corrects in his own *Historia* to 'como su madre los parió' or 'como sus madres los parieron'.

It is also important to bear in mind that not only was Columbus's Spanish that of a non-native speaker, but there was also a lapse of anything up to 30 years between the time when Columbus wrote and the time when Las Casas summarized and transcribed him. If the transcription is accurate, features of the language which were undergoing change at this time should be reflected differentially in the verbatim and summary sections of the text. An investigation of initial *f-* against initial *h-*, for example, shows this to be the case.[1]

One particular feature of Columbus's written style which survives in Las Casas's summary is his use of repetitive and what one might call formulaic description. One of the striking features of the digest is the way it repeatedly supplies information which Las Casas certainly knew, and which he in any case did not need to repeat because at the time he was writing for his own eyes alone. Ten times, for example, he tells himself that a 'canoa' is a boat made from a single piece of wood. Five times he reminds us that Martín Alonso Pinzón is the captain of the Pinta; indeed, the phrase becomes something of an epic epithet. Other small and relatively trivial examples of repetitive and formulaic description include his frequent comparison of the calm sea on the outward journey with the river at Seville:

> All those days he had a very calm sea, like the river at Seville. (18.9)

> They had a sea *like the river at Seville, thanks be to God,* the Admiral says; *the sweetest of breezes, like April in Seville, such that it is a pleasure to be in them, so fragrant are they.* (8.10)

> He says that it seems to him that the whole of that sea must always be calm like the river at Seville ... (29.10)

---

[1] See page xxxv.

> ... the breezes he says are very gentle and sweet, *as in Seville in April and May, and the sea*, he says, *is always calm, thanks be to God.* (20.1)

The allusion to the pleasant climate of Andalusia in April and May is also a formula which appears several times:

> Here the Admiral says that today and thenceforth they always encountered the most gentle breezes, that the enjoyment of the morning was a great pleasure, that all they needed was to hear nightingales, he says; and the weather was like April in Andalusia. (16.9)

> *During this time I wandered among those trees which were more beautiful to look at than anything else that has ever been seen; I saw as much greenery as in May in Andalusia* ... (17.10)

> *Here and in all the island everything is green and the vegetation is like April in Andalusia.* (21.10)

And there are many other examples. Compare, too, his account of the 'niames', the sweet potatoes which were an important part of the Indians' diet, which on three separate occasions (4.11, 13.12, 16.12) he says look like carrots and taste like chestnuts. If Las Casas were not summarizing fairly closely, he would have undoubtedly spared himself the effort of writing out the same thing several times.

As for Columbus himself, there are many reasons why the ways in which he describes places, events and impressions tend to be stereotyped. Undoubtedly he suffered from the limitations of vocabulary or range of expression which someone writing in a foreign language might be excused. But Columbus was not naive where language was concerned; for all his imperfect command of Spanish, Columbus understood what any writer understands - the power of language to constitute reality. Many times in the Journal Columbus comments on the importance of language in conquest, and the disadvantages under which he labours because he cannot understand the Indians and they cannot understand him. Columbus's initial impression of the docility of the Indians is like a closed door which requires only to be unlocked by the power of language for them to carry out the designs of the Spanish Crown:

> ... he says that the only thing needed is to know the language and give them orders ... (21.12)

Introduction

This task would, he says, be much easier in the Caribbean than in Guinea, for example, because here *'the language is one and the same in all these islands'* whereas in Guinea *'... there are a thousand different languages, with one not understanding the other.'* (12.11)

Columbus understands, too, the power of naming. He gives the islands, the headlands, the bays 'Christian' names, and he does so in the full knowledge of what the islands are 'really' called in the language of the inhabitants. When he baptises them he 'names' them, he does not 're-name' them.

This is not a picture of a linguistic novice, not least when he admits that language - or his poor command of it - cannot do justice to his achievement: *'... a thousand tongues would not suffice to give the Monarchs an account of what they had seen, and his hand could not write it ...'* (27.11) Rather, what it suggests is that the repetitive, somewhat formulaic language of the Journal is not just of use in evaluating the accuracy or otherwise of Las Casas's summary, but also gives us an important clue to the nature of Columbus's descriptive language and the way that he uses it. It also returns us to the key question of what Columbus's purpose was in writing his Journal.

The aim of the Journal

We are used to thinking of Columbus and the later generations of *conquistadores* as free agents, pioneers, driven by ideals and lusts of their own devising beyond the margins of the society they left behind. But this was almost never the case. Wherever they went, the *conquistadores* were constrained by a far-reaching network of controls administered with varying degrees of success by the Crown and the Church. Although they were always in conflict with that bureaucracy, they could not ignore it. When Columbus went ashore on the morning of Friday 12 October 1492 he had with him four individuals who embodied these forces in tension. On the one hand he had the brothers Martín Alonso and Vicente Yáñez Pinzón, captains of the Pinta and the Niña, archetypal adventurers, fractious and disobedient, always on the lookout for private gain. On the other, he had two Crown officials, the secretary of the expedition, Rodrigo de Escobedo, and the accountant, Rodrigo Sánchez de Segovia.

The presence of the two officials hardly seems to fit the popular image of the 1492 voyage as a do-or-die mission led by a hare-brained visionary. But they

were there because when Columbus sailed he did so under the auspices of what was fast becoming a very efficient, modern, bureaucratic state. The system of conciliar government which Ferdinand and Isabel were in the process of setting up would provide the newly-unified Spain with a powerful mechanism for administering a huge empire with a high degree of centralised control. The delegation of much of the work of discovery and conquest to private individuals like Columbus was not done without strict contractual obligations which were, in theory at least, closely monitored. The secretary and the accountant were there to keep tabs on progress, look after the Crown's interests and see that all the proper formalities were carried out. And when the first landing was made, it was they who officially witnessed the documents which formally constituted the act of possession.

The rate at which the central administration in Spain kept pace with territorial expansion in the New World is impressive indeed. By 1503, the enterprise of the Indies was being run by its own administrative unit in Seville, the Casa de Contratación. The head of this unit, Juan Rodríguez de Fonseca, Isabel's chaplain and later Bishop of Burgos, kept a remarkable degree of control over activities which were going on at the furthest edge of the known world. In 1524, as the network of governorships and tribunals grew in the Caribbean and the mainland, the Empire of the Indies acquired its own Council of State.

As the extent of the newly-discovered territories grew ever greater, there sprang up alongside the *conquistadores* a shadowy army of clerks and secretaries, recording the events for posterity and maintaining a discreet surveillance in the process. There was, it seems, no conquest without writing. As John Elliott has put it, 'Royal officials in the Indies, theoretically at large in the great open spaces of a great New World, in practice found themselves bound by chains of paper to the central government in Spain. Pen, ink and paper were the instruments with which the Spanish crown responded to the unprecedented challenges of distance implicit in the possession of a world-wide empire.'[1]

But the written records were not always created by civil servants and Crown officials. The *conquistadores* themselves often turned their own hands to writing, and between them they built up a huge volume of accounts of discovery and con-

---

[1] J.H. Elliott, 'Spain and America before 1700' in *Colonial Spanish America*, ed. Leslie Bethell, Cambridge: Cambridge University Press, 1987, p. 63.

quest which constitute an important chapter in Spanish and Latin-American literary history. In this they were following Columbus's own example. During the homeward journey, on Thursday 14 February, he records how, at the height of a terrible storm, fearing that if he were to perish Their Majesties would have no news of his voyage, he took a piece of parchment and wrote on it everything he could about everything he had found, beseeching whomsoever might find it to take it to the Monarchs. He then wrapped the parchment tightly in a waxed cloth and cast it afloat in a large wooden barrel.

Columbus's despair at the thought that everything he had achieved could easily go to the bottom of the ocean brought home to him how, in the end, words are much more important than deeds when one is working at the edge of the known world and the rewards are to be found at the centre. His writing, then, is characterised by two characteristic qualities which are often in tension in the Journal: the need to be accountable and the need to communicate effectively with the powerful people back in Spain. At times one feels a strong sense of the writer looking over his shoulder, fending off criticism and justifying his actions and decisions. At others he is desperately trying to get the people who hold the keys to reward and recognition to understand and re-live the problems he faces, the terrain, the culture, the sheer size of everything. And all this had to be done when the writer himself was often at a loss to understand the reality he was describing. Before attempting a comprehensive account of the city of Tenochtitlan, Cortés voices a characteristic complaint about the difficulties he faces as a narrator:

> Most powerful Lord, in order to give an account to Your Royal Excellency of the magnificence, the strange and marvellous things of this great city of Temixtitan and of the dominion and wealth of this Mutezuma, its ruler, and of the rites and customs of the people, and of the order there is in the government of the capital as well as in the other cities of Mutezuma's dominions, I would need much time and many expert narrators. I cannot describe one hundreth part of all the things which could be mentioned, but, as best I can, I will describe some of those I have seen which, although badly described, will, I well know, be so remarkable as not to be believed, for we who saw them with our own eyes could not grasp them with our understanding.[1]

---

[1] Hernán Cortés, *Letters from Mexico*, trans. Anthony Pagden, London: Oxford University Press, 1972, pp. 101-2.

Columbus was the first of a line of shrewd conquerors who learned not just to live with but to harness the power of the document and the written record, and to turn it to their advantage. They learned quickly and effectively how to set the record straight, using the written word to gain political and financial support in the pursuit of their aims. And they used writing to try to stamp political, linguistic and conceptual authority on the unknown. But the reality all too often rebelled.

The objectives of the 1492 voyage

In order to understand the problems Columbus faced in writing his Journal, it is important to understand his objectives. What was he trying to do, and to what extent did that first landfall confirm or confound his expectations? There are three main statements about Columbus's objectives in three different documents, and as one might expect, they all say different things. First there is the contract made between Columbus and the Crown and signed on 17 April 1492. This document, known as the *Capitulaciones*, is written in Spanish and sets out the terms of the agreement by which Columbus was to become viceroy and governor-general of any islands and mainland he might discover, the appointment to be hereditary in perpetuity; and, in exchange, the Crown would take 90% of all income from the territories under his jurisdiction.[1]

The second document is the passport issued to Columbus to ensure that he received maximum cooperation from any King, Prince, Duke, Marquis, Count, Viscount, Baron, Lord or Lady he might meet on his travels. This document, so that it might more readily be understood in the Far East, was written in Latin, and speaks of Columbus as engaged on matters concerning the service of God and the Catholic religion, 'necnon benefficium et utilitatem nostram'.[2]

The third statement about objectives comes from the prologue to the Journal itself. This is the longest and most detailed statement and it aims to put the 1492 voyage into a broad religious and diplomatic context. With the ending of the Reconquest in Spain, and the expulsion of the Moors and the Jews, the time was

---

[1] *Capitulaciones del Almirante don Cristóbal Colón y salvoconductos para el descubrimiento del nuevo mundo*, Madrid: Ministerio de Educación y Ciencia, 1970.
[2] *ibid.*, p. 23.

Introduction

ripe, it suggests, for a diplomatic mission to the lands of the Great Khan to promote the Catholic faith:

> *Your Highnesses, as Catholic Christians and princes devoted to the holy Christian faith and the furtherance of its cause, and enemies of the sect of Mohammed and of all idolatry and heresy, resolved to send me, Christopher Columbus, to the said regions of India to see the said princes and the peoples and lands and determine the nature of them and of all other things, and the measures to be taken to convert them to our holy faith; and you ordered that I should not go by land to the East, which is the customary route, but by way of the West, a route which to this day we cannot be certain has been taken by anyone else.*

The idea of a religious alliance with the Far East directed against Islam was a very long-standing one in the European mind; so long-standing, in fact, that the last Mongol Emperor of China, the Great Khan to which Columbus refers, had been deposed in 1368.

Clearly, if we take each of these documents at face value and assume that Columbus was trying to do all of those things, we get a mishmash of strategic objectives - scientific, economic, diplomatic and religious - which is so diffuse as to guarantee disaster. Columbus's objectives undoubtedly were unclear, but there was also, I believe, a firm sense of priorities underlying them. While the Capitulations speak entirely in terms of discovery and conquest, the terms used - 'descubrir' and 'ganar' (literally 'discover' and 'gain' or 'win') - are formulae which appear frequently in comparable documents licensing expeditions in the Atlantic. To that extent, the Capitulations need to be seen more as a pro-forma agreement drafted in very general terms to cover any eventuality than as a specific set of commands. For that reason, the more detailed statements of objectives which appear in the passport and the Journal appear to take priority. Columbus, then, was not primarily trying to discover anything at all. He was simply trying to get somewhere he had never been before, by a route no one had ever used, to make contact with a ruler who had been deposed 124 years earlier.

Now there is nothing inherently contradictory about each of the objectives as they have been stated - it is quite possible to be aiming for a known port of call, and to come across some previously unknown territory in the process; the Atlantic, everyone knew, was peppered with islands which Spain and Portugal had been busily identifying and colonising throughout the fifteenth century. But if

one is prepared for both the expected and the unexpected there will come a point in the voyage when the commander will have to decide: is this new phenomenon something he knows about and is expecting, or is it something unforeseen?

No one can blame Columbus for failing in his main objective; in failing to reach China he was wholly the victim of circumstance. But Columbus went on to compound his failure. At the first landfall and in the weeks that followed, he was apparently unable to make that crucial distinction between something foreseen and something unforeseen. In this, he was also a victim, but this time, perhaps, he was a victim of his preparation.

The preparations for the 1492 voyage

In terms of navigation, the preparation for the 1492 voyage was extraordinarily thorough. It had to be, for in aiming to reach a known destination by an unknown route, the very success of the enterprise depended on reducing unknown factors to a minimum. Planning was everything, not just because his life and those of his crews were at stake, but because Columbus had no means of his own, and if he was to obtain the funding for the expedition he had to convince his sponsors that there was a good chance of success, and a return on their investment. This was a particularly important consideration when the Portuguese voyages to Guinea were consistently self-financing and a much safer bet. The Catholic Monarchs were not in the business of funding disinterested research.

In planning his project Columbus did what anyone would do in the circumstances, that is, he tried to limit the number of unknown factors by thorough research. He made an extensive search of the available geographical literature, he consulted all the leading European geographers, and made sure that he got access to the best available maps, charts and guidebooks. His research told him what all the best geographers knew: that of course a western route to the east was a theoretical possibility and always had been. The difficulty was knowing if it was a practical proposition. There was a strong and growing body of opinion that the distances involved were not impossibly great, and as the true size of Africa became apparent throughout the 1480s, many were saying that the time had come to take a serious look at the western route. Columbus's reading and interpretation of the evidence of classical geographers was confirmed by a family of maps drawn by Henricus Martellus and Francesco Roselli in Florence, by Martin

Introduction

Behaim's globe made in Nuremberg, and by his own calculations based on first-hand observations made during extensive sailing experience in the Atlantic. All the evidence pointed to a transatlantic voyage from the Canaries to Japan of around 2,400 miles.

Columbus's presentation of his plan to the Portuguese coincided unhappily with the news of Bartolomeu Dias's rounding of Cape Horn in 1488, a success which revived faith in the viability of the southern route to the East. When Columbus turned to Spain, he was met by a cool response from a government which was still too preoccupied by the Reconquest to show any great interest in the rather remote possibility of scoring a point off their long-standing rivals. Nevertheless, Columbus lobbied with great vigour, his Genoese friends in Seville came up with some financial backing and the Crown contributed two caravels, the Pinta and the Niña, whose participation came as the result of a fine imposed on the town of Palos. The expedition left Palos on 3 August 1492, and on the morning of 12 October, 2,400 miles out into the Atlantic, just where he said it would be, he found land.

The landfall and its aftermath

The reality that confronted Columbus in the days following the landfall was, in some ways, a great disappointment, and the conflict between his expectations and the evidence of his eyes has been the object of a great deal of comment. Where he expected to find the sophisticated subjects of the Great Khan and the bustling ports of the Orient, he found naked innocents and little else. In a sense he was the victim of a cruel coincidence, but he was also unduly fixated by the written authority of charts and books, and for that he must take some of the blame. The days immediately following the landfall were therefore a period of crisis in Columbus's thinking, but he managed that crisis remarkably well. He was very resourceful, and he devised a number of strategies for coping with the mismatch between reality and expectation.

The most obvious one was closely tied in with his operational decision-making: what should he do now, where should he go next? While he could not admit that he was not in the Orient - to admit that was to admit the failed objective of the whole voyage - he could properly admit that he was not quite where he wanted to be. This strategy is a very effective one in terms of keeping spirits up, keeping the expedition going and giving it a sense of purpose. In explanatory terms it is

even more effective because the real objective is always constituted elsewhere, and writing is the perfect medium for doing just that, giving the products of the imagination substance in the text. Large parts of the Journal are designed to construct an alternative reality beyond the horizon. So while the characteristic gesture of the voyage is an out-stretched arm and a pointing finger - what we seek is on the next island - that gesture has a number of rhetorical equivalents in the Journal. One of the most commonly-used nouns in the Journal is 'gold' although no gold worth speaking of was found on the first voyage; and what was found is always referred to as 'samples'. Simply talking about gold often enough helps to create a strong impression of substance, or holds out the strong likelihood of substance.

By the same token, one of the most commonly-used groups of words in the Journal used to describe Columbus's impressions is that related to 'marvellous'. Columbus's use of this and related words is closely tied to another rhetorical strategy which also has a counterpart in operational terms. Operationally, if what he is looking for is not here, and is therefore somewhere else, he needs a means of deciding which way to go and whether he is making any progress. The first one is easy - just follow the signs marked 'gold' - but the second one involves finding a substitute for gold to which an incremental rhetoric can reasonably be applied. The substitute he uses most often is landscape, and Columbus's growing sense of the marvellous is an important element in the success of this strategy.

Landscape

In the early pages of the Journal, Columbus is very keen to make everything seem familiar. There are constant references back to the Spanish experience; everything is just like Spain, like spring in Andalusia, like the river in Seville, like the hills behind Córdoba. But as the voyage progresses, and particularly off the coasts of Cuba and Hispaniola, Columbus shows a much greater willingness to concede difference, to make things exotic. One can appreciate why he might want to refer back to common experience with the absent addressee in mind, and why on the outward voyage especially, he might want to give a strong sense of predictability almost, of a sense that everything is just as he expected. But once arrived, and in view of his limited success, he has to adopt a different posture. No one, having sailed to the other end of the earth wants to have to write back that 'it's just like Spain'.

Introduction

Columbus's response to the natural beauty of the islands is undoubtedly genuine, but it is also strategic. Each island is the most beautiful that eyes have ever seen. The trees are green, straight and tall, fragrant, and full of singing birds. The rivers are deep, and the harbours wide, wide enough to embrace all the ships of Christendom. His eyes never tired of looking nor his ears of listening. 'He praises all this very highly', Las Casas sums up at one point (25.11), evidently lacking Columbus's own stamina for hyperbole. On 25 November, Columbus assures Their Majesties that the reality is a hundred times greater than his description. By 5 January inflation had taken that to a thousand. And all the time Columbus's incremental rhetoric - this bay is more commodious than that, these people more intelligent than those, this island richer and more marvellous than that - is skilfully deployed to encourage the sense that he is getting warmer and warmer.

Morison has argued that Columbus's descriptions are not extravagant for the 1490s.[1] Undoubtedly the islands were heavily wooded and rich in exotic flora and fauna. But what we have in the Journal is not really a description, and to judge it in those terms is to misunderstand the genre to which this text belongs. For all Columbus's empiricism in the execution of the voyage, his account of it has more in common with travellers' tales than with a ship's log. Travellers' tales are supposed to be marvellous, and what Columbus describes is not so much what he saw, as the sense of wonder with which he saw it.

That is all very well, say the Crown officials, but beautiful views cannot be turned into cash. Columbus's answer appears to be: cut the trees down and turn them into ships, develop the natural resources for economic ends, and, of course, where there are such wonderful things, who can doubt that there are many more things of value yet to be discovered? Columbus anticipates in the Journal many of the forms of exploitation of both human and natural resources which will lead in a very short time to the total destruction of a whole way of life in the Caribbean. But, in privileging the landscape, even if for want of anything of more tangible value, Columbus inevitably calls up associations in the European mind with rural worth versus urban decadence, and in doing so he raises important questions about the nature of the inhabitants which point to a fundamental contradiction in Columbus's mind. Underlying what appears to be a systematic

---

[1] Samuel Eliot Morison, *Admiral of the Ocean Sea*, Boston: Little, Brown and Company, 1942, p. 234.

search for the epicentre of this oriental civilisation there is a network of contradictory behaviour and discourse which allows us to glimpse his sense of failure which is never explicitly articulated.

Native inhabitants

In an important and influential study of the origins of the cannibal mythology in the Journal, Peter Hulme has argued that it contains two conflicting discourses, of civilisation and savagery.[1] As the absence of cities, and therefore of gold, becomes more apparent, an alternative discourse emerges in which gold in the form of artifacts, to be traded for or plundered, is replaced with the idea of gold as an element to be dug from the earth. Marco Polo gives way to Herodotus. At the same time the docility of the natives - on which Columbus frequently comments, particularly in the early stages - is superceded by a growing fascination with the possibility that there may be another more aggressive and therefore more civilised tribe on a neighbouring island who prey on the inhabitants of Hispaniola. However, this conflict, between the native as a thing of value and a thing of no value, is there from the outset and is maintained throughout the first and subsequent voyages.

I have suggested that Columbus evolved some effective strategies for making the best of the reality which presented itself to him, and that he implemented these in the writing of the Journal with considerable skill. Although the landscape presented him with many opportunities to write up reality, the native inhabitants of the islands were more difficult. The Indians wore no clothes, in contrast to the rich robes described by Marco Polo, and this was a truth which was too naked to be covered up. But Columbus did his best. On 18 December he was visited on board the Santa María by a young chieftain and his entourage of 200 men, of whom four carried him on a litter. 'Your Highnesses would no doubt approve of the ceremony and respect with which they all treat him, although they all go naked', writes Columbus, and there follows a set-piece of savage nobility, an acting out by these two leaders of the kind of elaborate ceremonial which would be expected of men of their status in a sophisticated society.

---

[1] Peter Hulme, 'Columbus and the Cannibals' in *Colonial Encounters*, London and New York: Methuen, 1986, pp. 13-43.

Introduction

When the cacique comes aboard, Columbus is at table in the sterncastle. The Indian will not allow him to interrupt his meal or rise to greet him. Some food is brought for the visitor and the entourage is ordered outside, with the exception of two men whom Columbus judged to be his advisers and who sat at his feet. Of the food and the drink which are brought, the cacique takes just enough to taste, sending the rest to his men 'and all with an amazing gravity and with few words, and those he did speak, as far as I could understand, were very wise and considered, and those two men watched his mouth and spoke for him and with him and with great respect.'

Gifts and pleasantries are exchanged:

> *After he had eaten, a page brought a belt just like those from Castile in manufacture although the workmanship is different, which he took and gave to me, and two pieces of worked gold which were very thin, because I believe that they get very little of it here, although I hold that they are very close to its source and there is a great deal of it. I saw that he liked a tapestry which I had over my bed. I gave it to him with some very good amber beads which I had around my neck, and some red slippers, and a flask of orange-flower water with which he was so pleased that it was amazing. He and his advisers are very sad because they could not understand me nor I them. Nevertheless, I understood him to say that if I wanted anything from there, the whole island was at my disposal.* (18.12)

It takes very little to see in this awesome, well-mannered, softly-spoken and above all *generous* Indian a not too distant reflection of the Great Khan himself, attended by 12,000 liegemen in token of his power, surrounded by elaborate ritual and held in universal fear.

But though Columbus must find his Great Khan, one way or another, so much of what he says and does on the first voyage gainsays his praise of the land and its people, and that contradiction is evident from the very moment Columbus first goes ashore. If this is a diplomatic mission, why is Columbus's first act one of possession? He has a Latin passport and men aboard who speak Hebrew and Arabic and Chaldean so that he can present his credentials to one of the greatest princes and richest men in the world. Why, then, does he take twopenny trinkets - glass beads and hawks' bells - instead of something to impress the man who has everything? And if he is intent on conquering the lands of the Great Khan, why does he take such a small expedition, no soldiers and minimal weapons?

The answer to this question may well lie in the ceremony which took place on Guanahaní at the first landfall on 12 October. The Journal reads:

> ... they saw some naked people and the Admiral went ashore in the armed boat with Martín Alonso Pinzón and Vicente Yáñez, his brother, who was the captain of the Niña. The Admiral brought out the royal standard, and the captains unfurled two banners of the green cross, which the Admiral flew as his standard on all the ships, with an F and a Y, and a crown over each letter, one on one side of the + and one on the other. When they landed they saw trees, very green, many streams and a large variety of fruits. The Admiral called the two captains and the others who landed, and Rodrigo de Escobedo, secretary of the expedition, and Rodrigo Sánchez de Segovia, and made them bear witness and testimony that he, in their presence, took possession, as in fact he did take possession, of the said island in the names of the King and Queen, His Sovereigns, making the requisite declarations, as is more fully recorded in the statutory instruments which were set down in writing. (12.10)

The ceremony they enacted had many precedents in Roman and Germanic law and had been often used during the reconquest and the colonisation of the Canaries.[1] The act of possession always took a physical, symbolic form. Columbus would have taken a handful of earth, cut off the branch of a tree, drunk some water or eaten some fruit, or simply imprinted his footsteps on the soil. The mention of trees, water and fruit in the Journal may be an indication of the precise form the ceremony took. But that itself was not enough. Other elements had to be present for the act to be valid in law. There had to be witnesses (the Pinzón brothers); there had to be Crown representatives (the secretary and the accountant); and there had to be someone to give possession. Columbus knew about these formalities, because at the beginning of the prologue of the Journal he describes the handing over of the keys of the Alhambra to Their Majesties by the defeated Boabdil in a ceremony at which Columbus claims to have been present.

Now there were circumstances under which the third element could be dispensed with, that is when the lands being annexed were considered 'res nullius',

---

[1] F. Morales Padrón, 'Descubrimiento y toma de posesión', *Anuario de Estudios Americanos*, xii (1955), 321-80.

Introduction

when they belonged to no one. But these, surely, were the lands of the Great Khan; how could they be considered 'res nullius'? Clearly, the legal precedents put Columbus in some difficulty; either these lands belonged to someone, or they did not. Evidently, Columbus decided they did not. And if they did not, who were all these people who inhabited them?

Columbus's judgement in this legal matter clearly indicates that he had formed a view at a very early stage about the Taino inhabitants of the Caribbean. They were, it seems, nothing, a tabula rasa on which the Catholic faith and European civilisation had still to be inscribed. His chosen stylus was language, and the book in which this inscription would take place is the Journal. There is, however, an irony underlying Columbus's attempt at linguistic and cultural colonisation through language. We know that he made his first landfall on an island called Guanahaní, an island which he then (re)named 'San Salvador'. But to this day no-one knows for certain which island Guanahaní was. In suppressing the Indian name, Columbus has erased the site of his greatest triumph.

EDITORIAL NOTE

The purpose of this new edition of the Journal is to provide a clear, accurate and readable Spanish text which keeps faith as far as possible with the features of the original manuscript. Original orthography has been maintained, but all contractions have been resolved. Las Casas made over 1,000 corrections to the text as he was making the summary and no attempt has been made to document these, but all his marginal notes are retained, as footnotes tied to the nearest appropriate place in the Spanish text.

The punctuation of the original differs considerably from modern usage. Las Casas used three main punctuation marks, a slash and a point (/.), a colon (:), and a slash alone (/), in descending order of importance. An equivalent hierarchy has been used in the edition: a point (.), a semi-colon (;), and a comma (,). Very occasionally some punctuation has had to be added, but this is kept to a minimum.

Verbatim text is printed in *italic* on both the Spanish and the English pages. Explanatory notes are tied to the English text and follow it.

THE LANGUAGE OF CHRISTOPHER COLUMBUS

by R.J. Penny

Columbus's linguistic background

Columbus was born in Genoa in 1451, and lived there until 1473, when he was 22. Despite some opinions to the contrary, his family was in all probability Genoese,[1] and it is therefore reasonable to assume that his native language was the Genoese vernacular. Through his involvement in the wool trade, he may have become familiar with the commercial Latin of the time, and it is possible that he came into contact with Spanish and/or Portuguese speakers in the busy port (although this is a notion for which there is no direct evidence). What familiarity Columbus had with Tuscan is unknown; the idea that he was a student at Pavia has been discarded as a myth, created by Columbus, and the little that Columbus later wrote in Italian is heavily contaminated by Spanish.

Between the ages of 22 and 25 (1473-6), Columbus was employed as a commercial agent by the great Genoese shipping houses of Paolo di Negro and Ludovico Centurione, for one of whom he undertook a journey to the Greek-speaking island of Chios. The house of Centurione maintained agencies in Seville, Cádiz, and other Spanish ports, but there is no evidence that Columbus worked in or visited such offices.

At the age of 25, Columbus was shipwrecked off the coast of Portugal, and for the next nine years (1476 to the end of 1485) he made his home in Lisbon. During this time, he made voyages to England and Iceland, and to West Africa, as well as visits to Genoa and other Mediterranean ports, but for most of the period Columbus found himself in a Portuguese-speaking environment. Even before marrying a Portuguese wife in 1480, it can be assumed that he learned to speak Portuguese; after his marriage, it is a near certainty. At least from 1480, Columbus became involved in the social and intellectual life of Portugal, and it is

---

[1] Ramón Menéndez Pidal, *La lengua de Cristóbal Colón*, Buenos Aires: Austral, 1947, pp. 13-14.

probable that at the same time as he was formulating his projects for discovery he was also learning to write Spanish, in accordance with the practice of many educated Portuguese of the time.[1] In all probability, Spanish was the first language Columbus learned to write; there is no evidence that he ever learned to write Portuguese, and he could barely write Italian.

At the age of 34, Columbus moved to Spain and had his home there until his death. For most of this period (1485-1506) he was in the service of the Catholic Monarchs, and his various writings are almost exclusively in Spanish, even in the case of letters addressed to Italians. The few notes made by Columbus in Italian are, as we noted above, full of hispanisms.

Columbus's written Spanish

The evidence summarized in the previous section suggests that the only language Columbus learned to write was Spanish. He was at least 25 when he began this learning process, and it would be natural to assume that, as in the case of all adult language-learners, his native speech (i.e., Genoese, not Italian) would have interfered with and distorted his written Spanish. Furthermore, because of the fact that he was learning to write Spanish after learning to speak Portuguese and in a milieu where the native language was Portuguese, it would be unsurprising to find that the language he learned to speak in Portugal should have influenced the way he wrote Spanish. There are some instances where these two outside influences (Genoese and Portuguese) can be expected to conspire; that is, there are features of development which are common to Genoese and Portuguese which are not shared by Spanish. On other occasions, namely where Genoese and Portuguese differ in their development both from each other and from Spanish, it is in theory possible to identify which of the two vernaculars concerned is responsible for a given non-native feature in Columbus's Spanish.

The language of the 1492 Journal

It should be noted at the outset that, since the journal only survives in Las Casas's summary (although with extensive verbatim quotation), it is to be expected that at least some non-native features of Columbus's Spanish would have been filtered out by copyists of the Journal and by Las Casas himself. Such

---

[1] Menéndez Pidal, *La lengua de Cristóbal Colón*, p. 6.

modifications are most likely at the level of spelling, possible at the level of morphology and syntax, and perhaps least likely in the case of lexis and semantics. In order to minimize the effect of such standardization, the following discussion is based entirely on those sections of Las Casas's text in which it is evident that he is directly quoting Columbus's words.

Influence exercised jointly by Genoese and Portuguese

- The absence of diphthong /ue/, /ie/ in cases like *al longo de* (20.10; vs. *luengo* [13.10]), *aviamento* (26.12), *pagamento* (16.10), may be a case of joint Genoese-Portuguese influence on Columbus's Spanish. This is certainly claimed by Milani.[1] However, Rohlfs claims that the graphs *e*, *o* are used in 13th-century Genoese texts to represent diphthongs, which have today receded to remoter parts of Liguria.[2] It is possible (but not proven) that such diphthongs had already been lost from the Genoese vernacular of the 1450s, so that their occasional absence from Columbus's Spanish may indeed be due to Genoese as well as Portuguese influence.

- The form *gavilano* (22.10) (for *gavilán*) may be due to awareness on Columbus's part that Genoese *-ā, -an*, Portuguese *-āo* often corresponded to Spanish *-ano* (e.g., Genoese *mā*,[3] Portuguese *mão*, Spanish *mano*), although such cross-linguistic comparisons, if they are at work here, have led in this case to an erroneous result.

- Use of the form *non* with final /n/ (20.10: *una de limpio y otra de non*), unusual at this stage in Spanish, may argue for combined Genoese and Portuguese influence, since in these varieties the negative particle ended in a nasal (e.g., Old Portuguese *nom*).

- Columbus's preference for /r/ in the forms *temperada* (17.10, 12.11), *temperadas* (23.10), *temperança* (27.11), rather than *templar* and its

---

[1] Virgil I. Milani, *The Written Language of Christopher Columbus* (supplement to *Forum Italicum*), Buffalo: State University of New York at Buffalo, 1973, p. 137.
[2] Gerhard Rohlfs, *Grammatica Storica della Lingua Italiana e dei suoi Dialetti*, (three vols., I Fonetica, II Morfologia, III Sintassi e Formazione delle Parole, Turin: Einaudi, 1966, 1968, 1969), I pp. 112, 139.
[3] Rohlfs, I, p. 427.

derivatives, which were becoming normal in Spanish at the end of the 15th century, perhaps reveals both Genoese and Portuguese influence, since both these varieties continued (and continue) to use forms with /r/. Additionally, absence of syncope may be due to Genoese, where syncope is less frequent than in Hispano-Romance.[1]

- Columbus's use of monosyllabic *nos*, rare in late 15th-century Spanish, to the exclusion of *nosotros* (e.g. *porque dé buenas nuevas de nos* [15.10, 16.10]; *vinieron a nos* [17.10]) is perhaps due to the fact that contemporary Genoese and Portuguese used monosyllabic forms of the corresponding pronoun.

- The sense 'steal, seize' for the verb *prender* was probably obsolete in Spanish by the end of the 15th century. Its use by Columbus in this sense (12.11) is arguably due to the fact that in both Genoese and Portuguese, the verb *prender* continued to be used with this value.

Influence exercised by Genoese alone

Such evidence is hard to come by, owing to the scarcity of sources of information on 15th-century Genoese, so that the following cases must be viewed with caution. Evidence is often available from medieval Italian (i.e., Tuscan) sources, but it goes without saying that such data by no means necessarily imply that a given form was used in contemporary Genoese. Caution is all the more necessary in that we have seen that it cannot be established that Columbus was a fluent user of 'standard' Italian.

- The otherwise unprecedented form *símplice(s)* 'simple', used by Columbus in 14.10, may owe its form to interference from a Genoese cognate of Italian *sémplice*. Likewise, the final vowel of *doblo* (26.12) may be accounted for in similar manner (cf. Italian *doppio*). *Doblo* does not elsewhere appear in Spanish until 1640, and then only as a legal term.[2]

- Columbus uses the words *diforme* and *disforme* in the sense 'different' (*muy diformes de los nuestros* [16.10], *disformes de los nuestros* [16.10,

---

[1] Rohlfs, I, pp. 169-72.
[2] Juan Corominas and J.A. Pascual, *Diccionario crítico etimológico castellano e hispánico*, 6 vols., Madrid: Gredos, 1980-, s.v. *dos*.

17.10], with the same meaning as in *tan diversas de las nuestras* [19.10]). Such a meaning is associated with late medieval and Renaissance Italian *disforme*[1] and may conceivably have been attached to a cognate Genoese term, introduced unconsciously by Columbus into his Spanish. Similar arguments can be applied to Columbus's use of *estimulados* 'excited, worried' (22.9), *infra* (*infra la tierra* 'inland' [27.11]), and to *temporejar* 'to delay, stand off (a coast)' (15.10, 20.10). For the latter verb Milani (pp. 155-6) quotes cases of late medieval and Renaissance Italian *temporeggiare*, 'id.', a Genoese cognate of which Columbus may have introduced into his Spanish; the Spanish verb does not otherwise appear until the late 19th century, when it is borrowed from Catalan or Portuguese.

- Columbus uses the verb *ser* in impersonal constructions, as the equivalent of impersonal *haber*. Thus: *es* [= *hay*] *en estas tierras grandíssima suma de oro* (12.11); *adonde es* [= *hay*] *mill maneras de lenguas* (12.11); *es* [= *hay*] *tanto* (29.12); *que más mejor gente ni tierra puede ser* [= *haber*] (24.12). Since such a construction does not occur in Spanish or Portuguese, we may be dealing with a case of interference from Genoese, if it can be shown that 15th-century Genoese was like Tuscan in using the verb 'to be' in this way. It should be noted that Columbus also uses *aver* (modern *haber*) in this role.

Influence exercised by Portuguese alone

Evidence of such interference in Columbus's Spanish is abundant and, in some instances, has long been known.[2]

- The verb *sufrir* 'suffer' appears spelt with *ç*- (*çufriré* [9.1]). It is not inconceivable that this spelling reveals that the Portuguese Columbus learned was subject to incipient merger of /s/ and /ts/ (since *seseo* had begun in Southern Portuguese in the 13th century, even if it did not become fully acceptable in educated usage until the mid-16th century.[3]

---

[1] Milani, *Written Language*, pp. 49-50.
[2] Menéndez Pidal, *La lengua de Cristóbal Colón*, pp. 17-23.
[3] Serafim da Silva Neto, *História da língua portugêsa*, Rio de Janeiro: Livros de Portugal, 1952, pp. 484-7.

However, such a hypothesis is weakened by the fact that this verb is also spelt with ç- (five instances) in the non-verbatim sections of Las Casas's summary.

- Raising of atonic /o/ to /u/, not unknown in non-standard varieties of Castilian, is regular in Portuguese, and may account for Columbus's use of *cudiçia* 'greed' (25.12, 26.12).

- The form *convertería* (11.10; vs. *convertirán* [6.11, 27.11]), as well as reflecting vocalic uncertainty similar to the preceding case, may reveal interference in Columbus's Spanish of the Portuguese infinitive *converter*.

- *Cogujos* 'buds (?)' (4.11) is conceivably a falsely castilianized form of a Portuguese word. Latin CUCULLIO, -ONIS 'hood' might be expected to provide Portuguese **cogulhão*, or conceivably by back-formation **cogulho*. The latter form may have been the one learned by Columbus, for which he invented a non-existent Castilian cognate.

- *Multidumbre* 'multitude' (12.11) appears to be a blend created by Columbus from separate components of Spanish *muchedumbre* and Portuguese *multidão* 'id.'.

- Columbus confuses the pronouns *el* and *lo*, using *lo* as a masculine (*lexos de lo uno y de lo otro* [= *del uno y del otro*], with reference to geographical locations [1.11]). This confusion is likely to be due to the dual masculine and neuter function of the Portuguese pronoun *o*.

- The masculine gender of *nariz* (11.10, 15.10 17.10 22.10; vs. fem. twice at 13.10) and of *señal* 18.12 (vs. fem. at 1.11 and 12.11) probably reveals Portuguese influence, since the Portuguese cognates of these words are masculine. However, in the second case, Genoese may have conspired with Portuguese, if the Genoese cognate was masc., as is the Italian *segnale*.

- The verb *tener* is used as an auxiliary to form the perfect and other compound tenses with some frequency in Golden-Age Spanish. However, the consistency with which Columbus uses this auxiliary (rather than *haber*) argues for considerable Portuguese influence on his Spanish. E.g., *aquellos hombres que yo tenía tomado* (14.10), *como tenía determinado*

(23.10); *tengo determinado de la rodear* (16.10); *desnuda como dicho tengo* (4.11); *como hasta aquí tienen fecho* (6.11); *tengo hablado del sitio* (27.11); *menos de lo que yo tengo dicho* (24.12).

- It has long been known that Columbus's Spanish contains items of Portuguese vocabulary not elsewhere attested in Spanish, or not attested there until later. Among such items we find: *angla* 'inlet of the sea' (19.10; probably Portuguese *angra* 'id.', castilianised by Columbus; Castilian *angra* is attested only from 1573,[1] *arambel* 'bed-cover' (18.12) (< Portuguese *alambel* 'id.', otherwise attested only from 1527, *corredíos* 'straight, smooth (hair)' (13.10), *fugir* 'to flee' (*fugir, fugió, fugeron, se avía fugido* [15.10], *se avían fugido* [21.10], *fugir* [21.10, 27.11]; vs. *fuyen* [12.11], *huyr* [16.12]).

Other evidence of Columbus's imperfect learning of Spanish

In the following cases it can be argued that we are witnessing errors typical of those made by an adult learner of a second or subsequent language. In the absence of detailed information on 15th-century Genoese, these departures from the Castilian norm are not here assigned to interference from Columbus's native language (or from any previously learned language), although subsequent investigation may reveal that they are interference errors.

- Columbus twice uses *ningúnd* (27.11), perhaps modifying *ningún* in imitation of *según*, which genuinely alternated with *segúnd* in medieval and early modern Spanish.[2] Columbus uses the form *segúnd* at 12.12.

- The verb-form *andássemos* (19.10), for standard *andoviéssemos* or *anduviéssemos*, although not totally unprecedented in the history of Spanish, is likely to be an adult language-learner's error, perhaps ultimately due to the regular nature (if this is indeed the case) of the cognate verb in Genoese.

- The relative pronoun *qui* was ousted by *quien* in the 13th century.[3] Columbus nevertheless uses this pronoun (*un cabo a qui yo llamé el Cabo*

---

[1] Corominas-Pascual, s.v. *angra*.
[2] Corominas-Pascual, s.v. *seguir*.
[3] Corominas-Pascual, s.v. *que*.

Hermoso [19.10]; *este, a qui yo digo Cabo Fermoso* [19.10]; *uno se llegó a qui yo di unos cascaveles* [21.10]); he must either have picked up this archaism from his reading of Spanish, or it is due to some (unidentified) outside interference.

- Columbus's use of the Spanish personal pronouns is notoriously confused. Like some speakers (but few writers) of Spanish, he uses *le* with plural value (additionally confusing direct with indirect object function): *y si se le* [= *los*] *trastorna, luego se echan todos a nadar* (13.10); *de siete que yo hize tomar para le* [= *los/les*] *llevar y deprender nuestra lengua* (14.10); *muchas vezes le* [= *les*] *entiendo una cosa por otra* (27.11); *porque ... le* [= *les*] *obedezcan* (26.12); *yo no le* [= *les*] *dexé tocar* (21.10); *como le* [= *les*] *amuestran* (1.11). He also uses *le* (for standard *lo*) in reference to non-animate (including mass) nouns: *yo no le falle* [sc. *oro*] (15.10); *sin le* [sc. *algodón*] *llevar a España* (12.11); *que todos le* [sc. *acatamiento*] *tienen* (18.12). In the following case, *le* is used as a feminine direct object form (i.e., for standard *la*): *nos le* [sc. *a la sierpe*] *seguimos dentro* (21.10). Similarly, as in some of the phrases listed first in this paragraph, Columbus confuses plural *los* with *les*: *los pareçe a ellos* (19.10). Finally, his use of tonic *ello* to refer to mass nouns, although non-standard, is similar to the present-day usage of northern and north-western dialect areas:[1] *tenía grandes vasos de ello* [sc. *de oro*] (13.10), *topar en ello* [sc. *oro*] (19.10). However, *él* and *ella* also appear in this role: *aquí alcançan poco de él* [sc. *oro*] (18.12), *no e podido aver de ella* [sc. *resina*], *salvo muy poquita* (*sic*) (12.11), while *ello* on one occasion has a count-noun as its referent: *la entrada de ello* [sc. *del puerto*] (14.10).

- Like many non-standard speakers of Spanish today, Columbus sometimes pluralizes finite forms of the verb *haber* when they are used with impersonal value: *an en ella 5 leguas* (15.10), *an en ella más de diez leguas* (15.10). However, both *ay* and *(h)a* are also found with plural complements.

- Unless we are dealing with an error of transcription by Las Casas, Columbus confuses indicative and subjunctive mood in the following case: *no me pareçe que las puede aver* (27.11).

---

[1] R.J. Penny, 'Mass nouns and metaphony in the dialects of north-western Spain', *Archivum Linguisticum*, n.s., 1 (1970), 21-30.

- One or two items of Columbus's vocabulary may be due to imperfect learning. I find no corroborative trace of the verb *asensar* (*asensar la ánima* [14.2]), which is conceivably an error for *assentar* 'to calm'. *Oppósito* '(personal) opposition' (15.3) likewise appears to lack documentation in Spanish as a noun; on its rare appearances, it functions as a participle, alternating with *opuesto*. In the realm between lexis and syntax, Columbus entirely conflates Spanish *salvo* and *sino*, using only *salvo* (e.g., 23.10, 30.10 12.11, *no falta salvo assiento* [16.12]). Although other writers occasionally use *salvo* where the modern language prefers *sino*, Columbus stands out from his contemporaries by never using *sino*.

Aspects of Columbus's language which are in keeping with late 15th-century practice

The language of Columbus's Journal is, in a majority of its features, typical, unsurprisingly, of the language used by other late 15th-century writers. Among such features, there are of course a good number which differ from those of the modern standard, and it is worthwhile to note here the most important.

- We have noted above that the spelling used by Columbus is very likely to have been 'standardized' either by the copyists of the original Journal or by Las Casas himself. However, it is interesting to observe one aspect of Spanish spelling which underwent substantial change between the time of the composition of the journal and its publication in summary form. In 1492, the letter *f* was still used with two values, that of /f/ (as in *favor*, *fortaleza*) and that of the aspirate /h/, then the normal educated pronunciation appropriate to words like *fablar*, *fazer*, *fijo*, etc. However, some writers, led by Antonio de Nebrija[1] were beginning to use *h* to indicate /h/ (*hablar*, *hazer*, *hijo*, etc.). In those parts of Las Casas's text which are evidently verbatim quotations of Columbus's words, we find 88 cases in which *f* arguably represents Castilian /h/, and 144 cases in which this phoneme is written *h*. By contrast, in those parts of Las Casas's summary where he is not directly quoting, use of the letter *f* for the phoneme /h/ is rare. It is likely that Columbus used *f* spellings in all cases like *fablar*,

---

[1] Antonio Nebrija, *Gramática de la lengua castellana* [1492], ed. Antonio Quilis, Madrid: Editora Nacional, 1980, p. 121.

fazer, etc., and that in the majority of cases his spelling was replaced by *hablar*, *hazer*, etc., but leaving a substantial number of original spellings intact.

- Within the word, cases of /h/ were relatively rare in 15th-century Castilian. Columbus spells with *f* the verb *refe[r]tar* (*rehertar* 'to dispute, haggle over' [16.10]), and we find both the spellings *bofío* and *bohío* for a word, borrowed from Arawak with sense 'hut', which almost certainly contained /h/ in that language and therefore also in the receptor language, Castilian.

- What these spellings cannot tell us is how Columbus pronounced such words. Had he learned the Castilian pronunciation /h-/, or, having learned his Spanish in Portugal, where the Portuguese words cognate with Spanish *fazer/hazer*, etc., were pronounced with /f/, did he pronounce some or all of the Spanish words with /f/? The latter pronunciation would have been foreign in his period, but cannot be entirely excluded, since we have independent testimony that Columbus spoke Spanish with a foreign accent.

- The morphology of certain words (for verbs, see below) differs from that of their modern counterparts. As mentioned above, we find *segúnd* (12.12); similarly, one notes *vidro* 'glass' (11.10, etc.), still common in the Golden Age beside *vidrio*, and *peçe* 'fish' (11.10), the only form used by Nebrija.[1]

- Verbal morphology still allowed considerable free variation in Columbus's time. In the stem of -*ir* verbs, variation between /o/ and /u/ continued to be common, irrespective of the structure of the verbal ending; thus Columbus uses forms like *sorgí* beside *surgí*, *descobrir* beside *descubrir*, *descubrí*, *descubrirán*, etc. Similarly, in the 1st pers. sing. pres. ind. and throughout the pres. subj. of inceptive verbs, forms in -*sc*- compete with those in -*zc* (e.g., *cognosco* [15.3], *aclaresca* [17.10], by contrast with five cases of *cognozco* and one each of *cognozca* and *cognozcan*), and the corresponding forms of the verbs *caer*, *oír*, *traer* may still lack analo-

---

[1] Antonio de Nebrija, *Vocabulario de romance en latín* [1516], ed. Gerald J. Macdonald, Madrid: Castalia, 1973, s.v.

gical /g/ (thus *oyo* 'I hear' [21.10], vs. *traygo* 'I take' [21.10] and 14 other cases with /g/).

- In the preterite of the verb *ver* 'to see', the 1st and 3rd pers. sing. forms may appear with or without /d/. On 11 occasions Columbus uses *vi*, against 29 cases of *vide*; for the 3rd pers., he uses only *vido* (two cases). In the case of the verbs *ser* and *ir*, the 1st pers. sing. preterite form hesitated between *fue* (the only form recommended for both verbs by Nebrija[1] and *fui/fuy*. However, in Columbus's use of these forms, he appears to use *fue* as the preterite of *ser* (e.g., 15.10, 17.10) and *fui/fuy* as that of *ir* (e.g., 17.10, 18.10). In the case of the verb *traer*, Columbus uses the commonest medieval preterite form, one which was still frequent in the Golden Age: *truxeron* (15.10, 12.12).

- The imperfect of *ver* 'to see' is given its more usual Golden Age form *vía* (15.10, 18.12), while, more unusually, the participle of *ser* appears in its medieval guise of *seydo* (14.1), rather than the by then usual form *sido* (Nebrija, *Gramática*, pp. 238-45).

- The verb *llevar* 'to take, carry' could, in the 15th century, appear with initial /l/ in those forms in which the word-stress did not fall on the first syllable. Thus Columbus is able to use *levar* (23.10), *levaré* (11.10) (against six cases with *ll-*: *llevar, llevamos, llevasen, llevava, llevávamos, llevé*).

- In the field of syntax, it should be noted that until about the middle of the 16th century the auxiliary used to form the compound tenses of intransitive verbs (especially verbs of motion) was frequently *ser*, although *haber* was becoming dominant in this role. Columbus uses both constructions: *si éramos venido (sic) del çielo* (14.10), *éramos venidos del çielo* (22.10), *todo es venido mucho a pelo* (26.12), against *nosotros avemos venido del çielo* (12.11).

- In normal medieval and frequent Golden Age usage, the 'personal a' construction is only required where it is otherwise unclear whether a given noun functioned as the subject or object of its clause. Columbus can therefore write, in accordance with contemporary syntactical usage, *así truxeron la muger* (12.12).

---

[1] *Gramática*, p. 238.

- Medieval partitive expressions, based on *de* + noun or pronoun, continued in use in the early Golden Age. We find cases like *buscar del oro* 'search for some gold' (6.11) which exemplify this usage.

- In the late 15th century, the semantic range of words was naturally sometimes different from their present range. Thus the verb *ser* could still indicate location, as in the following cases: *que en ella era* (15.10); *adonde es el oro* (17.10); *fue acerca* 'I was nearby' (17.10); *por ser en ella más presto* 'in order to arrive there' (17.10); *aquí ... no es la poblaçión* (19.10); *adonde entendí ... que era la poblaçión* (20.10); *es ella en esta comarca* (24.10); *las otras que son entremedio* (21.10). The same verb can be used to indicate non-permanent attributes, fulfilling a role currently fulfilled by the verb *estar*: *[sus casas] eran de dentro muy barridas y limpias* (17.10).

- The verb *aver* (= modern Spanish *haber*) could still in the 15th century indicate 'possession': *aya lengua con este rey* (19.10, 21.10), *para aver lengua con este rey* (23.10); *y ver si puedo aver de él el oro* (21.10); *aquí se avría grande suma de algodón* (12.11); *avrán en dicha servir* 'they will consider themselves fortunate to serve' (12.11); *aviendo mugeres* (12.11); *el benefiçio de que aquí se pueda aver* (27.11). However, *tener* is already found with this value: *teniendo sus mugeres* (12.11).

- The expression *después que* can mean 'since', as in *después que en estas Yndias estoy* (17.10).

- Some of the vocabulary used by Columbus represents the earliest attestation in Spanish of the words concerned. Corominas-Pascual list only later examples of *restinga/restringa* '(underwater) rock' (14.10, 19.10, 26.12), and the noun *tomo* (*de tanto tomo* 'of such importance' [31.12]). Other words no longer current (or current only in modified form) were normal in Columbus's time: *alfilel* (21.10; cf. Nebrija: *alhilel/alfilel*), *aviamento* (for *aviamiento*) 'supplies (of food, etc.)' (26.12), *enxeridos* (now *injertados*) 'grafted' (16.10), *estima* 'esteem' (15.10), *mareantes* 'sailors' (21.10), *refe[r]tar* 'to dispute, haggle over' (16.10, whence the noun *refierta-/rehierta*, now spelt *reyerta* 'quarrel'), *roquedos* (15.10, 24.10), now replaced by the infrequent *roqueda* 'rocky ground', *hazer la salva* 'to taste food, in case of poison (before a king, etc., eats)' (18.12), *ventar* (e.g., *no ventavan... vientos* [22.9], *vienta* [23.10], *ventar muy amoroso* [24.10]) 'to blow', now *ventear*.

Amerindian words borrowed by Columbus

It is hardly surprising that Columbus uses few amerindianisms, since his Journal is only intermittently concerned with description of the life and customs of the territories he discovered. He does make attempts at verbal communication with the islands' inhabitants (usually in an effort to gain information on the availability of gold and other commodities), but it appears from his account that such attempts had only limited success. Novel concepts are therefore labelled, for the most part, with the Spanish vocabulary available to Columbus. Thus, the dug-out canoes of the islanders are generally (on 16 occasions) referred to as *almadías*, while the borrowing *canoa* appears only four times (all at 17.12). The only other amerindianisms used by Columbus are *cacique* 'Indian chieftain' (17.12), and the disputed *ajes* 'yam' (21.12), which is described by Corominas-Pascual as a 'voz de origen antillano', but which may be an arabism.[1] The same plants are referred to as *mames* (4.11), a variant of (or perhaps a misreading of) *niames*, a form found in the non-verbatim part of Las Casas's digest of the Journal, later *ñames*, a word which is possibly of W. African origin.[2]

Idiosyncrasies of Columbus's language

Columbus's Spanish sometimes suffers from overcomplexity of syntax, seen in its most opaque form in the prologue of the Journal. On other occasions, one identifies less acute infelicities of style, as in the entry where Columbus is describing the bargaining abilities of different groups of natives: *cositas que saben mejor refe[r]tar el pagamento que no hazían los otros* (16.10). Elsewhere it is difficult to distinguish clumsiness from imperfect learning of Spanish: *es en esto mucho de aver gran diligencia* (16.10), *para otra isla grande mucho* (21.10), *en todos tres los navíos* (27.11), *yo he visto solos tres de estos marineros* (16.12), *más gente al doblo* 'twice the population' (26.12), *no pudiera errar de ver alguna* 'I could not have failed to see one' (16.10), *para pujar a rodear toda la ysla* (16.10). The last case is a strange instance of the use of the verb *pujar*, which usually means 'to raise'; the sentence is still odd even if *pujar* is an error for *puxar* 'to push', since the writer's intended meaning seems to be 'to try to sail around the whole island'.

---

[1] Professor L.P. Harvey, by personal communication.
[2] Corominas-Pascual, s.v. *ñame*.

Conclusion

This study of Columbus's language has been based exclusively upon the sections of Las Casas's summary of the Journal in which he explicitly quotes Columbus's words. However, there is no reason to think that the observations made on this portion of Columbus's output are not relevant to his other writings. We have seen that Columbus's native language, Genoese, probably influenced the Spanish he later learned, but that it is easier to identify interference from Portuguese, the language he learned to speak in adulthood before learning to write (and speak) Castilian. Other non-standard features of his language can be put down to inadequate learning of Spanish, while in other ways his language does not depart from the late 15th-century norm. There are few cases of borrowing of Amerindian terms and we have noted certain infelicities of Columbus's style.

BIBLIOGRAPHY

There is a vast bibliography relating to Columbus and his age. The following list is restricted to important editions and translations of the Journal, and a small number of major studies of Columbus. The text of Las Casas's digest was unknown until 1790, when Martín Fernández de Navarrete discovered it in the library of the Duque del Infantado. Robert H. Fuson discusses the history of the Journal and its reliability in 'The *Diario de Colón*: A legacy of poor transcription, translation, and interpretation', de Vorsey and Parker, 51-75.

Editions

Manuel Alvar, *Cristóbal Colón, Diario del descubrimiento*, 2 vols., Madrid: La Muralla, 1976.
Joaquín Arce and Manuel Gil Esteve, *Diario de a bordo de Cristóbal Colón*, Turin, 1971.
Luis Arranz Márquez, *Cristóbal Colón, Diario de a bordo*, Madrid, 1985.
Oliver Dunn [partial], 'The Diario, or Journal, of Columbus's First Voyage: A New Transcription of the Las Casas Manuscript for the Period October 10 through December 6, 1492', in de Vorsey and Parker, 173-231.
Oliver Dunn and James E. Kelley, Jr., *The 'Diario' of Christopher Columbus's First Voyage to America 1492-1493*, Norman and London: University of Oklahoma Press, 1989.
Martín Fernández de Navarrete, *Colección de los viajes y descubrimientos que hicieron por mar los españoles desde fines del siglo quince con varios documentos inéditos*, 5 vols., (Madrid, 1825-37), 2, 1-197. [Edited by Carlos Seco Serrano in *Obras de D. Martín Fernández de Navarrete*, Biblioteca de Autores Españoles, Madrid, 1954, vol. 75.]
Julio F. Guillén y Tato, *El primer viaje de Cristóbal Colón*, Madrid, 1943.
Cesare de Lollis, *Raccolta di documenti et studi*, 14 vols., Rome, 1892-96.
Vicente Muñoz Puelles, ed., *Cristóbal Colón, Diario de a bordo*, Madrid: Anaya, 1985.
Carlos Sanz, *Diario de Colón: Libro de la primera navegación y descubrimiento de las Indias*, 2 vols., Madrid, 1962.
Consuelo Varela, *Cristóbal Colón, Textos y documentos completos*, Madrid: Alianza, 1982, 2nd edition, 1984.

Consuelo Varela, *Diario del primer y tercer viaje de Cristóbal Colón*, Madrid: Alianza, 1989 [vol. 14 of the *Obras completas* of Fray Bartolomé de las Casas].

Translations

J.M. Cohen, *The Four Voyages of Christopher Columbus*, Harmondsworth, 1952.
Robert H. Fuson, *The Log of Christopher Columbus*, Southampton: Ashford Press, 1987.
Cecil Jane, *The Voyages of Christopher Columbus*, London, 1930. [Revised and annotated by L.A. Vigneras with an appendix by R.A. Skelton, Hakluyt Society, Extra series, 38, London, 1960.]
Clements R. Markham, *The Journal of Christopher Columbus (during his first voyage, 1492-93), and Documents Relating to the Voyages of John Cabot and Gaspar Corte Real*, Hakluyt Society, London, 1893.
Samuel Eliot Morison, *Journals and Other Documents on the Life and Voyages of Christopher Columbus*, New York, 1963.

Studies

R.H. Major, trans. and ed., *Select Letters of Christopher Columbus, with other Original Documents, Relating to his Four Voyages to the New World*, 1st ed. Hakluyt Society 1st series, 2, London 1847; 2nd ed. Hakluyt Society, 1st series, 43, London, 1870. [Re-edited with additional material by Cecil Jane, 2 vols., Hakluyt Society, 2nd series, 65, 70, London, 1930, 1933; reprinted by Kraus Reprint Co., 1967]
Alain Milhou, *Colón y su mentalidad mesiánica en el ambiente franciscanista español*, Valladolid: Casa-Museo de Colón, 1983.
Samuel Eliot Morison, *Admiral of the Ocean Sea: A Life of Christopher Columbus*, 2 vols, Boston, 1942.
John Boyd Thacher, *Christopher Columbus: His Life, His Work, His Remains*, 3 vols., New York, 1903-04, reprinted Kraus, 1962.
Tzvetan Todorov, *La conquête de l'Amérique. La question de l'autre*, Paris: Seuil, 1982.
Henry Vignaud, *Toscanelli and Columbus*, New York, 1902.
Louis de Vorsey, Jr. and John Parker, *In the Wake of Columbus. Islands and Controversy*, Detroit: Wayne State University Press, 1985.

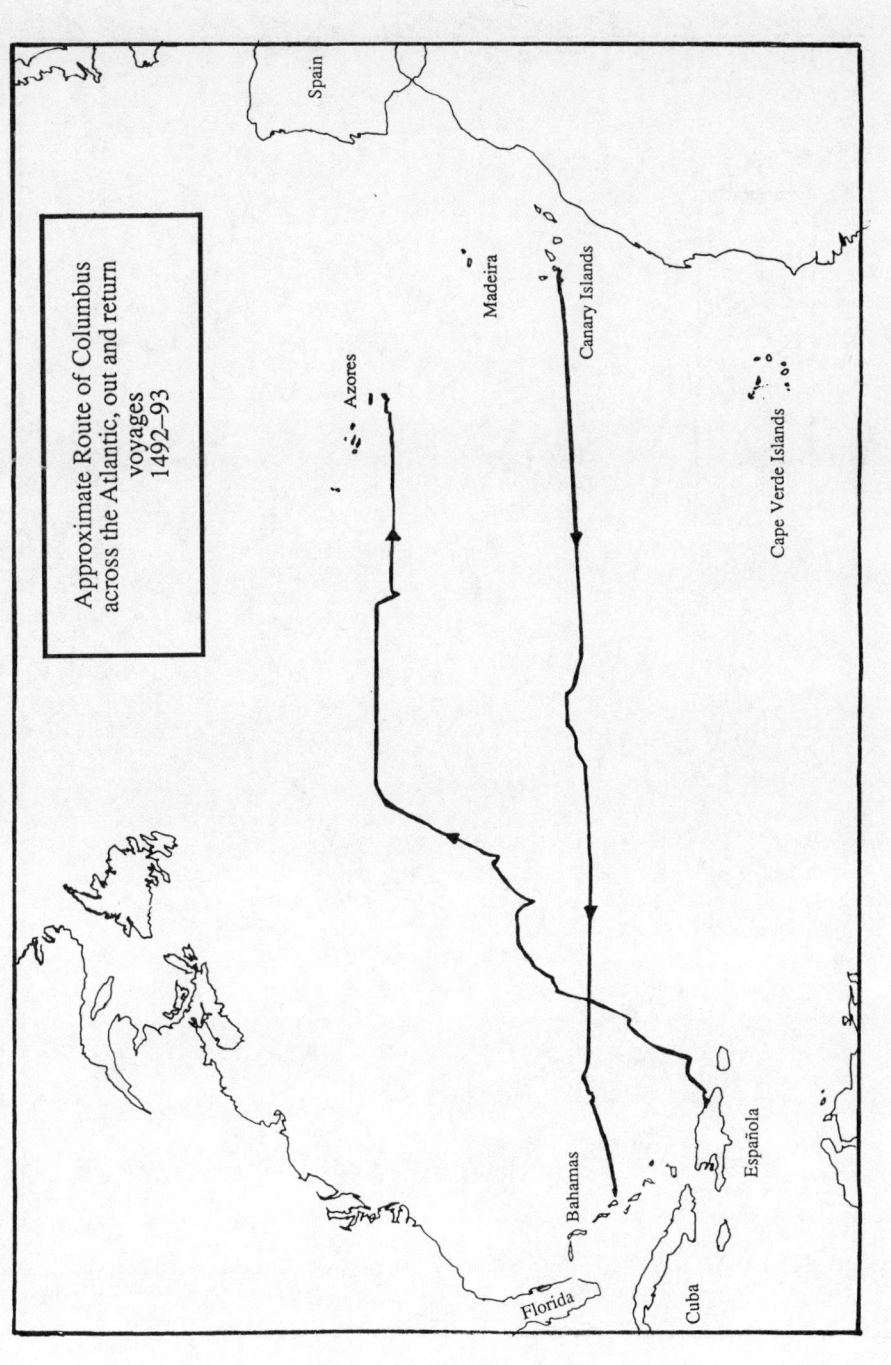

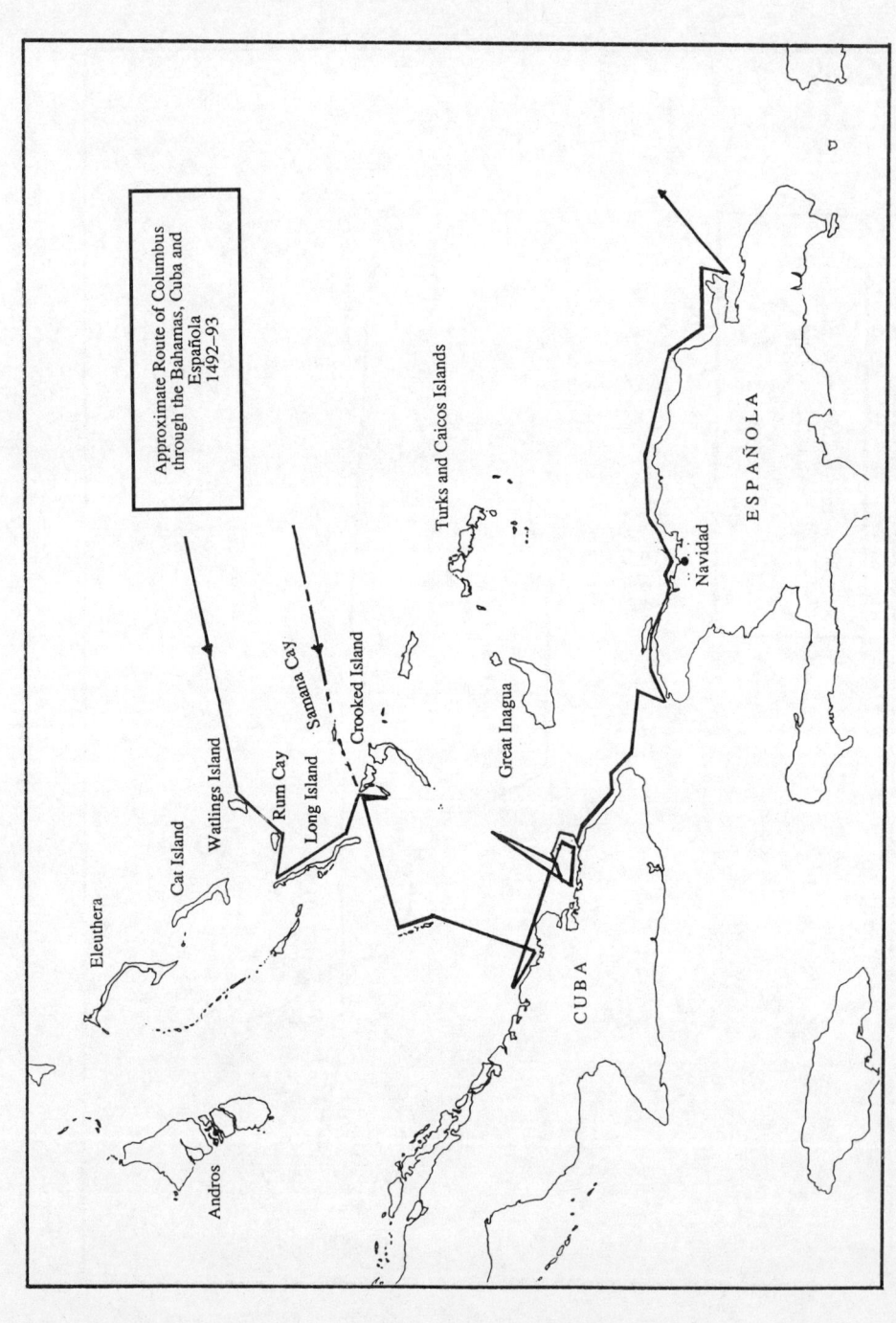

DIARIO DE COLON

JOURNAL OF COLUMBUS

Este es el primer viaje y las derrotas y camino que hizo el Almirante don Cristóval Colón quando descubrió las yndias, puesto sumariamente, sin el prólogo que hizo a los Reyes que va a la letra y comiença desta manera:

YN NOMINE DOMINI NOSTRI IHESU CHRISTI

Porque, cristianíssimos y muy altos y muy excelentes y muy poderosos príncipes Rey e Reyna de las Españas y de las yslas de la mar Nuestros Señores, este presente año de 1492, después de Vuestras Altezas aver dado fin a la guerra de los moros que reynavan en Europa y aver acabado la guerra en la muy grande çiudad de Granada adonde este presente año a dos días del mes de enero por fuerça de armas vide poner las vanderas reales de Vuestras Altezas en las torres de la Alfambra, que es la fortaleza de la dicha çiudad, y vide salir al rey moro a las puertas de la çiudad y besar las reales manos de Vuestras Altezas y del Príncipe mi Señor; y luego en aquel presente mes, por la información que yo avía dado a Vuestras Altezas de las tierras de Yndia y de un prínçipe que es llamado Gran Can, que quiere dezir en nuestro romançe Rey de los Reyes, cómo muchas vezes él y sus anteçessores avían enbiado a Roma a pedir doctores en nuestra sancta fe porque le enseñasen en ella y que nunca el Sancto Padre le avía proveydo, y se perdían tantos pueblos cayendo en ydolatrías e resçibiendo en sí sectas de perdiçión; y Vuestras Altezas como cathólicos cristianos y prínçipes amadores de la sancta fe cristiana y acreçentadores della, y enemigos de la secta de Mahoma y de todas ydolatrías y heregías, pensaron de embiarme a mí, Cristóval Colón, a las dichas partidas de Yndia para ver los dichos prínçipes y los pueblos y las tierras y la disposiçión dellas y de todo y la manera que se pudiera tener para la conversión dellas a nuestra sancta fe; y ordenaron que yo no fuese por tierra al Oriente por donde se costumbra de andar, salvo por el camino de Occidente, por donde hasta oy no sabemos por çierta fe que aya passado nadie. Así que después de aver echado fuera todos los judíos de todos vuestros reynos y señoríos, en el mismo mes de enero, mandaron Vuestras Altezas a mí que con armada suffiçiente me fuese a las dichas partidas de Yndia. Y para ello me hizieron grandes merçedes y me anobleçieron que dende en adelante yo me llamase Don y fuesse Almirante Mayor de la Mar Occéana y Visorey e Governador perpetuo de todas las yslas y tierra firme que yo descubriese y ganasse, y de aquí adelante se descubriesen y ganasen en la mar Occéano y así sucediese mi hijo mayor y él así de grado en grado para siempre jamás. Y partí yo de la çiudad de Granada a doze días del mes de mayo del mesmo año de 1492[1] *en sábado, y vine a la villa de Palos, que es puerto de*

---

[1] Margin: Quando salió despachado de la çiudad de Granada el Almirante Colón para yr a descubrir las yndias.

Journal of Columbus

This is the first voyage with the courses and route which the Admiral don Christopher Columbus took when he discovered the Indies, set down in an abbreviated form, except for the prologue to the Monarchs which is given in full and begins:

IN THE NAME OF OUR LORD JESUS CHRIST

Most Christian and most exalted and most excellent and most powerful Princes, King and Queen of the Spains and of the islands of the sea, Our Sovereigns: Forasmuch as in this present year of 1492, after Your Highnesses had brought to an end the war with the Moors who reigned in Europe and had concluded the war in the great city of Granada where this same year on the second day of January I saw Your Highnesses' royal banners placed by force of arms on the towers of the Alhambra, which is the fortress of the said city, and I saw the Moorish king come out to the city gates and kiss Your Highnesses' royal hands and those of My Lord the Prince;[1] and then in the same month from information which I had given Your Highnesses about the lands of India and a prince called the Great Khan, which means in our language King of Kings, and how he and his ancestors had many times sent to Rome for learned men to instruct him in our holy faith, and how the Holy Father had never provided them, and how so many people were being lost, falling into idolatry and embracing doctrines of perdition;[2] and Your Highnesses, as Catholic Christians and princes devoted to the holy Christian faith and the furtherance of its cause, and enemies of the sect of Mohammed and of all idolatry and heresy, resolved to send me, Christopher Columbus, to the said regions of India to see the said princes and the peoples and lands and determine the nature of them and of all other things, and the measures to be taken to convert them to our holy faith; and you ordered that I should not go by land to the East, which is the customary route, but by way of the West, a route which to this day we cannot be certain has been taken by anyone else: So then, after having expelled all the Jews from all your kingdoms and domains, in the same month of January,[3] Your Highnesses commanded me to take sufficient ships and sail to the said regions of India. And in consideration you granted me great favours and honoured me thenceforth with the title 'Don' and the rank of Admiral of the Ocean Sea and Viceroy and Governor in perpetuity of all the islands and mainland that I should discover and take possession of and which should hereafter be discovered and occupied in the Ocean Sea, and that my eldest son should succeed in turn, and so on from generation to generation for ever.[4] And I left the city of Granada on the twelfth day of May of the same year 1492, a Saturday, and came to the town of Palos, which is a seaport, where I

mar adonde yo armé tres navíos muy aptos para semejante fecho, y partí del dicho puerto muy abasteçido de muy muchos mantinimientos, y de mucha gente de la mar a tres días del mes de agosto del dicho año[1] en un viernes antes de la salida del sol con media ora, y llevé el camino de las yslas de Canaria de Vuestras Altezas que son en la dicha mar Occéana para de allí tomar mi derrota y navegar tanto que yo llegase a las yndias y dar la embaxada de Vuestras Altezas a aquellos prínçipes y complir lo que así me avían mandado. Y para esto pensé de escrevir todo este viaje muy puntualmente de día en día todo lo que yo hiziese y viese y passasse como adelante se veyrá. También, Señores Prínçipes, allende de escrevir cada noche lo que el día passare, y el día lo que la noche navegare, tengo propósito de hazer carta nueva de navegar; en la qual situaré toda la mar e tierras del mar Occéano en sus proprios lugares debaxo su viento y más componer un libro y poner todo por el semejante por pintura por latitud del equinocial y longitud del Occidente; y sobre todo cumple mucho que yo olvide el sueño y tiente mucho el navegar porque así cumple, las quales serán gran trabajo.

Viernes 3 de agosto

Partimos viernes 3 días de agosto de 1492 años de la barra de Saltés a las ocho oras. Anduvimos con fuerte virazón hasta el poner del sol hazia el sur sesenta millas que son 15 leguas; después al sudueste y al sur quarta del sudueste que era el camino para las Canarias.

El sábado 4º de agosto

Anduvieron al sudueste quarta del sur.

Domingo 5 de agosto

Anduvieron su vía entre día y noche más de quarenta leguas.

Lunes 6 de agosto

Saltó, o desencasóse el governario a la caravela Pinta, donde yva Martín Alonso Pinçón a lo que se creyó, o sospechó por industria de un Gómez Rascón y Cristóval Quintero, cuya era la caravela, porque le pesava yr aquel viaje; y dize el

---

[1] Margin: Quando partió el Almirante del puerto de Palos para su descubrimiento.

prepared three ships suitable for such an undertaking, and I set out from the said port well stocked with supplies and with many seamen[5] on the third day of August of the same year, a Friday, half an hour before sunrise, and I took the route to the Canary Islands, Your Highnesses' possessions in the said Ocean Sea, thence to set my course and navigate until I should reach the Indies, and deliver Your Highnesses' embassy to those princes and comply with what you had ordered.[6] And for this purpose I decided to write down the whole of this voyage in detail, day by day, everything that I should do and see and undergo, as will be seen in due course.[7] Furthermore, My Lords, besides writing down each night whatever the day should bring, and each day the course taken during the night, I propose to make a new navigational chart, on which I shall note all the sea and land in the Ocean Sea in their proper places with their bearings and also keep a book in which, in the same way, I shall record them by latitude from the equator and by longitude to the west; above all, I must give no thought to sleep, and must work diligently at my navigation, because such is my duty; all of which will require great effort.[8]

Friday 3 August

We left the bar of Saltés at eight o'clock on Friday 3 August 1492. We sailed with a strong sea breeze until sunset, S for 60 miles, which is 15 leagues, then SW and S by W, which was the route to the Canaries.[9]

Saturday 4 August

They sailed SW by S.

Sunday 5 August

They made more than 40 leagues during the day and night.

Monday 6 August

The rudder of the caravel Pinta, in which Martín Alonso Pinzón was sailing, came adrift or became dislodged and it was believed or suspected that this was the work of one Gómez Rascón and Cristóbal Quintero, the owner of the caravel, because he was annoyed at having to make the voyage; and the Admiral

Almirante que antes que partiesen avían hallado en çiertos reveses y grisquetas, como dizen, a los dichos. Vídose allí el Almirante en gran turbaçión por no poder ayudar a la dicha caravela sin su peligro y dize que alguna pena perdía con saber que Martín Alonso Pinçón era persona esforçada y de buen ingenio. En fin, anduvieron entre día y noche veynte y nueve leguas.

Martes 7 de agosto

Tornóse a saltar el governalle a la Pinta, y adobáronlo y anduvieron en demanda de la ysla de Lançarote, que es una de las yslas de Canaria, y anduvieron entre día y noche xxv leguas.

Miércoles 8 de agosto

Ovo entre los pilotos de las tres caravelas opiniones diversas dónde estavan y el Almirante salió más verdadero; y quisiera yr a Gran Canaria por dexar la caravela Pinta porque yva mal acondiçionada del governario y hazía agua, y quisiera tomar allí otra si la hallara; no pudieron tomarla aquel día.

Jueves 9 de agosto

Hasta el domingo en la noche no pudo el Almirante tomar la Gomera y Martín Alonso quedóse en aquella costa de Gran Canaria por mandado del Almirante porque no podía navegar. Después tornó el Almirante a Canaria[1] y adobaron muy bien la Pinta con mucho trabajo y diligençia del Almirante, de Martín Alonso y de los demás y al cabo vinieron a la Gomera. Vieron salir gran huego de la sierra de la ysla de Tenerife que es muy alta en gran manera. Hizieron la Pinta redonda porque era latina. Tornó a la Gomera domingo a dos de setiembre con la Pinta adobada. Dize el Almirante que juravan muchos hombres honrrados españoles que en la Gomera estavan con doña Inés Peraça, madre de Guillén Peraça que después fue el primer Conde de la Gomera, que eran vezinos de la ysla del Hierro, que cada año vían tierra al vueste de las Canarias, que es al poniente; y otros de la Gomera afirmavan otro tanto con juramento. Dize aquí el Almirante que se acuerda que estando en Portogal el año de 1484 vino uno de la ysla de la Madera al Rey a le pedir una caravela para yr a esta tierra que vía, el qual jurava que cada año la vía y siempre de una manera. Y

---

[1] Margin: o a Tenerife.

says that before they left they had been obstructive and always putting a spoke in, as they say.[10] The Admiral was very worried at not being able to assist the Pinta without danger to himself and says that his fears were somewhat allayed by the thought that Martín Alonso Pinzón was a courageous and resourceful man. They eventually made 29 leagues during the the day and night.

Tuesday 7 August

The Pinta's rudder again came adrift and they repaired it and made for the island of Lanzarote, which is one of the Canary Islands, and they made 25 leagues during the day and night.

Wednesday 8 August

There were differing opinions among the pilots of the three caravels about their position and the Admiral turned out to be nearest the truth. He wished to go to Gran Canaria to leave the caravel Pinta there because her rudder was in bad condition and she was taking in water, and he wished to obtain another if he could find one. They were unable to obtain one that day.

Thursday 9 August

The Admiral could not make Gomera until Sunday night and Martín Alonso remained off the coast of Gran Canaria at the Admiral's orders because he could not navigate. Then the Admiral returned to Canaria[11] and with painstaking effort from the Admiral, Martín Alonso and the rest the Pinta was put in very good repair and eventually they reached Gomera. They saw a great fire issuing from the peak of the island of Tenerife, which is extremely high.[12] They fitted the Pinta[13] with square sails because she was lateen-rigged. He returned to Gomera on Sunday 2 September with the Pinta repaired. The Admiral says that many trustworthy Spaniards from the neighbouring island of Ferro who were on Gomera with doña Inés Peraza, mother of Guillén Peraza, later to be the first Count of Gomera, swore that every year they could see land to the W of the Canaries, that is, under the setting sun. And others from Gomera affirmed as much on oath. Here the Admiral says he remembers that, when in Portugal in 1484, a man from the island of Madeira[14] came to the King to ask for a caravel to go to this land which he had sighted, and which he swore he sighted every year and always in the same way.[15] And he also says that he remembers the same

también dize que se acuerda que lo mismo dezían en las yslas de los Açores, y todos estos en una derrota y en una manera de señal y en una grandeza. Tomada, pues, agua y leña y carnes y lo demás que tenían los hombres que dexó en la Gomera el Almirante quando fue a la ysla de Canaria a adobar la caravela Pinta, finalmente se hizo a la vela de la dicha ysla de la Gomera con sus tres caravelas jueves a seys días de setiembre.

Jueves 6 de setiembre

Partió aquel día por la mañana del puerto de la Gomera y tomó la buelta para yr su viaje; y supo el Almirante de una caravela que venía de la ysla del Hierro que andavan por allí tres caravelas de Portugal para lo tomar; devía de ser de enbidia que el Rey tenía por averse ydo a Castilla. Y anduvo todo aquel día y noche en calma, y a la mañana se halló entre la Gomera y Tenerife.

Viernes 7 de setiembre

Todo el viernes y el sábado hasta tres oras de noche estuvo en calmas.

Sábado 8 de setiembre

Tres oras de noche sábado començó a ventar nordeste y tomó su vía y camino al güeste; tuvo mucha mar por proa que le estorvava el camino y andaría aquel día nueve leguas con su noche.

Domingo 9 de setiembre

Anduvo aquel día 15 leguas y acordó contar menos de las que andava porque si el viaje fuese luengo no se espantase y desmayase la gente; en la noche anduvo çiento y veynte millas a diez millas por ora que son 30 leguas. Los marineros governavan mal decayendo sobre la quarta del norueste y aun a la media partida, sobre lo qual les riñó el Almirante muchas vezes.

Lunes 10º de setiembre

En aquel día con su noche anduvo sesenta leguas a diez millas por ora que son dos leguas y media pero no contava sino quarenta y ocho leguas porque no se asombrase la gente si el viaje fuese largo.

thing being said in the Azores, with everyone seeing land in the same direction with the same aspect and the same size. Having taken on water and firewood and meat and the other things obtained by the men the Admiral left on Gomera when he went to the island of Canaria to repair the caravel Pinta, he finally set sail from the said island of Gomera with his three caravels on Thursday 6 September.

Thursday 6 September

He left that day in the morning from the port of Gomera and rounded the island to proceed on his voyage. The Admiral learned from a caravel which came from the island of Ferro that three caravels from Portugal were cruising in the area with the intention of detaining him; it must have been due to the envy the King felt at his having gone to Castile. He was becalmed that day and night and in the morning lay between Gomera and Tenerife.

Friday 7 September

He was becalmed all of Friday and until three o'clock on Saturday morning.

Saturday 8 September

At three o'clock on Saturday morning the wind began to blow from the NE and he set his course and sailed W; he had heavy sea over the bows which made progress hard going and he made about 9 leagues that day and night.

Sunday 9 September

He made 15 leagues that day and decided to reckon fewer than he was making so that if the journey were long the men should not be afraid and discouraged.[16] That night he made 120 miles at 10 miles an hour, which is 30 leagues. The sailors steered badly, drifting W by N and even WNW. The Admiral several times rebuked them about this.

Monday 10 September

He made 60 leagues that day and night at 10 miles an hour, which is 2½ leagues, but he only reckoned 48 leagues so that the men should not be afraid if the journey were long.

Martes 11º de setiembre

Aquel día navegaron a su vía que era el güeste y anduvieron 20 leguas y más y vieron un gran troço de mástel de nao de çiento y veynte toneles y no lo pudieron tomar. La noche anduvieron çerca de veynte leguas y contó no más de diez y seys por la causa dicha.

Miércoles 12 de setiembre

Aquel día yendo su vía anduvieron en noche y día 33 leguas, contando menos por la dicha causa.

Jueves 13 de setiembre

Aquel día con su noche yendo a su vía que era el güeste anduvieron xxxiii leguas y contava tres o quatro menos. Las corrientes le eran contrarias. En este día al comienço de la noche las agujas noruesteavan y a la mañana nordesteavan algún tanto.

Viernes 14 de setiembre

Navegaron aquel día su camino al güeste con su noche y anduvieron xx leguas; contó alguna menos. Aquí dixeron los de la caravela Niña que avían visto un garxao y un rabo de junco, y estas aves nunca se apartan de tierra quando más xxv leguas.

Sábado 15 de setiembre

Navegó aquel día con su noche xxvii leguas su camino al güeste y algunas más, y en esta noche al principio della vieron caer del çielo un maravilloso ramo de huego en la mar lexos dellos quatro o çinco leguas.

Domingo 16 de setiembre

Navegó aquel día y la noche a su camino el güeste; andarían xxxviiii leguas pero no contó sino 36. Tuvo aquel día algunos ñublados; lloviznó. Dize aquí el Almirante que oy y siempre de allí adelante hallaron ayres temperatíssimos, que era plazer grande el gusto de las mañanas, que no faltava sino oyr rruyseñores, dize él; y era el tiempo como por abril en el Andaluzía. Aquí començaron a ver

Tuesday 11 September

That day they steered their course W and made 20 leagues or more and saw a great portion of the mast of a ship of 120 tons, but they could not catch it. During the night they made nearly 20 leagues and he counted no more than 16 for the reason stated.

Wednesday 12 September

Proceeding on their course that day they made 33 leagues during the night and day, reckoning fewer for the reason given.

Thursday 13 September

That day and night, following their course W, they made 33 leagues and he reckoned three or four fewer. The currents were against him. On this day at nightfall the needles pointed NW and in the morning somewhat to the NE.[17]

Friday 14 September

That day and night they steered their course W and made 20 leagues; he reckoned somewhat fewer. Here the crew of the caravel Niña said that they had seen a tern and a reed-tail, and these birds never go more than 25 leagues from land.

Saturday 15 September

That day and night he steered his course W making a little over 27 leagues, and at nightfall they saw a marvellous sheet of fire fall from the sky into the sea 4 or 5 leagues away.

Sunday 16 September

That day and night he steered his course W; they made about 39 leagues but he reckoned no more than 36. He had an overcast sky and some drizzle. Here the Admiral says that today and thenceforth they always encountered the most gentle breezes, that the enjoyment of the mornings was a great pleasure, that all they needed was to hear nightingales, he says; and the weather was like April in

muchas manadas de yerva muy verde que poco avía (según le pareçía) que se avía desapegado de tierra por la qual todos juzgavan que estavan çerca de alguna ysla pero no de tierra firme segúnd el Almirante, que dize: *porque la tierra firme hago más adelante.*

Lunes 17 de setiembre

Navegó a su camino al güeste y andarían en día y noche çinquenta leguas y más; no asentó sino 47. Ayudávales la corriente. Vieron muchas yerva[s] y muy a menudo y era yerva de peñas y venían las yerva[s] de hazia poniente. Juzgavan estar çerca de tierra. Tomaron los pilotos el norte marcándolo y hallaron que las agujas noruesteavan una gran quarta, y temían los marineros y estavan penados y no dezían de qué. Cognosciólo el Almirante, mandó que tornasen a marcar el norte en amaneçiendo y hallaron que estavan buenas las agujas. La causa fue porque la estrella que pareçe haze movimiento y no las agujas. En amaneçiendo aquel lunes vieron muchas más yervas y que pareçían yervas de ríos en las quales hallaron un cangrejo bivo el qual guardó el Almirante, y dize que aquellas fueron señales çiertas de tierra, porque no se hallan ochenta leguas de tierra. El agua de la mar hallavan menos salada desde que salieron de las Canarias. Los ayres siempre más suaves. Yvan muy alegres todos y los navíos quien más podía andar andava por ver primero tierra. Vieron muchas toninas y los de la Niña mataron una. Dize aquí el Almirante que aquellas señales eran del poniente, *donde espero en aquel alto Dios en cuyas manos están todas las victorias que muy presto nos dará tierra.* En aquella mañana dize que vido una ave blanca que se llama rabo de junco que no suele dormir en la mar.

Martes 18 de setiembre

Navegó aquel día con su noche y andarían más de çinquenta y çinco leguas; pero no asentó sino 48. Llevava en todos estos días mar muy bonanço como en el río de Sevilla. Este día Martín Alonso con la Pinta que era gran velera no esperó porque dixo al Almirante desde su caravela que avía visto gran multitud de aves yr hazia el poniente y que aquella noche esperava ver tierra y por eso andava tanto. Apareçió a la parte del norte una gran çerrazón que es señal de estar sobre la tierra.

Miércoles 19 de setiembre

Navegó su camino y entre día y noche andaría xxv leguas porque tuvieron calma. Escrivió xxii. Este día a las diez oras vino a la nao un alcatraz y a la tarde vieron

Andalusia. Here they began to see great clumps of deep green seaweed which (so it seemed to him) had only recently been torn from land.[18] On account of this they all thought that they were near some island, but not the mainland, according to the Admiral, who says: *Because I make the mainland further on*.

Monday 17 September

He steered his course W and they would have made more than 50 leagues during the day and night; he only reckoned 47. The current was assisting them. They saw a good deal of weed, very frequently, and it was a rock weed and came from westward. They reckoned they were near land. The pilots took the north and marked it and found that the compass needles veered NW by a full point, and the sailors were fearful and anxious and did not say why. The Admiral sensed this and ordered them to mark north again at dawn and they found that the compasses were true. The reason was that it is the star that appears to move, not the needles. At dawn on that Monday they saw much more weed which seemed to be a river weed and in which they found a live crab which the Admiral kept. He says that these were sure signs of land, because they are not found 80 leagues from land. They found the seawater less salty since they had left the Canaries and the breezes more and more gentle. Everyone was very happy and the ships sailed as fast as they could to be the first to sight land. They saw many tunny fish and the crew of the Niña killed one. Here the Admiral says that those indications came from the west, *where I hope that Almighty God, in whose hands all victories are found, will soon grant us land*. That morning he says that he saw a white bird called a reed-tail which does not usually sleep on the sea.

Tuesday 18 September

He proceeded on his course during that day and night and they made about 55 leagues; but he only put down 48. All those days he had a very calm sea, like the river at Seville. On this day Martín Alonso with the Pinta, which was very fast, did not wait, for he told the Admiral from his caravel that he had seen a large flock of birds flying westward and that he expected to sight land that night and for that reason was making such speed. A mass of dark cloud appeared to the N which is a sign that one is near land.

Wednesday 19 September

He steered his course and during the day and night he made about 25 leagues, since it was calm. He put down 22. On this day at ten o'clock a gannet came to

otro, que no suelen apartarse xx leguas de tierra. Vinieron unos llovizneros sin viento, lo que es señal çierta de tierra. No quiso detenerse barloventeando el Almirante para averiguar si avía tierra, más de que tuvo por çierto que a la vanda del norte y del sur avía algunas yslas, como en la verdad lo estavan, y él yva por medio dellas; porque su voluntad era de seguir adelante hasta las yndias, *y el tiempo es bueno porque, plaziendo a Dios, a la buelta todo se vería.* Estas son sus palabras. Aquí descubrieron sus puntos los pilotos: el de la Niña se hallava de las Canaria[s] 440 leguas; el de la Pinta, 420; el de la donde yva el Almirante, 400 justas.

Jueves 20 de setiembre

Navegó este día al güeste quarta del norueste y a la media partida, porque se mudaron muchos vientos con la calma que avía. Andarían hasta siete o ocho leguas. Vinieron a la nao dos alcatraçes y después otro, que fue señal de estar çerca de tierra, y vieron mucha yerva aunque el día passado no avían visto della. Tomaron un páxaro con la mano que era como garjao; era páxaro de río y no de mar; los pies tenía como gaviota. Vinieron al navío en amaneçiendo dos o tres paxaritos de tierra cantando y después antes del sol salido desapareçieron. Después vino un alcatraz; venía del güesnorueste; yva al sueste, que era señal que dexava la tierra al güesnorueste, porque estas aves duermen en tierra y por la mañana van a la mar a buscar su vida y no se alexan xx leguas.

Viernes 21 de setiembre

Aquel día fue todo lo más calma y después algún viento. Andarían entre día y noche, dello a la vía y dello no, hasta 13 leguas. En amaneçiendo hallaron tanta yerva que pareçía ser la mar quajada della y venía del güeste. Vieron un alcatraz. La mar muy llana como un río y los ayres los mejores del mundo. Vieron una vallena que es señal que estavan çerca de tierra porque sienpre andan çerca.

Sábado 22 de setiembre

Navegó al güesnorueste más o menos acostándose a una y a otra parte; andarían xxx leguas. No vían quasi yerva. Vieron unas pardelas y otra ave. Dize aquí el Almirante: *Mucho me fue neçessario este viento contrario, porque mi gente andavan muy estimulados, que pensavan que no ventavan en estos mares vientos*

the ship and they saw another during the afternoon; they do not usually go 20 leagues from land. There was some drizzle without wind, which is a sure sign of land. The Admiral did not wish to delay by beating to windward in order to see if there were any land, especially as he was certain that to the N and S there were some islands, as in fact there were,[19] and he was sailing between them, because his intention was to press on to the Indies, *and the weather is fine, because, God willing, everything could be seen on the way back.* These are his words. Here the pilots compared their reckonings: the Niña's pilot made it 440 leagues from the Canaries; the Pinta's, 420; and the Admiral's pilot, 400 exactly.

Thursday 20 September

On this day he steered W by N and WNW because, with the prevailing calm, the winds were variable. They made about 7 or 8 leagues. Two gannets came to the ship and then another, which was a sign that they were near land, and they saw a lot of seaweed although on the previous day they had not seen any. They caught by hand a bird which was like tern; it was a river bird and not a sea bird; it had feet like a gull's. At dawn two or three small land birds came singing to the ship and then disappeared before the sun got up. Then came a gannet; it came from the WNW and was flying SE, which was an indication that it was leaving land to the WNW, because these birds sleep on land and in the morning fly out to sea to look for food and do not go further than 20 leagues.

Friday 21 September

That day was mostly calm and later there was some wind. They made about 13 leagues during the day and night, some of it on course, some not. At dawn they found so much seaweed that the sea seemed clotted with it and it came from the W. They saw a gannet. The sea was as flat as a river and the breezes the best in the world. They saw a whale, which is a sign that they were near land because they always stay close by.

Saturday 22 September

He steered WNW more or less, sometimes inclining one way, sometimes another. They made about 30 leagues. They saw hardly any weed. They saw some petrels and another bird. At this point the Admiral says: *I was in great need of this head wind because my men were very agitated and thought that no winds*

para bolver a España.[1] Por un pedaço de día no ovo yerva; después muy espessa.

Domingo 23 de setiembre

Navegó al norueste y a las vezes a la quarta del norte y a las vezes a su camino que era el güeste y andaría hasta xxvii leguas. Vieron una tórtola, y un alcatraz y otro paxarito de río y otras aves blancas. Las yervas eran muchas y hallavan cangrejos en ellas. Como la mar estuviese mansa y llana,[2] murmurava la gente diziendo que pues por allí no avía mar grande que nunca ventaría para bolver a España. Pero después alçóse mucho la mar y sin viento que los asombrava; por lo qual dize aquí el Almirante: *Así que muy neçessario me fue la mar alta que no pareçió salvo el tiempo de los judíos quando salieron de Egipto contra Moysén que los sacava de captiverio.*[3]

Lunes 24 de setiembre

Navegó a su camino al güeste día y noche y andarían quatorze leguas y media; contó doze. Vino al navío un alcatraz y vieron muchas pardelas.

Martes 25 de setiembre

Este día ovo mucha calma y después ventó, y fueron su camino al güeste hasta la noche. Yva hablando el Almirante con Martín Alonso Pinçón, capitán de la otra caravela Pinta, sobre una carta[4] que le avía enbiado tres días avía a la caravela donde según pareçe tenía pintadas el Almirante ciertas yslas por aquella mar. Y dezía el Martín Alonso que estavan en aquella comarca y respondía el Almirante que así le pareçía a él; pero puesto que no oviesen dado con ellas lo devía[n] de aver causado las corrientes que siempre avían echado los navíos al nordeste y que no avían andado tanto como los pilotos dezían. Y estando en esto díxole el Almirante que le enbiase la carta dicha, y enbiada con alguna cuerda començó el Almirante a cartear en ella con su piloto y marineros. Al sol puesto, subió el Martín Alonso en la popa de su navío y con mucha alegría llamó al

---

[1] Margin: Aquí comiençan a murmurar la gente del largo viaje, etc.
[2] Margin: Murmurava la gente.
[3] Margin: No[ta].
[4] Margin: Nota sobre esta carta.

blew on these seas that would get them back to Spain.[20] For part of the day there was no weed; then it was very thick.

Sunday 23 September

He steered NW and at times NW by N and at others kept his course which was W, and he made about 27 leagues. They saw a pigeon and a gannet and another river bird and other white birds. There was much seaweed and they found crabs in it. As the sea was so calm and flat the men began muttering, saying that as there was no heavy sea in that area, there would never be enough wind to return to Spain. But later the sea stirred a great deal and without wind, which surprised them; for which reason the Admiral says at this point: *So I had great need of that high sea the like of which has not been seen since the time of the Jews when they were leaving Egypt and murmured against Moses who was delivering them from captivity.*[21]

Monday 24 September

He steered his course W during the day and night and they made about 14½ leagues; he reckoned 12. A gannet came to the ship and they saw many petrels.

Tuesday 25 September

On this day it was very calm and then the wind got up and they proceeded on course to the W until nightfall. The Admiral had talks with Martín Alonso Pinzón, captain of the other caravel Pinta, about a chart[22] which he had sent to the caravel three days previously and on which it seems the Admiral had certain islands depicted in that area of the sea. And Martín Alonso said that they were in that vicinity and the Admiral replied that he thought the same; but since they had not come across them the reason must have been that the currents had been continually pushing the ships NE and that they had not sailed as far as the pilots said. Such being the case, the Admiral asked him to send back the said chart and when it had been sent back on a line the Admiral began to plot a course on it with his pilot and crew. At sunset Martín Alonso went up to the poop of his ship and with great joy called to the Admiral, claiming a reward because he could see

Almirante pidiéndole albriçias que vía tierra.[1] Y quando se lo oyó dezir con afirmación el Almirante dize que se echó a dar gracias a Nuestro Señor de rodillas, y el Martín Alonso dezía *Gloria in excelsis Deo* con su gente. Lo mismo hizo la gente del Almirante y los de la Niña. Subiéronse todos sobre el mástel y en la xarçia y todos affirmaron que era tierra y al Almirante así pareçió y que avría a ella 25 leguas. Estuvieron hasta la noche affirmando todos ser tierra. Mandó el Almirante dexar su camino que era el güeste y que fuesen todos al sudueste adonde avía pareçido la tierra. Avrían andado aquel día al güeste 4 leguas y media y en la noche al sud[u]este 17 leguas, que son xxi, puesto que dezía a la gente 13 leguas, porque siempre fingía a la gente que hazía poco camino porque no les pareçiese largo; por manera que escrivió por dos caminos aquel viaje: el menor fue el fingido; y el mayor el verdadero. Anduvo la mar muy llana por lo qual se echaron a nadar muchos marineros. Vieron muchos dorados y otros peçes.

Miércoles 26 de setiembre

Navegó a su camino al güeste hasta después de mediodía; de allí fueron al sudueste hasta cognosçer que lo que dezían que avía sido tierra no lo era sino çielo. Anduvieron día y noche 31 leguas, y contó a la gente 24. La mar era como un río, los ayres dulçes y suavíssimos.

Jueves 27 de setiembre

Navegó a su vía al güeste. Anduvo entre día y noche 24 leguas; contó a la gente 20 leguas. Vinieron muchos dorados; mataron uno. Vieron un rabo de junco.

Viernes 28 de setiembre

Navegó a su camino al güeste. Anduvieron día y noche con calmas 14 leguas; contó treze. Hallaron poca yerva; tomaron dos peçes dorados, y en los otros navíos más.

Sábado 29 de setiembre

Navegó a su camino al güeste. Anduvieron 24 leguas; contó a la gente xxi. Por calmas que tuvieron anduvieron entre día y noche poco. Vieron un ave que se

---

[1] Margin: Alegrón de tierra por Martín Alonso pero no lo era.

land. And when he heard it said so positively, the Admiral says that he fell on his knees to give thanks to God and Martín Alonso said the *Gloria in excelsis Deo* with his men. The Admiral's crew and that of the Niña did the same. They all climbed the mast and the rigging and they all said that it was definitely land and the Admiral thought so too and that it must be about 25 leagues away. They were all still positive that it was land until the evening. The Admiral ordered them to change course from W to SW where the land had appeared. They sailed that day about 4 and a half leagues W and in the evening 17 leagues SW, which is 21, although he told the men 13 leagues because he always pretended to the men that he was making little headway so that the voyage should not seem long; so that he kept two reckonings for that voyage: the shorter was the false one and the longer was the true one.[23] They had very calm sea and many sailors went swimming. They saw many dorados and other fish.

Wednesday 26 September

He steered his course W until after midday; then they went SW until they realized that what they had said was land was merely sky. They sailed 31 leagues day and night, and he told the men 24. The sea was like a river, the breezes sweet and very gentle.

Thursday 27 September

He steered his course W. He sailed 24 leagues during the day and night; he told the men 20 leagues. Many dorados came along and they killed one. They saw a reed-tail.

Friday 28 September

He steered his course W. They made 14 leagues day and night between periods of calm; he reckoned 13. They saw little weed; they caught two dorados, and the crew of the other ships caught more.

Saturday 29 September

He steered his course W. They made 24 leagues; he told the men 21. Because they were becalmed they made little progress during the day and night. They saw

llama rabiforçado que haze gomitar a los alcatraçes lo que comen para comerlo ella y no se mantiene de otra cosa. Es ave de la mar pero no posa en la mar ni se aparta de tierra 20 leguas. Ay destas muchas en las yslas de Cabo Verde. Después vieron dos alcatraces. Los ayres eran muy dulçes y sabrosos que dizque no faltava sino oyr el ruyseñor y la mar llana como un río. Pareçieron después en tres vezes tres alcatraçes y un forçado. Vieron mucha yerva.

Domingo 30 de setiembre

Navegó su camino al güeste. Anduvo entre día y noche por las calmas 14 leguas; contó onze. Vinieron al navío quatro rabos de junco que es gran señal de tierra; porque tantas aves de una naturaleza juntas es señal que no andan desmandadas ni perdidas. Viéronse quatro alcatraçes en dos vezes; yerva mucha. Nota que *las estrellas que se llaman las Guardias quando anocheçe están junto al braço de la parte del poniente; y quando amaneçe están en la línea debaxo del braço al nordeste, que pareçe que en toda la noche no andan salvo tres líneas, que son 9 oras, y esto cada noche*; esto dize aquí el Almirante. También en anocheçiendo las agujas noruestean una quarta,[1] y en amaneçiendo están con la estrella justo. Por lo qual pareçe que la estrella haze movimiento, como las otras estrellas; y las agujas piden siempre la verdad.

Lunes 1º de otubre

Navegó su camino al güeste. Anduvieron 25 leguas; contó a la gente 20 leguas. Tuvieron grande aguaçero. El piloto del Almirante tenía oy en amaneçiendo que avían andado desde la ysla del Hierro hasta aquí 578 leguas al güeste. La cuenta menor que el Almirante mostrava a la gente eran 584, pero la verdadera que el Almirante juzgava y guardava eran 707.

Martes 2 de otubre

Navegó a su camino al güeste noche y día 39 leguas; contó a la gente obra de 30 leguas. *La mar llana y buena siempre, a Dios muchas graçias sean dadas*, dixo aquí el Almirante. Yerva venía de leste a güeste por el contrario de lo que solía. Pareçieron muchos peçes; matóse uno. Vieron un[a] ave blanca que pareçía gaviota.

---

[1] Margin: No[ta].

a bird called a frigate-bird which makes gannets regurgitate what they have eaten and then eats the vomit itself and lives off nothing else. It is a sea bird but does not settle on the sea nor does it fly more than 20 leagues from land. There are many of these in the Cape Verde islands. Later they saw two gannets. The breezes were very gentle and sweet and he says that all that was missing was the sound of the nightingale, and the sea was as flat as a river. On three occasions there appeared three gannets and a frigate-bird. They saw a lot of weed.

Sunday 30 September

He steered his course W. He made 14 leagues during the day and night between periods of calm; he reckoned 11. Four reed-tails came to the ship, which is a sure sign of land, because so many birds together is a sign that they had not strayed or were lost. Four gannets were seen on two occasions; much weed. He notes that *when night falls the stars which are called the Guards are beside the western arm; and at sunrise they are on a line below the arm to the NE, so that it appears that throughout the night they only move three lines, that is 9 hours, and this is the case every night;*[24] this is what the Admiral says at this point. Furthermore, at nightfall the compass needles point a quarter NW, and at sunrise they are right on the star. From which it seems that the star appears to move, like the other stars;[25] and the needles always seek the truth.

Monday 1 October

He steered his course W. They made 25 leagues; he told the men 20 leagues. They had a downpour. This morning at dawn the Admiral's pilot held that they had made 578 leagues W from the island of Ferro. The Admiral's shorter reckoning which he showed to the men was 584, but the true one which the Admiral calculated and kept to himself was 707.

Tuesday 2 October

He steered his course W night and day for 39 leagues; he told the men a matter of 30 leagues. *The sea is still good and flat, thanks be to God*, the Admiral says at this point. The weed floated from E to W contrary to what it usually did. Many fish appeared; one was killed. They saw a white bird which looked like a seagull.

Miércoles 3 de otubre

Navegó su vía ordinaria. Anduvieron 47 leguas; contó a la gente 40 leguas. Aparecieron pardelas; yerva mucha, alguna muy vieja y otra muy fresca y traya como fruta. No vieron aves algunas y creya el Almirante que le quedavan atrás las yslas que traya pintadas en su carta.[1] Dize aquí el Almirante que no se quiso detener barloventeando la semana passada y estos días que vía tantas señales de tierra aunque tenía noticia de çiertas yslas en aquella comarca, por no se detener, pues su fin era passar a las yndias y si se detuviera dize él que no fuera buen seso.

Jueves 4º de otubre

Navegó a su camino al güeste. Anduvieron entre día y noche 63 leguas; contó a la gente 46 leguas. Vinieron al navío más de quarenta pardelas juntos y dos alcatraçes y al uno dio una pedrada un moço de la caravela. Vino a la nao un rabiforçado y una blanca como gaviota.

Viernes 5º de otubre

Navegó a su camino. Andarían onze millas por ora; por noche y día andarían 57 leguas porque afloxó la noche algo el viento. Contó a su gente 45. *La mar bonança y llana, a Dios*, dize, *muchas gracias sean dadas*. El ayre muy dulçe y temprado; yerva ninguna; aves, pardelas muchas; peces golondrinos volaron en la nao muchos.

Sábado 6 de otubre

Navegó su camino al vueste o güeste que es lo mismo. Anduvieron 40 leguas entre día y noche; contó a la gente 33 leguas. Esta noche dixo Martín Alonso que sería bien navegar a la quarta del güeste a la parte del sudueste; y al Almirante pareçió que no. Dezía esto Martín Alonso por la ysla de Çipango, y el Almirante vía que si la erravan que no pudieran tan presto tomar tierra, y que era mejor una vez yr a la tierra firme y después a las yslas.

---

[1] Margin: No[ta].

Wednesday 3 October

He steered his usual course. They made 47 leagues; he told the men 40 leagues. Some petrels appeared; much weed, some of it old and some very fresh, with what looked like fruit.[26] They saw no birds at all and the Admiral believed that they had left behind the islands depicted on his chart. Here the Admiral says that he did not wish to delay by beating to windward during the past week and on those days when he saw so many signs of land although he knew of certain islands in the vicinity, so as not to delay matters, because his aim was to reach the Indies and he says that if he were to have delayed it would not have been very sensible.

Thursday 4 October

He steered his course W. They made 63 leagues during the day and night; he told the men 46 leagues. More than forty petrels came to the ship in a flock, and two gannets, and one of the ship's boys hit one with a stone. A reed-tail came to the ship and a white bird like a seagull.

Friday 5 October

He proceeded on his course. They made about 11 miles an hour; during the night and day they made about 57 leagues because during the night the wind eased somewhat. He told the men 45. *The sea is beautifully calm*, he says, *thanks be to God*. The air very sweet and temperate; no weed; birds, many petrels; many flying fish flew onto the ship.

Saturday 6 October

He steered his course W.[27] They made 40 leagues during the day and night; he told the men 33 leagues. That night Martín Alonso said that it would be a good idea to steer SW by W; and the Admiral thought not. Martín Alonso suggested this with the island of Cipangu[28] in mind, and as the Admiral saw it, if they missed it, they would not be able to reach land so quickly, and that it was better to make for the mainland first and then for the islands.

Domingo 7 de otubre

Navegó a su camino, el güeste. Anduvieron 12 millas por ora dos oras y después 8 millas por ora y andaría hasta una ora de sol 23 leguas; contó a la gente 18. En este día al levantar del sol la caravela Niña que yva delante por ser velera y andavan quien más podía por ver primero tierra por gozar de la merced que los Reyes a quien primero la viese avía[n] prometido, levantó una vandera en el topo del mástel y tyró una lombarda por señal que vían tierra, porque así lo avía ordenado el Almirante. Tenía también ordenado que al salir del sol y al ponerse se juntasen todos los navíos con él, porque estos dos tiempos son más proprios para que los humores den más lugar a ver más lexos. Como en la tarde no viesen tierra la que pensavan los de la caravela Niña que avían visto, y porque passavan gran multitud de aves de la parte del norte al sudueste, por lo qual era de creer que se yvan a dormir a tierra o huyan quiçá del invierno que en las tierras de donde venían devía de querer venir, porque sabía el Almirante que las más de las yslas que tienen los Portugueses, por las aves las descubrieron; por esto el Almirante acordó dexar el camino del güeste, y pone la proa hazia güesu[du]este con determinación de andar dos días por aquella vía. Esto començó antes una ora del sol puesto. Andaría en toda la noche obra de çinco leguas, y xxiii del día. Fueron por todas veynte y ocho leguas noche y día.

Lunes 8 de otubre

Navegó al güesudueste y andarían entre día y noche onze leguas y media o doze y a ratos parece que anduvieron en la noche quinze millas por ora, si no está mentirosa la letra. Tuvieron la mar *como el río de Sevilla, gracias a Dios*, dize el Almirante; *los ayres muy dulces, como en abril en Sevilla, que es plazer estar a ellos, tan olorosos son*. Pareció la yerva muy fresca; muchos paxaritos de campo, y tomaron uno, que yvan huyendo al sudueste; grajaos y ánades y un alcatraz.

Martes 9 de otubre

Navegó al sudueste. Anduvo 5 leguas; mudóse el viento y corrió al güeste quarta al norueste y anduvo 4 leguas. Después con todas xi leguas de día y a la noche xx leguas y media. Contó a la gente 17 leguas. Toda la noche oyeron passar páxaros.

Sunday 7 October

He steered his course W. They made 12 miles an hour for two hours and then 8 miles an hour and he made about 23 leagues by an hour after sunrise; he told the men 18. On this day at sunrise the caravel Niña, which went on ahead as she was fast and as they were all sailing for all their worth to be the first to sight land and claim the reward which the Monarchs had promised to the first person to sight it, ran a flag up to the mast-head and fired a lombard shot as a sign that they could see land, because that is what the Admiral had ordered. He had also ordered that at sunrise and sunset all the ships should join him because at those times the atmosphere was such as to allow them to see furthest. Since in the evening the crew of the Niña could not see the land which they thought they had seen, and because a large flock of birds passed from the N to the SW, which led them to believe that they were going to roost on land or were perhaps fleeing the winter which was approaching in the lands from which they came, and because the Admiral knew that most of the islands which the Portuguese held had been discovered because of birds; for these reasons the Admiral agreed to leave the course to the W and steer WSW with the intention of following that course for two days. This he began an hour before sunset. Throughout the night he made a matter of 5 leagues, and 23 in the day, a total of 28 during the night and day.

Monday 8 October

He steered WSW and they made about 11 and a half or 12 leagues during the day and night and it seems that at times during the night they were making 15 miles an hour, if the text is to be believed. They had a sea *like the river at Seville, thanks be to God*, the Admiral says; *the sweetest of breezes, like April in Seville, such that it is a pleasure to be in them, so fragrant are they*. The seaweed seemed very fresh; many land birds fleeing to the SW - they caught one; terns and ducks and a gannet.

Tuesday 9 October

He steered SW. He made 5 leagues; the wind veered and he ran W by N for 4 leagues; thereafter, a total of 11 leagues by day and 20 and a half by night. He told the men 17 leagues. All night they heard birds passing.

Miércoles 10 de otubre

Navegó al güesudueste; anduvieron a diez millas por ora y a ratos 12 y algún rato a 7 y entre día y noche 59 leguas; contó a la gente 44 leguas no más. Aquí la gente ya no lo podía çufrir;[1] quexávase del largo viaje; pero el Almirante los esforçó lo mejor que pudo dándoles buena esperança de los provechos que podrían aver. Y añidía que por demás era quexarse, pues que él avía venido a las yndias y que así lo avía de proseguir hasta hallarlas, con el ayuda de Nuestro Señor.

Jueves 11º de otubre

Navegó al güesudueste. Tuvieron mucha mar, más que en todo el viaje avían tenido. Vieron pardelas y un junco verde junto a la nao. Vieron los de la caravela Pinta una caña y un palo, y tomaron otro palillo labrado a lo que pareçía con hyerro y un pedaço de caña, y otra yerva que naçe en tierra, y una tablilla. Los de la caravela Niña también vieron otras señales de tierra y un palillo cargado de escaramojos; con estas señales respiraron y alegráronse todos. Anduvieron en este día hasta puesto el sol 27 leguas. Después del sol puesto navegó a su primer camino al güeste; andarían doze millas cada ora y hasta dos oras después de media noche andarían 90 millas que son 22 leguas y media.[2] Y porque la caravela Pinta era más velera e yva delante del Almirante halló tierra y hizo las señas que el Almirante avía mandado. Esta tierra vido primero un marinero que se dezía Rodrigo de Triana, puesto que el Almirante a las diez de la noche, estando en el castillo de popa, vido lumbre aunque fue cosa tan çerrada que no quiso affirmar que fuese tierra. Pero llamó a Pero Gutiérrez, repostero de estrados del Rey e díxole que parecía lumbre, que mirasse él, y así lo hizo y vídola; díxolo también a Rodrigo Sánches de Segovia que el Rey y la Reyna enbiavan en el armada por veedor el qual no vido nada porque no estava en lugar do la pudiese ver. Después que el Almirante lo dixo, se vido una vez o dos, y era como una candelilla de cera que se alçava y levantava, lo qual a pocos pareçiera ser indiçio de tierra; pero el Almirante tuvo por çierto estar junto a la tierra. Por lo qual quando dixeron la Salve, que la acostumbran dezir e cantar a su manera todos los marineros, y se hallan todos, rogó y amonestólos el Almirante que hiziesen buena guarda al castillo de proa y mirasen bien por la

---

[1] Margin: No[ta].
[2] Margin: Hallan ya tierra.

Wednesday 10 October

He steered WSW; they made 10 miles an hour and at times 12 and at others 7, and during the day and night 59 leagues; he told the men no more than 44 leagues. Here the men could stand it no longer; they complained of the long journey; but the Admiral encouraged them as best he could, holding out good hope of the rewards they could gain. And he added that there was no point in complaining, because he had set out for the Indies and that he intended to persist until he found them, with the help of Our Lord.

Thursday 11 October

He sailed WSW. They had a rough sea, rougher than any they had had throughout the voyage. They spotted some petrels and a green reed near the flagship. The crew of the caravel Pinta spotted a cane and a twig and they fished out another piece of stick, carved with iron by the looks of it, and a piece of cane and other vegetation that grows on land, and a small plank. The crew of the caravel Niña also saw signs of land and a branch covered in barnacles. At these signs they breathed again and all took heart. They sailed 27 leagues today, until sunset. After sunset he set his former course due W. They were making about twelve miles an hour and until two in the morning they made about 90 miles, that is 22 leagues and a half. And because the caravel Pinta was faster and sailed ahead of the Admiral, she sighted land and gave the signals that the Admiral had commanded. The first man to see this land was a sailor by the name of Rodrigo de Triana,[29] although the Admiral had seen a light at ten in the evening on the poop deck, but it was so indistinct that he would not swear that it was land.[30] But he called Pero Gutiérrez, His Majesty's chamberlain, told him that it seemed to be a light and asked him to look, which he did, and he did see it. He also called Rodrigo Sánchez de Segovia, whom the King and Queen had sent as comptroller, and he saw nothing as he was not in a position from which he could see it. After the Admiral had spoken, the light was spotted a couple of times, and it was like a small wax candle being raised and lowered, which struck very few people as being a sign of land, but the Admiral was certain that he was near land. So when they had said the Salve, which all sailors are accustomed to say or chant in their own way, and they were all gathered together, the Admiral urged them to keep a good lookout from the forecastle and watch for land, saying that he would

tierra; y que al que le dixese primero que vía tierra le daría luego un jubón de seda, sin las otras merçedes que los Reyes avían prometido, que eran diez mill maravedís de juro a quien primero la viese. A las dos oras después de media noche pareçió la tierra de la qual estarían dos leguas. Amaynaron todas las velas y quedaron con el treo, que es la vela grande sin bonetas y pusiéronse a la corda temporizando hasta el día viernes que llegaron a una ysleta de los Lucayos que se llamava en lengua de yndios Guanahaní. Luego vieron gente desnuda, y el Almirante salió a tierra en la barca armada y Martín Alonso Pinçón y Viçeynte Anes, su hermano, que era capitán de la Niña.[1] Sacó el Almirante la vandera real; y los capitanes con dos vanderas de la cruz verde que llevava el Almirante en todos los navíos por seña, con una F y una Y, ençima de cada letra su corona, una de un cabo de la + y otra de otro. Puestos en tierra vieron árboles muy verdes, y aguas muchas y frutas de diversas maneras. El Almirante llamó a los dos capitanes y a los demás que saltaron en tierra y a Rodrigo de Escobedo, escrivano de toda el armada, y a Rodrigo Sánches de Segovia, y dixo que le diesen por fe y testimonio como él por ante todos tomava como de hecho tomó possessión de la dicha ysla por el Rey e por la Reyna sus señores, haziendo las protestaciones que se requirían, cómo más largo se contiene en los testimonios que allí se hizieron por escripto. Luego se ayuntó allí mucha gente de la ysla. Esto que se sigue son palabras formales del Almirante en su libro de su primera navegaçión y descubrimiento destas yndias. *Yo*, dize él, *porque nos tuviesen mucha amistad, porque cognoscí que era gente que mejor se libraría y convertería a nuestra sancta fe con amor que no por fuerça, les di a algunos dellos unos bonetes colorados y unas cuentas de vidro que se ponían al pescueço y otras cosas muchas de poco valor con que ovieron mucho plazer y quedaron tanto nuestros que era maravilla. Los quales después venían a las barcas de los navíos adonde nos estávamos, nadando; y nos trayan papagayos y hylo de algodón en ovillos y azagayas y otras cosas muchas y nos las trocavan por otras cosas que nos les dávamos como cuentezillas de vidro y cascaveles. En fin, todo tomavan y davan de aquello que tenían de buena voluntad. Mas me pareçió que era gente muy pobre de todo. Ellos andan todos desnudos como su madre los parió, y también las mugeres, aunque no vide más de una farto moça. Y todos los que yo vi eran todos mancebos, que ninguno vide de edad de más de xxx años, muy bien hechos de muy fermosos cuerpos y muy buenas caras; los cabellos gruessos quasi como sedas de cola de cavallos e cortos. Los cabellos traen por encima de las çejas salvo unos*

---

[1] Margin: Saltó el Almirante y los demás en la primera tierra de las yndias viernes de mañana a 12 de otubre de 1492.

give the first man to tell him that he could see land a silk doublet, quite apart from the other rewards which the King and Queen had promised, such as the annual payment of ten thousand maravedís to the first man to see land. Two hours after midnight land appeared at a distance of about two leagues. They shortened all sail, kept the mainsail without the bonnets and lay to, waiting for Friday to dawn, the day on which they finally reached a small island of the Lucayos which was called in the language of the Indians Guanahaní.[31] Then they saw some naked people and the Admiral went ashore in the armed boat with Martín Alonso Pinzón and Vicente Yáñez, his brother, who was the captain of the Niña. The Admiral brought out the royal standard, and the captains unfurled two banners of the green cross, which the Admiral flew as his standard on all the ships, with an F and a Y, and a crown over each letter, one on one side of the + and one on the other. When they landed they saw trees, very green, many streams and a large variety of fruits. The Admiral called the two captains and the others who landed, and Rodrigo de Escobedo, secretary of the expedition, and Rodrigo Sánchez de Segovia, and made them bear witness and testimony that he, in their presence, took possession, as in fact he did take possession, of the said island in the names of the King and Queen, His Sovereigns, making the requisite declarations, as is more fully recorded in the statutory instruments which were set down in writing. Then, many islanders gathered round. What follows are the Admiral's own words from the journal of his first voyage and discovery of these Indies. *In order to win their good will,* he says, *because I could see that they were a people who could more easily be won over and converted to our holy faith by kindness than by force, I gave some of them red hats and glass beads that they put round their necks, and many other things of little value, with which they were very pleased and became so friendly that it was a wonder to see. Afterwards they swam out to the ships' boats where we were and brought parrots and balls of cotton thread and spears and many other things, and they bartered with us for other things which we gave them, like glass beads and hawks' bells. In fact they took and gave everything they had with good will, but it seemed to me that they were a people who were very poor in everything.*[32] *They go as naked as their mothers bore them, even the women, though I only saw one girl, and she was very young. All those I did see were young men, none of them more than thirty years old. They were well built, with handsome bodies and fine features. Their hair is thick, almost like a horse's tail, but short; they wear it down over their eyebrows except for a few strands behind which they wear long and never cut. Some of them paint themselves black, though they are naturally the colour of Canary Islanders,*[33] *neither black nor white; and*

pocos detrás que traen largos que jamás cortan. Dellos se pintan de prieto, y ellos son de la color de los canarios, ni negros ni blancos; y dellos se pintan de blanco; y dellos de colorado; y dellos de lo que fallan. Y dellos se pintan las caras; y dellos todo el cuerpo; y dellos solos los ojos; y dellos sólo el nariz. Ellos no traen armas ni las cognosçen, porque les amostré espadas y las tomavan por el filo, y se cortavan con ignorançia. No tienen algún fierro; sus azagayas son unas varas sin fierro y algunas dellas tienen al cabo un diente de peçe y otras de otras cosas. Ellos todos a una mano son de buena estatura de grandeza y buenos gestos, bien hechos. Yo vide algunos que tenían señales de feridas en sus cuerpos y les hize señas qué era aquello, y ellos me amostraron cómo allí venían gente de otras yslas que estavan açerca y los querían tomar y se defendían. Y yo crey e creo que aquí vienen de tierra firme a tomarlos por captivos. Ellos deven ser buenos servidores y de buen ingenio, que veo que muy presto dizen todo lo que les dezía.[1] Y creo que ligeramente se harían cristianos, que me pareçió que ninguna secta tenían. Yo, plaziendo a Nuestro Señor, levaré de aquí al tiempo de mi partida seys a Vuestras Altezas para que deprendan fablar. Ninguna bestia de ninguna manera vide salvo papagayos en esta ysla. Todas son palabras del Almirante.

Sábado 13 de otubre

Luego que amaneçió vinieron a la playa muchos destos hombres, todos mançebos como dicho tengo; y todos de buena estatura, gente muy fermosa; los cabellos no crespos salvo corredíos y gruessos como sedas de cavallo; y todos de la frente y cabeça muy ancha, más que otra generaçión que fasta aquí aya visto, y los ojos muy fermosos y no pequeños. Y ellos ninguno prieto, salvo de la color de los canarios; ni se deve esperar otra cosa pues está lestegüeste con la ysla del Fierro en Canaria so una línea.[2] Las piernas muy derechas, todos a una mano; y no barriga salvo muy bien hecha. Ellos vinieron a la nao con almadías[3] que son hechas del pie de un árbol como un barco luengo y todo de un pedaço y labrado muy a maravilla según la tierra y grandes en que en algunas venían 40 y 45 hombres; y otras más pequeñas fasta aver dellas en que venía un solo hombre. Remavan con una pala como de fornero y anda a maravilla, y si se le trastorna luego se echan todos a nadar y la endereçan y vazían con calabaças que traen ellos. Trayan ovillos de algodón filado y papagayos y azagayas y otras cositas que

---

[1] Margin: No[ta].
[2] Margin: La isleta de guanahaní está en el altura que la ysla del Hierro.
[3] Margin: Canoas.

some paint themselves white, some red and some whatever colour they can find; some paint their faces, some their whole bodies, some only the eyes and some only the nose. They do not carry arms and do not know of them because I showed them some swords and they grasped them by the blade and cut themselves out of ignorance. They have no iron: their spears are just shafts without a metal tip, and some have a fish tooth at the end, and some have other things. They are all fairly tall, good looking and well proportioned. I saw some who had signs of wounds on their bodies and in sign language I asked them what they were, and they indicated that other people came from other islands nearby and tried to capture them, and they defended themselves.[34] *I believed then and still believe that they come here from the mainland to take them as slaves. They ought to make good slaves for they are of quick intelligence since I notice that they are quick to repeat what is said to them, and I believe that they could very easily become Christians, for it seemed to me that they had no religion of their own. God willing, when I come to leave I will bring six of them to Your Highnesses so that they may learn to speak. I have seen no animals of any kind on this island, except parrots.* These are all the Admiral's own words.

Saturday 13 October

At sunrise many of these men, all youths, as I have said, came to the shore. They were all of good stature, very handsome people, with hair which is not curly but thick and flowing like a horse's mane. They all have very wide foreheads and heads, wider than those of any race I have seen before; their eyes are very beautiful and not small. None of them is black, rather the colour of the Canary islanders, which is to be expected since this island lies E-W with the island of Ferro in the Canaries on the same latitude.[35] *They all have very straight legs; they are not potbellied, but very well formed. They came to the ship in dugouts*[36] *which are made out of a tree-trunk, like a long boat, and all in one piece and wonderfully well carved in the local manner. Some are large, to the extent that 40 or 45 men came in some of them, and others are smaller, so small that there were some with only one man in them. They paddle with a kind of baker's peel and it goes along wonderfully, and if one overturns they all swim around and right it and bale it out with gourds which they carry. They brought balls of cotton thread and parrots and spears*

sería tedio de escrevir y todo davan por qualquiera cosa que se los diese. Y yo estava atento y trabajava de saber si avía oro, y vide que algunos dellos trayan un pedaçuelo colgado en un agujero que tienen a la nariz. Y por señas pude entender que yendo al sur o bolviendo la ysla por el sur que estava allí un rey que tenía grandes vasos dello y tenía muy mucho. Trabajé que fuesen allá, y después vide que no entendían en la yda. Determiné de aguardar fasta mañana en la tarde y después partir para el subdueste que según muchos dellos me enseñaron dezían que avía tierra al sur y al sudueste y al norueste, y que estas del norueste les venían a combatir muchas vezes; y así yr al sudueste a buscar el oro y piedras preciosas. Esta ysla es bien grande y muy llana y de árboles muy verdes y muchas aguas y una laguna en medio muy grande sin ninguna montaña y toda ella verde que es plazer de mirarla. Y esta gente farto mansa y por la gana de aver de nuestras cosas y temiendo[1] que no se les a de dar sin que den algo y no lo tienen, toman lo que pueden y se echan luego a nadar. Mas todo lo que tiene[n] lo dan por qualquiera cosa que les den, que fasta los pedaços de las escudillas y de las taças de vidro rotas rescatavan fasta que vi dar 16 ovillos de algodón por tres çeotís de Portugal que es una blanca de Castilla, y en ellos avría más de un arrova de algodón filado. Esto defendiera y no dexara tomar a nadie salvo que yo lo mandara tomar todo para Vuestras Altezas si oviera en cantidad. Aquí nace en esta ysla mas por el poco tiempo no pude dar así del todo fe; y también aquí nace el oro que traen colgado a la nariz, mas por no perder tiempo quiero yr a ver si puedo topar a la ysla de Çipango. Agora como fue noche todos se fueron a tierra con sus almadías.

Domingo 14 de otubre

En amaneçiendo mandé adereçar el batel de la nao y las barcas de las caravelas, y fui al luengo de la ysla en el camino del nornordeste para ver la otra parte, que era de la parte del leste, qué avía; y también para ver las poblaçiones, y vide luego dos o tres, y la gente que venía todos a la playa llamándonos y dando gracias a Dios. Los unos nos trayan agua; otros, otras cosas de comer; otros, quando veyan que yo no curava de yr a tierra, se echavan a la mar nadando y venían y entendíamos que nos preguntavan si éramos venido del çielo. Y vino uno viejo en el batel dentro y otros a bozes grandes llamavan todos, hombres y mugeres: 'Venid a ver los hombres que vinieron del çielo; traedles de comer y de bever'.[2] Vinieron muchos y muchas mugeres cada uno con algo dando gracias a Dios, echándose al suelo, y

---

[1] MS: teniendo.
[2] Margin: No[ta].

and other things which it would be tedious to list and they gave anything in exchange for whatever was given to them. I watched intently and tried to find out if there was any gold and I saw that some of them wore a small piece hanging from a hole in the nose. By sign language I gathered that to the south, or rounding the southern end of the island, there was a king who had great quantities of it in large pots. I tried to get them to go there but I subsequently saw that they were not interested in going.[37] I decided to wait until tomorrow afternoon and then leave for the SW, where from what many of them pointed out to me, they said there was land to the S and SW and NW, and that from these islands to the NW they very often came to attack them; and so to the SW, to look for the gold and precious stones. This island is very large and very flat, the trees are very green, and there is much water; there is a very large lake[38] in the centre. There are no mountains and it is all so green that it is a pleasure to see. And these people are so gentle, and out of a desire to have some of our things and fear that they will not be given anything without their giving something in exchange, and they have nothing to give, they grab whatever they can and set off swimming. They give everything they have for whatever we give them, even pieces of broken bowls and glass cups they will barter for. I even saw 16 balls of cotton given for three Portuguese coins worth a few farthings in Castile, but there must have been more than 25 pounds of cotton thread there. I would forbid this and not allow anyone to take any of it, but order it all to be kept for Your Highnesses, if there were plenty of it. It grows here on this island but for lack of time I could not make a full investigation; the gold which they wear hanging from their noses is also found here, but, in order not to waste time I want to go and see if I can find the island of Cipangu. Now since it is night they have all gone ashore in their canoes.

Sunday 14 October

At dawn I ordered the ship's boat and the caravels' boats to be got ready and went NNE along the island to see what there was on the other side, the eastern side, and also to see the villages; and I saw two or three, and all the people came to the shore calling us and giving thanks to God. Some brought us water; others, other things to eat; others, when they saw that I was not intending to land, jumped into the sea and swam out and we understood them to be asking us if we had come from heaven.[39] One old man got into the boat[40] and all the others, men and women, called in loud voices 'Come and see the men who have come from heaven; bring them something to eat and drink'. Many came, and many women, each with something and giving thanks to God, throwing themselves on the ground and lifting their hands to the sky

levantavan las manos al çielo y después a bozes nos llamavan que fuésemos a tierra mas yo temía de ver una grande restinga de piedras que çerca toda aquella ysla alrededor. Y entremedias queda hondo y puerto para quantas naos ay en toda la cristiandad, y la entrada dello muy angosta. Es verdad que dentro desta çintha ay algunas baxas, mas la mar no se mueve más que dentro en un pozo. Y para ver todo esto me moví esta mañana porque supiese dar de todo relación a Vuestras Altezas, y también adónde pudiera hazer fortaleza. Y vide un pedaço de tierra que se haze como ysla, aunque no lo es,[1] en que avía seys casas; el qual se pudiera atajar en dos días por ysla, aunque yo no veo ser neçessario, porque esta gente es muy símplice en armas, como verán Vuestras Altezas de siete que yo hize tomar[2] para le llevar y deprender nuestra fabla y bolvellos, salvo que Vuestras Altezas quando mandaren puédenlos todos llevar a Castilla o tenellos en la misma ysla captivos, porque con çinquenta hombres los terná[n] todos sojuzgados, y les hará[n] hazer todo lo que quisiere[n]. Y después, junto con la dicha ysleta están güertas de árboles las más hermosas que yo vi e tan verdes y con sus hojas como las de Castilla en el mes de abril y de mayo, y mucha agua. Yo miré todo aquel puerto y después me bolví a la nao y di la vela y vide tantas yslas que yo no sabía determinarme a quál yría primero. Y aquellos hombres que yo tenía tomado me dezían por señas que eran tantas y tantas que no avía número; y anombraron por su nombre más de çiento. Por ende yo miré por la más grande y [a] aquella determiné andar y así hago y será lexos desta de Sant Salvador çinco leguas, y las otras, dellas más, dellas menos. Todas son muy llanas sin montañas y muy fértiles y todas pobladas y se hazen guerra la una a la otra, aunque estos son muy símplices y muy lindos cuerpos de hombres.

Lunes 15 de otubre

Avía temporejado esta noche con temor de no llegar a tierra a sorgir antes de la mañana, por no saber si la costa era limpia de baxas, y en amaneçiendo cargar velas. Y como la ysla fuese más lexos de çinco leguas, antes será siete, y la marea me detuvo, sería mediodía quando llegué a la dicha ysla. Y fallé que aquella haz que es de la parte de la ysla de San Salvador se corre norte sur y an en ella 5 leguas; y la otra que yo seguí se corría leste güeste, y an en ella más de diez leguas. Y como desta ysla vide otra mayor al güeste, cargué las velas por andar todo aquel día fasta la noche, porque aún no pudiera aver andado al cabo del güeste, a la

---

[1] Margin: Península.
[2] Margin: Siete personas tomó el Almirante de guanahaní.

and then calling out to us to go ashore, but I was fearful of a great reef of rock which surrounds the whole of that island. Inside it is deep and there is a harbour for as many ships as there are in the whole of christendom, and the entrance to it is very narrow. It is true that inside this ribbon there are some shoals, but the sea is as still as a well. I made an effort to see all this this morning so that I would be able to give a full account of it to Your Highnesses and inform you also where a fortress could be built. I saw a piece of land shaped like an island although it is not one, on which there were six houses; this could be cut off and made into an island in two days,[41] except that I do not see it as necessary because these people are unfamiliar with weapons, as Your Highnesses will see from seven[42] of them whom I have had captured to take with me to learn our language before returning. Alternatively they can all be taken to Castile when Your Highnesses wish or they can be held captive on the island, because with fifty men one could keep them all in subjection and make them do whatever one might wish. Later, next to the said islet, there are the most beautiful groves of trees I have seen, and so green, and with leaves like those in Castile in April and May, and much water. I looked around the whole of that harbour and then went back to the flagship and set sail and saw so many islands that I could not decide which I would go to first. The men whom I had taken told me in sign language that there were so many of them that they were without number and they mentioned by name more than a hundred. So I looked for the biggest and decided to make for that, which I am doing; it must 5 leagues from this island of San Salvador, and of the others, some are nearer, some further away. They are all very flat, without mountains and very fertile and all populated, and they make war with each other, although these are very simple people and very attractive.

Monday 15 October

I stood off that night for fear of reaching land and having to anchor before morning, not knowing if the coast were free of shoals, and intending to put on sail at dawn. And since the island was more than 5 leagues off, more like 7, and the current held me up, it would have been midday when I reached the said island.[43] And I found that the side facing the island of San Salvador runs N-S for 5 leagues, and the other, which I followed, ran E-W for more than 10 leagues. And since from this island I saw another larger one to the W, I put on full sail to proceed all that day until the evening because, otherwise, I would not have been able to reach the

qual puse nombre la ysla de Sancta María de la Concepçión. Y quasi al poner del sol sorgí acerca del dicho cabo por saber si avía allí oro, porque estos que yo avía hecho tomar en la ysla de San Salvador me dezían que ay trayan manillas de oro muy grandes a las piernas y a los braços. Yo bien crey que todo lo que dezían era burla para se fugir. Con todo, mi voluntad era de no passar por ninguna ysla de que no tomase possessión, puesto que, tomado de una, se puede dezir de todas. Y sorgí e estuve hasta oy martes, que en amaneçiendo fue a tierra con las barcas armadas; y salí y ellos que eran muchos así desnudos y de la misma condiçión de la otra ysla de San Salvador nos dexaron yr por la ysla y nos davan lo que les pedía. Y porque el viento cargava a la traviesa sueste, no me quise detener y partí para la nao. Y una almadía grande estava a bordo de la caravela Niña, y uno de los hombres de la ysla de Sant Salvador que en ella era se echó a la mar y se fue en ella. Y la noche de antes a medio echado el otro y fue atrás la almadía, la qual fugió, que jamás fue barca que le pudiese alcançar, puesto que le teníamos grande avante. Con todo, dio en tierra y dexaron la almadía y alguno[s] de los de mi compañía salieron en tierra tras ellos, y todos fugeron como gallinas. Y la almadía que avían dexado la llevamos a bordo de la caravela Niña, adonde ya de otro cabo venía otra almadía pequeña con un hombre que venía a rescatar un ovillo de algodón, y se echaron algunos marineros a la mar porque él no quería entrar en la caravela y le tomaron. Y yo, que estava a la popa de la nao, que vide todo, enbié por él y le di un bonete colorado y unas cuentas de vidro verdes pequeñas que le puse al braço, y dos cascaveles que le puse a las orejas, y le mandé bolver su almadía que también tenía en la barca y le enbié a tierra. Y di luego la vela para yr a la otra ysla grande que yo vía al güeste. Y mandé largar también la otra almadía que traya la caravela Niña por popa. Y vide después en tierra al tiempo de la llegada del otro a quien yo avía dado las cosas susodichas y no le avía querido tomar el ovillo de algodón, puesto que él me lo quería dar. Y todos los otros se llegaron a él y tenía a gran maravilla e bien le pareçió que éramos buena gente, y que el otro que se avía fugido nos avía hecho algún daño y que por esto lo llevávamos. Y a esta razón usé esto con él de le mandar alargar y le di las dichas cosas porque nos tuviese en esta estima, porque otra vez quando Vuestras Altezas aquí tornen a enbiar no hagan mala compañía. Y todo lo que yo le di no valía quatro maravedís. Y así partí que serían las diez oras con el viento sueste y tocava de sur para passar a estotra ysla, la qual es grandíssima y adonde todos estos hombres que yo traygo de la de San Salvador hazen señas que ay muy mucho oro, y que lo traen en los braços en manillas y a las piernas y a las orejas y al nariz y al

western cape. I called the island Santa María de la Concepción. And almost at sunset I anchored off the said cape to see if there were any gold there, because the men I had had captured on the island of San Salvador said that there they wore very large gold bracelets on their legs and arms. I thought that everything they said was a ruse to escape. However, I did not wish to pass by any island without taking possession of it, although it might be said that once one had been taken, they all were. I anchored and stayed there until today, Tuesday, when I went ashore at dawn with the armed boats. I disembarked and they, who were many and naked and of the same type as those on the other island of San Salvador, allowed us to go about the island and gave us what was asked of them. And as the wind was blowing strongly across from the SE, I did not wish to wait and left for the ship. A large canoe was alongside the caravel Niña, and one of the men from the island of San Salvador who was on board jumped into the sea and went off in the canoe. And before midnight another threw himself overboard[44] and went after the canoe, which raced off and there was never a boat which could catch it, since we were a long way behind it. It eventually reached land and they abandoned the canoe and some of my crew went ashore after them, and they all ran off like hens. We brought the canoe which they had abandoned alongside the Niña, to which there came another small canoe from another direction with a man who wanted to barter a ball of cotton, and some sailors jumped into the sea because he would not come aboard the caravel and they captured him. I was on the poop of the flagship and saw it all and sent for him and gave him a red bonnet and some small green glass beads which I put on his arm and two hawks' bells which I put on his ears and I ordered his canoe which had also been brought aboard the ship's boat to be returned to him and I sent him ashore. I then set sail to make for the other large island which I could see to the W. I also ordered the other canoe which the Niña was towing astern to be set free. And later, on land, at the time of our arrival, I saw the man to whom I had given the things I mentioned and from whom I refused to take the ball of cotton although he wished to give it to me, and all the others clustered around him and he was ecstatic and clearly thought that we were good people and that the other man who had escaped had done us some wrong and for that reason we were taking him away. And it was for that reason that I treated him as I did, ordering him to be released and giving him the things I said, so that they should think highly of us in this way so that the next time Your Highnesses should send someone here they would not be badly received. And everything I gave him was not worth more than 4 maravedís. And so I left at what must have been 10 o'clock with a SE wind that veered southerly, to go across to this other island which is very large and where all these men I have brought from San Salvador make signs that there is a great deal of gold, and that they wear it in bracelets on their arms and legs, in their ears

pescueço. Y avía desta ysla de Sancta María a esta otra nueve leguas leste güeste, y se corre toda esta parte de la ysla norueste sueste. Y se pareçe que bien avría en esta costa más de veynte ocho leguas en esta faz. Y es muy llana sin montaña ninguna, así como aquella de Sant Salvador y de Sancta María, y todas playas sin roquedos, salvo que a todas ay algunas peñas açerca de tierra debaxo del agua, por donde es menester abrir el ojo quando se quiere surgir e no surgir mucho açerca de tierra aunque las aguas son siempre muy claras y se vee el fondo. Y desviado de tierra dos tyros de lombarda, ay en todas estas yslas tanto fondo que no se puede llegar a él. Son estas yslas muy verdes y fértiles y de ayres muy dulçes, y puede aver muchas cosas que yo no sé porque no me quiero detener por calar y andar muchas yslas para fallar oro. Y pues éstas dan así estas señas, que lo traen a los braços y a las piernas, y es oro porque les amostré algunos pedaços del que yo tengo; no puedo errar con el ayuda de Nuestro Señor que yo no le falle adonde naçe. Y estando a medio golpho destas dos yslas, es de saber de aquella de Sancta María y desta grande a la qual pongo nombre la Fernandina,[1] fallé un hombre solo en una almadía que se passava de la ysla de Sancta María a la Fernandina y traya un poco de su pan que sería tanto como el puño y una calabaça de agua y un pedaço de tierra bermeja hecha en polvo y después amassada, y unas hojas secas, que deve ser cosa muy apreçiada entre ellos, porque ya me truxeron en San Salvador dellas en presente. Y traya un çestillo a su guisa en que tenía un ramalejo de cuentezillas de vidro y dos blancas, por las quales cognosçí que él venía de la ysla de Sant Salvador y aví[a] passado a aquella de Sancta María y se passava a la Fernandina. El qual se llegó a la nao; yo le hize entrar, que así lo demandava él, y le hize poner su almadía en la nao y guardar todo lo que él traya y le mandé dar de comer pan y miel y de bever. Y así le passaré a la Fernandina y le daré todo lo suyo, porque dé buenas nuevas de nos por a Nuestro Señor aplaziendo, quando Vuestras Altezas enbíen acá, que aquellos que vinieren resçiban honrra y nos den de todo lo que oviere.

Martes y miércoles 16 de otubre

Partí de las ysla[s] de Sancta María de Concepçión, que sería ya çerca de mediodía, para la ysla Fernandina, la qual amuestra ser grandíssima al güeste y navegué todo aquel día con calmería. No pude llegar a tiempo de poder ver el fondo para surgir en limpio porque es en esto mucho de aver gran diligençia por no perder las anclas. Y así temporizé toda esta noche hasta el día que vine a una

---

[1] Margin: Fernandina.

and noses and around their necks. Between the island of Santa María and this one there were 9 leagues E-W and all this part of the island tends NW-SE. And it looks as if the coast may run for more than 28 leagues[45] on this side of the island. It is very flat, with no mountains, just like San Salvador and Santa María, and all the beaches are free of rocks except that they all have some offshore reefs under the water, which makes it necessary to keep one's eyes open when anchoring and not to anchor very close to shore, although the water is always very clear and the bottom can be seen. And two lombard shots offshore from all these islands the water is so deep that it cannot be sounded. These islands are very green and fertile and with very gentle breezes and there may be many things on them which I do not know about because I do not want to delay by searching the islands for gold. And since signs are such that they wear it on their arms and legs, and it is gold because I showed them some pieces which I have, I cannot fail with the help of Our Lord to find the place where it originates. And being midway between these two islands, that is, between Santa María and this large island to which I have given the name Fernandina,[46] I found a man alone in a canoe who was crossing from Santa María to Fernandina and he had with him a piece of their bread about the size of a fist and a gourd of water and a piece of brown earth powdered and kneaded into a mass and some dried leaves, which must be something they value highly because I was brought some on San Salvador as a present.[47] And he was carrying one of their baskets in which he had a string of glass beads and two blancas, from which I realised that he came from San Salvador and had crossed to Santa María and was on his way to Fernandina. He came alongside the flagship; I brought him aboard at his request and told him to bring his canoe aboard also and ordered that everything he was carrying should be kept safely, and that he should be given bread and honey to eat and something to drink. And so I shall take him across to Fernandina and give him back all his belongings, so that he will spread good news about us, and when, God willing, Your Highnesses send others here, those who come will be received with honour and the Indians will give us everything there is.

Tuesday and Wednesday 16 October

I left the islands of Santa María de la Concepción at what must have been midday for the island of Fernandina which shows signs of being very large, to the W, and I navigated all that day in a calm. I was not able to arrive in time to see the bottom to anchor in safe water, for it is essential to take great care not to lose the anchors. And so I stood off all that night until the morning when I saw a village where I

poblaçión adonde yo surgí e adonde avía venido aquel hombre que yo hallé ayer en aquella almadía a medio golfo. El qual avía dado tantas buenas nuevas de nos,[1] que toda esta noche no faltó almadías a bordo de la nao que nos trayan agua y de lo que tenían. Yo a cada uno le mandava dar algo, es a saber, algunas contezillas, diez o doze dellas de vidro en un filo, y algunas sonajas de latón destas que valen en Castilla un maravedí cada una, y algunas agujetas, de que todo tenían en grandíssima exçelençia, y también los mandava dar para que comiesen quando venían en la nao, y miel de açúcar. Y después a oras de tercia embié el batel de la nao en tierra por agua, y ellos de muy buena gana le enseñavan a mi gente adónde estava el agua y ellos mesmos trayan los barriles llenos al batel y se folgavan mucho de nos hazer plazer. Esta ysla es grandíssima y tengo determinado de la rodear, porque según puedo entender en ella o açerca della ay mina de oro. Esta ysla está desviada de la de Sancta María 8 leguas quasi leste güeste y este cabo adonde yo vine y toda esta costa se corre nornorueste y sursudueste, y vide bien veynte leguas della mas ay no acabava. Agora, escriviendo esto, di la vela con el viento sur para pujar a rodear toda la ysla, y trabajar hasta que halle Samaot que es la ysla o ciudad adonde es el oro, que así lo dizen todos estos que aquí vienen en la nao y nos lo dezían los de la ysla de San Salvador y de Sancta María. Esta gente es semejante a aquella de las dichas yslas y una fabla y unas costumbres salvo que estos ya me pareçen algún tanto más doméstica gente y de tracto y más sotiles. Porque veo que an traydo algodón aquí a la nao y otras cositas que saben mejor refetar el pagamento que no hazían los otros. Y aun en esta ysla vide paños de algodón fechos como mantillos, y la gente más dispuesta, y las mugeres traen por delante su cuerpo una cosita de algodón que escassamente les cobija su natura. Ella es ysla muy verde y llana y fertilíssima y no pongo duda que todo el año siembran panizo y cogen y así todas otras cosas. Y vide muchos árboles muy diformes de los nuestros, y dellos muchos que tenían los ramos de muchas maneras y todo en un pie, y un ramito es de una manera y otro de otra; y tan disforme, que es la mayor maravilla del mundo quánta es la diversidad de la una manera a la otra. Verbigracia: un ramo tenía las fojas de manera de cañas, y otro de manera de lantisco, y así en un solo árbol de çinco o seys destas maneras, y todos tan diversos. Ni estos son enxeridos porque se pueda dezir que el enxerto lo haze; antes son por los montes, ni cura dellos esta gente. No le cognozco secta ninguna y creo que muy presto se tomarían cristianos,[2] porque ellos son de muy buen entender. Aquí son los peçes tan disformes de los nuestros que es maravilla.

---

[1] Margin: No[ta].
[2] Margin: No[ta].

anchored and to which the man I met yesterday in the canoe in the middle of the gulf had come. He had given such good news about us that all night there were nothing but canoes alongside the ship bringing water and whatever they had. I ordered them all to be given something, that is, some beads, ten or a dozen glass ones on a string, and some little brass bells of the type that cost a maravedí each in Castile, and some thongs, all of which they were delighted with, and I also ordered them to be fed when they came aboard the ship and given molasses. Later, at terce,[48] I sent the ship's boat ashore to fetch water and they very gladly showed my men where the water was and themselves carried the full barrels to the boat and took great delight in obliging us. This island is very large and I have decided to round it, because, as far as I can understand, on it or near it there is a gold mine. This island is 8 leagues almost E-W of Santa María and the cape to which I have come and the whole of this coast runs NNW and SSW, and I saw a good 20 leagues of it but it did not end there.[49] Now, as I write this, I have set sail with a southerly wind to push on round the whole island, and make an effort to find Samaot,[50] which is the island or city where the gold is, for that is what all those who have come to the ship say, and what they told us on the islands of San Salvador and Santa María. These people are similar to those on the other islands, with the same language and customs except that these people already seem to me to be rather more civilised in manner and more intelligent, because I notice that they have brought cotton to the ship and other things for which they drive a harder bargain than the others did. Further, I have seen on this island cotton cloth made into headdresses, and the people are better looking and the women wear on the front of their bodies a little cotton garment which barely covers their genitals. This is a very green and flat and fertile island and I have no doubt that they sow and reap corn[51] and all other things all year round. And I saw many trees of a very different kind from ours, many of which had several different kinds of branches on one trunk; one branch is of one kind and one of another and they are so unlike each other that it is the greatest wonder of the world, so great is the difference. For example, one branch had leaves like a cane, one like a mastic tree, and so on, so that there were five or six kinds, all very diverse from each other.[52] And they are not grafted, for it might be said to be the result of grafting; but they are also on the scrubland where these people do not cultivate them. I do not think they have any religion and I believe that they would quickly become Christians because they are very intelligent. Here there are fish of such different kinds from ours that it is a

Ay algunos hechos como gallos, de las más finas colores del mundo, azules, amarillos, colorados y de todas colores, y otros pintados de mill maneras; y las colores son tan finas, que no ay hombre que no se maraville y no tome gran descanso a verlos. También ay vallenas. Bestias en tierra no vide ninguna de ninguna manera, salvo papagayos y lagartos. Un moço me dixo que vido una grande culebra. Ovejas ni cabras ni otra ninguna bestia vide, aunque yo e estado aquí muy poco, que es medio día; mas si las oviese, no pudiera errar de ver alguna. El çerco desta ysla escriviré después que yo la oviere arrodeada.

Miércoles 17 de otubre

A mediodía partí de la poblaçión adonde yo estava surgido y adonde tomé agua para yr rodear esta ysla Fernandina y el viento era sudueste y sur. Y como mi voluntad fuese de seguir esta costa desta ysla adonde yo estava al sueste porque así se corre toda nornorueste y sursueste, y quería llevar el dicho camino del sur y sueste, porque aquella parte todos estos yndios que traygo y otro de quien ove señas en esta parte del sur a la ysla a que ellos llaman Samoet adonde es el oro, y Martín Alonso Pinçón, capitán de la caravela Pinta, en la qual yo mandé a tres destos yndios, vino a mí y me dixo que uno dellos muy çertificadamente le avía dado a entender que por la parte del nornorueste muy más presto arrodearía la ysla. Yo vide que el viento no me ayudava por el camino que yo quería llevar y era bueno por el otro, di la vela al nornorueste y quando fue açerca del cabo de la ysla a dos leguas, hallé un muy maravilloso puerto con una boca, aunque dos bocas se le puede dezir, porque tiene un ysleo en medio y son ambas muy angostas y dentro muy ancho para çien navíos si fuera fondo y limpio y fondo al entrada. Parecióme razón del ver bien y sondear y así surgí fuera de él y fuy en él con todas las barcas de los navíos y vimos que no avía fondo. Y porque pensé quando yo le vi que era boca de algún río, avía mandado llevar barriles para tomar agua y en tierra hallé unos ocho o diez hombres que luego vinieron a nos y nos amostraron ay çerca la poblaçión adonde yo enbié la gente por agua, una parte con armas, otros con barriles, y así la tomaron. Y porque era lexuelos me detuve por espaçio de dos oras. En este tiempo anduve así por aquellos árboles que eran la cosa más fermosa de ver que otra que se aya visto, veyendo tanta verdura en tanto grado como en el mes de mayo en el Andaluzía. Y los árboles todos están tan disformes de los nuestros como el día de la noche, y así las frutas y así las yervas y las piedras y todas las cosas. Verdad es que algunos árboles eran de la naturaleza de otros que ay en Castilla; por ende avía muy gran diferençia, y los otros árboles de otras maneras eran tantos que no ay persona que lo pueda dezir ni asemejar a otros de Castilla. La gente toda era una con los otros ya dichos, de las mismas condiçiones y así

marvel. There are some crested like dories, of the finest colours in the world, blue, yellow, red and all colours, and others of a thousand different hues. And the colours are so subtle that there is no man who would not marvel at them and delight in seeing them. There are also whales. I have seen no land animals of any kind except parrots and lizards. A boy told me that he saw a large snake. I have seen no sheep or goats or any other beast, although as it is midday I have not been here long, but if there were any, I could not fail to see them. I will describe the circuit of this island when I have rounded it.

Wednesday 17 October

At midday I left the village, where I had anchored and took on water, to make the circuit of this island of Fernandina and the wind was SW and S. My intention was to follow the coast of this island to the SE because the coast runs NNW and SSE and I wanted to take the route SSE because that was where all the Indians[53] I have with me, and another from whom I got information, said the gold was, to the south on an island they call Samoet.[54] Martín Alonso Pinzón, captain of the caravel Pinta, in which I sent three of these Indians, came to me and said that one of them had very definitely given him to believe that I would circle the island much more quickly to the NNW. I saw that the wind was not favourable for the route I wanted to take but was favourable for the other one, and I set sail to the NNW and when I was 2 leagues from the cape of the island I found a marvellous harbour with a mouth, or rather two mouths one might say, because it has an islet in the middle and both are very narrow, and it is wide enough inside for a hundred ships if it is deep enough and if it is clear and there is sufficient depth at the mouth. It seemed a good idea to inspect it and take soundings and so I anchored outside and went in with all the ships' boats and we saw that it was shallow. And because when I saw it I thought that it was the mouth of a river, I had ordered barrels to be brought to fetch water and onshore I found eight or ten men who then came to us and showed us the village nearby, where I sent the men for water, some of them armed, the others with barrels, and they fetched it. And because it was some distance away I waited for a space of two hours. During this time I wandered among those trees which were more beautiful to look at than anything else that has ever been seen; I saw as much greenery as in May in Andalusia, and the trees are as different from ours as the day from the night, and the fruits too and the grass and the stones and everything else. It is true that some of the trees were like some found in Castile, yet they were still very different and there were so many other types of trees that they could not be said to be comparable with those in Castile. The people were the same as those already mentioned, of the same type and as naked and of the same stature

desnudos y de la misma estatura, y davan de lo que tenían por qualquiera cosa que les diesen. Y aquí vide que unos moços de los navíos les trocaron azagayas [por] unos pedaçuelos de escudillas rotas y de vidro, y los otros que fueron por el agua me dixeron cómo avían estado en sus casas, y que eran de dentro muy barridas y limpias, y sus camas y paramentos de cosas que son como redes de algodón.[1] Ellas, las casas, son todas a manera de alfaneques y muy altas y buenas chimeneas,[2] mas no vide entre muchas poblaçiones que yo vide ninguna que passasse de doze hasta quinze casas. Aquí fallaron que las mugeres casadas trayan bragas de algodón; las moças no, sino salvo algunas que eran ya de edad de diez y ocho años. Y ay avía perros mastines y branchetes, y ay fallaron uno que avía al nariz un pedaço de oro que sería como la mitad de un castellano, en el qual vieron letras. Reñí yo con ellos porque no se lo resgataron y dieron quanto pedía por ver qué era y cúya esta moneda era, y ellos me respondieron que nunca se lo osó resgatar. Después de tomada la agua bolví a la nao y di la vela y salí al noruesté tanto que yo descubrí toda aquella parte de la ysla hasta la costa que se corre leste güeste. Y después todos estos yndios tornaron a dezir que esta ysla era más pequeña que no la ysla Samoet, y que sería bien bolver atrás por ser en ella más presto.[3] El viento allí luego nos calmó y començó a ventar güesnorueste, el qual era contrario para donde avíamos venido,[4] y así tomé la buelta y navegué toda esta noche passada al leste sueste, y quando al leste todo, quando al sueste, y esto para apartarme de la tierra porque hazía muy gran çerrazón y el tiempo muy cargado; el [viento] era poco y no me dexó llegar a tierra a surgir. Así que esta noche llovió muy fuerte después de medianoche hasta quasi el día y aun está nublado para llover y nos al cabo de la ysla de la parte del sueste adonde espero surgir fasta que aclaresca para ver las otras yslas adonde tengo de yr. Y así todos estos días después que en estas Yndias estoy a llovido poco o mucho. Crean Vuestras Altezas que es esta tierra la mejor e más fértil y temperada y llana y buena que aya en el mundo.

---

[1] Margin: Hamacas.
[2] Margin: Estas chimeneas no son para humeros sino unas coronillas que tienen encima las casas de paja los indios. Por esto lo dize, puesto que dexan abierto por arriba algo para que salga el humo.
[3] Margin: No[ta].
[4] Margin: Buélvese a la Española.

and they gave what they had for whatever was given to them. And here I saw some
of the ships' boys exchanging pieces of broken crockery and glass for spears. The
others who went for water told me how they had been in their houses, and that
inside they were swept very clean and that their beds and coverings were like cotton
nets.[55] They, the houses, were all like tents, very high and with good chimneys,[56]
but of all the villages I have seen, I have not seen one which had more than 12 or
15 houses. Here they found that the married women wore cotton breeches, but not
the young girls, except some who were already about eighteen years old. And there
were mastiffs and small dogs there,[57]*Historia ü »* (I.42) *that they were more like
hounds than mastiffs, and that they did not bark but emitted a grunt through the
windpipe. The Indians bred them for food, and their popularity with the Spaniards
seems to have led to their extinction. ü »* and they found a man who had in his
nose a piece of gold about the size of half a castellano, on which they saw letters. I
rebuked them for not bartering for it, and giving whatever he asked, in order to see
what it was and whose money this was, and they replied that he dared not barter it
with them. After the water was taken on, I returned to the ship and set sail and
steered NW until I had investigated the whole of that part of the island until the
coast runs E-W. And afterwards all these Indians again said that this island was
smaller than the island of Samoet, and that it would be better to turn back in order
to be there sooner. Then the wind dropped and began to blow from the WNW,
which was the wrong way for the course we had been following, and so I turned
back and steered the whole of last night ESE, sometimes due E, sometimes SE, to
keep clear of the land because the clouds were very dark and the sky very heavy;
there was little wind and it did not allow me to reach land and anchor. So that last
night it rained very heavily from after midnight until just before dawn and it is still
cloudy with a threat of rain. We are off the SE cape of the island where I hope to
anchor until the weather clears so that I can see the other islands I intend to visit.
And so every day since I have been in the Indies it has rained more or less. Believe
me, Your Highnesses, that this is the best and most fertile and temperate and flat
and good land that there is in the world.

Jueves 18 de otubre

Después que aclaresçió seguí el viento y fui en derredor de la ysla quanto pude y surgí al tiempo que ya no era de navegar; mas no fui en tierra y en amaneçiendo di la vela.

Viernes 19 de otubre

En amaneçiendo levanté las anclas y envié la caravela Pinta al leste y sueste, y la caravela Niña al sursueste, y yo con la nao fui al sueste, y dado orden que llevasen aquella buelta fasta mediodía y después que ambas se mudasen las derrotas y se recogieron para mí. Y luego, antes que andássemos tres oras, vimos una ysla al leste sobre la qual descargamos y llegamos a ella todos tres los navíos antes de mediodía a la punta del norte, adonde haze un isleo y un[a] restinga de piedra fuera de él al norte y otro entre él y la ysla grande, la qual anombraron estos hombres de San Salvador que yo traygo la ysla Saomete, a la qual puse nombre la Ysabela.[1] El viento era norte y quedava el dicho ysleo en derrota de la ysla Fernandina de adonde yo avía partido leste güeste, y se corría después la costa desde el ysleo al güeste, y avía en ella doze leguas fasta un cabo y aquí yo llamé el Cabo Hermoso que es de la parte del güeste. Y así es fermoso, redondo y muy fondo, sin baxas fuera de él y al comienço es de piedra y baxo y más adentro es playa de arena como quasi la dicha costa es. Y ay surgí esta noche viernes hasta la mañana. Esta costa toda, y la parte de la ysla que yo vi es toda quasi playa, y la ysla, la más fermosa cosa que yo vi, que si las otras son muy hermosas, esta es más. Es de muchos árboles y muy verdes y muy grandes. Y esta tierra es más alta que las otras yslas falladas, y en ella alguno altillo, no que se le pueda llamar montaña, mas cosa que afermosea lo otro y parece de muchas aguas. Allá al medio de la ysla desta parte al nordeste haze una grande angla y a muchos arboledos y muy espessos y muy grandes. Yo quise yr a surgir en ella para salir a tierra y ver tanta fermosura, mas era el fondo baxo y no podía surgir salvo largo de tierra y el viento era muy bueno para venir a este cabo adonde yo surgí agora; al qual puse nombre Cabo Fermoso, porque así lo es. Y así no surgí en aquella angla y aun porque vide este cabo de allá tan verde y tan fermoso, así como todas las otras cosas y tierras destas yslas que yo no sé adónde me vaya primero, ni me sé cansar los ojos de ver tan fermosas verduras y tan diversas de las nuestras. Y aun creo que a en ellas muchas yervas y muchos árboles que valen mucho en España

---

[1] MS: Yslabela.

Thursday 18 October

When the weather cleared I sailed before the wind as far as I could around the island, and anchored when I could no longer navigate; but I did not land and set sail at dawn.

Friday 19 October

At dawn I weighed anchor and sent the caravel Pinta to the ESE, and the caravel Niña to the SSE, and I went SE with the flagship, having given orders that they should steer those courses until midday and that they should then both change course and rejoin me. And then, before we had been sailing for three hours, we saw an island to the east which we made for and before midday all three ships reached the northern point where it becomes an islet and there is a reef to seaward to the N and another between it and the main island. The men I have brought from San Salvador call this island Saomete, and I gave it the name of Isabela.[58] The wind was northerly and the said islet lay on a line E-W with the island of Fernandina from which I had set out, and the coast then ran from the islet to the W for 12 leagues to a cape which I called Cabo Hermoso which is at the western end.[59] It is indeed beautiful, round and with deep water and with no shoals offshore, and at first it is low and stony and further on it is a sandy beach, as almost all the coast is. I anchored there last night, Friday, until this morning. The whole of this coast, and the part of the island I have seen is almost all beach, and the island is the most beautiful thing I have seen, for if the others are very beautiful, this one is more so. It has many trees, very green and tall. This land is higher than the other islands I have discovered, and it has a high point, not such that it could be called a mountain, but something which enhances the rest, and there appears to be much water. The centre of this side of the island, from here to the NE, forms a great bay and it is thickly wooded with tall trees. I wanted to anchor off it to go ashore and see all these beautiful things, but the water was shallow and I could only anchor a long way offshore and the wind was favourable for reaching this cape where I am now anchored; I called it Cabo Hermoso, because that is what it is. So I did not anchor in that bay because I saw this cape from there, so green and so beautiful, like all the other things and lands in these islands, that I do not know where to go first nor do my eyes tire of seeing such beautiful greenery and so different from our own. I believe that there are on the islands many plants and trees which would be of great

para tinturas y para medicinas de espeçería, mas yo no los cognozco de que llevo grande pena. Y llegando yo aquí a este cabo vino el olor tan bueno y suave de flores o árboles de la tierra que era la cosa más dulçe del mundo. De mañana antes que yo de aquí vaya yré en tierra a ver qué es aquí en el cabo. No es la población salvo allá más dentro adonde dizen estos hombres que yo traygo que está el rey y que trae mucho oro. Y yo de mañana quiero yr tanto avante que halle la población y vea o aya lengua con este rey que, según éstos dan las señas, él señorea todas estas yslas comarcanas, y va vestido y trae sobre sí mucho oro; aunque no doy mucha fe a sus dezires, así por no los entender yo bien, como en cognosçer que ellos son tan pobres de oro que qualquiera poco que este rey trayga los pareçe a ellos mucho. Este a qui yo digo Cabo Fermoso creo que es ysla apartada de Saometo y aun ay ya otra entremedias pequeña. Yo no curo así de ver tanto por menudo, porque no lo podría fazer en çinquenta años, porque quiero ver y descubrir lo más que yo pudiere para bolver a Vuestras Altezas, a Nuestro Señor aplaziendo, en abril. Verdad es que, fallando adónde aya oro o espeçería en cantidad, me determé fasta que yo aya dello quanto pudiere. Y por esto no fago sino andar para ver de topar en ello.

Sábado 20 de otubre

Oy al sol salido levanté las anclas de donde yo estava con la nao surgido en esta ysla de Saometo al cabo del sudueste adonde yo puse nombre el Cabo de la Laguna y a la ysla, la Ysabela, para navegar al nordeste y al leste de la parte del sueste y sur, adonde entendí destos hombres que yo traygo que era la población y el rey della, y fallé todo tan baxo el fondo que no pude entrar ni navegar a ella, y vide que siguiendo el camino del sudueste era muy gran rodeo. Y por esto determiné de me bolver por el camino que yo avía traydo del nornordeste de la parte del güeste y rodear esta isla para [a]y. Y el viento me fue tan escasso que yo no nunca pude aver la tierra al longo de la costa salvo en la noche. Y porque es peligro surgir en estas yslas salvo en el día que se vea con el ojo adónde se echa el ancla porque es todo manchas, una de limpio y otra de non, yo me puse a temporejar a la vela toda esta noche del domingo. Las caravelas surgieron porque [se] hallaron en tierra temprano y pensaron que a sus señas que eran costumbradas de hazer, yría a surgir, mas no quise.

value in Spain as dyes and medicinal spices, but I do not recognise them, which I much regret. When I came to this cape such a fine sweet aroma of flowers or tree came from the land that it was the most delightful thing in the world. In the morning before I leave I shall go ashore to see what there is on this cape. The village where the men I have brought with me say the king who wears much gold is to be found is further inland. Tomorrow I want to press on to find the village and see or speak to this king who, according to the signs these men make, rules over all the neighbouring islands, wears clothes and much gold, although I do not give much credence to their stories, as much because I do not understand them very well as because I realize that they are so poor in gold that however little this king may have will seem a great deal to them. This cape which I call Cabo Hermoso I believe is a separate island from Saometo and that there is even another small island in between. I am not attempting to look at everything in great detail, because I could not do so in fifty years, and because I want to see and discover as much as possible to be able to return to Your Highnesses, God willing, in April. It is true that, if I find gold or spices in quantity, I shall delay until I have as much of it as possible. And for this reason I am simply pushing on to see if I can locate it.

Saturday 20 October

Today at sunrise I weighed anchor from where I was with the ship, anchored off the SW cape of this island of Saometo, the cape which I called Cabo de la Laguna[60] and the island, Isabela, to steer NE and E from the SE and S, where I understood from these men I have brought with me that the village and the king are to be found. I found the water so shallow that I could not enter nor steer towards it, and I could see that the SW would be a very roundabout route. For this reason I decided to return by the way I had come, NNE from the W, and to round the island that way. There was so little wind that I could not make any headway along the coast except during the night. And because it is dangerous to anchor off these islands except in daytime when it is possible to see with the eye where the anchor is dropping, since it is all patchy, clear in one place and not in the next, I decided to stand off on the alert all Sunday night. The caravels anchored because they found themselves near land earlier and they thought that, from the signals which they are accustomed to make, I would drop anchor, but I did not wish to do so.

Domingo 21 de otubre

A las diez oras llegué aquí a este Cabo del Ysleo y surgí y asimismo las caravelas. Y después de aver comido[1] fui en tierra, adonde aquí no avía otra población que una casa, en la qual no fallé a nadie, que creo que con temor se avían fugido, porque en ella estavan todos sus adereços de casa. Yo no le dexé tocar nada salvo que me salí con estos capitanes y gente a ver la ysla, que si las otras ya vistas son muy fermosas y verdes y fertibles, ésta es mucho más y de grandes arboledos y muy verdes. Aquí es unas grandes lagunas y sobre ellas y a la rueda es el arboledo en maravilla. Y aquí y en toda la ysla son todos verdes y las yervas como en el abril en el Andaluzía. Y el cantar de los paxaritos que pareçe que el hombre nunca se querría partir de aquí. Y las manadas de los papagayos que ascureçen el sol, y aves y paxaritos de tantas maneras y tan diversas de las nuestras que es maravilla. Y después ha árboles de mill maneras y todos [dan] de su manera fruto, y todos güelen que es maravilla; que yo estoy el más penado del mundo de no los cognosçer, porque soy bien çierto que todos son cosa de valía y dellos traygo la demuestra, y asimismo de las yervas. Andando así en çerco de una destas lagunas, vide una sierpe, la qual matamos y traygo el cuero a Vuestras Altezas. Ella como nos vido se echó en la laguna, y nos le seguimos dentro, porque no era muy fonda, fasta que con lanças la matamos. Es de siete palmos en largo.[2] Creo que destas semejantes ay aquí en estas lagunas muchas. Aquí cognosçí del lignáloe y mañana e determinado de hazer traer a la nao diez quintales porque me dizen que vale mucho. También andando en busca de muy buena agua, fuimos a una población aquí çerca adonde estoy surto media legua, y la gente della como nos sintieron dieron todos a fugir y dexaron las casas y escondieron su ropa y lo que tenían por el monte. Yo no dexé tomar nada ni la valía de un alfilel. Después se llegaron a nos unos hombres dellos y uno se llegó aquí. Yo di unos cascaveles y unas cuentezillas de vidro y quedó muy contento y muy alegre. Y porque la amistad creciese más y los requiriese algo le hize pedir agua. Y ellos después que fui en la nao vinieron luego a la playa con sus calabaças llenas y folgaron mucho de dárnosla. Y yo les mandé dar otro ramalejo de cuentezillas de vidro y dixeron que de mañana vernían acá. Yo quería hinchar aquí toda la vasija de los navíos de agua; por ende si el tiempo me da lugar luego me partiré a rodear esta isla fasta que yo aya lengua con este rey y ver si puedo aver de él el oro que oyo que trae, y después partir para otra isla grande mucho que creo que deve ser Cipango según

---

[1] MS: comigo.
[2] Margin: Yuana devió de ser ésta.

Sunday 21 October

At ten o'clock I arrived here at the Cabo del Isleo[61] and anchored and the caravels did likewise. And after eating I went ashore, where there was no more habitation than a house, in which I found no-one, for I think that they had fled in fear, because all their household goods were there. I allowed nothing to be touched, but left with the captains and the men to see the island. If the others we have seen are very beautiful and green and fertile, then this is much more so, with great groves of trees, very green. There are some large lakes here and around and overlooking them there are marvellous woods. Here and in all the island everything is green and the vegetation is like April in Andalusia. And the singing of the birds is such that it would seem that a man would never wish to leave here. And the flocks of parrots that darken the sun, and birds of so many kinds so different from our own that it is a marvel! And then there are trees of a thousand kinds all producing their own kind of fruit, and all wonderfully aromatic; I am the saddest man in the world at not recognising them, because I am certain that they are all of value, and I am bringing samples of them, and of the herbs. While walking around one of these lakes, I saw a snake, which we killed and I am bringing the skin back for Your Highnesses. When it saw us it jumped into the lake and we followed it, because it was not very deep, until we killed it with our lances. It is seven palms in length.[62] I believe that there are many similar to these in these lakes. Here I recognised some aloe[63] and have decided to have ten quintales brought to the ship tomorrow because they tell me that it is very valuable. Further, going in search of good water we went to a village near here about half a league from where I am anchored, and when the people there heard us they all fled and left their houses and hid their clothes and belongings in the scrub. I allowed nothing to be taken, not even to the value of a pin. Then some men approached us and one came nearer. I gave him some hawks' bells and some little glass beads and he was very happy and contented. And to make them more friendly and to ask something in return, I had them ask him for water. And after I had returned to the ship they came to the beach with their gourds full and were very pleased to give it to us. I ordered them to be given another string of glass beads and they said they would come here in the morning. I wanted to fill all the ships' casks with water here; and if the weather permits I shall soon set out to sail round the island until I can speak to the king and see if I can get from him the gold which I hear he carries, and then go to another very large island which I believe must be Cipangu according to the signs which the Indians I have brought with me

las señas que me dan estos yndios que yo traygo, a la qual ellos llaman Colba, en la qual dizen que a naos y mareantes muchos y muy grandes. Y desta ysla [a] otra que llaman Bofío que también dizen que es muy grande. Y a las otras que son entremedio veré así de passada, y según yo fallare recaudo de oro o espeçería, determinaré lo que e de fazer. Mas todavía tengo determinado de yr a la tierra firme y a la çiudad de Quisay y dar las cartas de Vuestras Altezas al Gran Can[1] y pedir respuesta y venir con ella.

Lunes 22 de otubre

Toda esta noche y oy estuve aquí aguardando si el rey de aquí o otras personas traherían oro o otra cosa de sustançia y vinieron muchos desta gente semejantes a los otros de las otras yslas así desnudos y así pintados dellos de blanco, dellos de colorado, dellos de prieto, y así de muchas maneras. Trayan azagayas y algunos ovillos de algodón a resgatar, el qual trocavan aquí con algunos marineros por pedaços de vidro, de taças quebradas, y por pedaços de escudillas de barro. Algunos dellos trayan algunos pedaços de oro colgado al nariz, el qual de buena gana davan por un cascavel destos de pie de gavilano, y por cuentezillas de vidro, mas es tan poco que no es nada, que es verdad que qualquier poca cosa que se les dé; ellos también tenían a gran maravilla nuestra venida y creyan que éramos venidos del çielo. Tomamos agua para los navíos en una laguna que aquí está açerca del Cabo del Ysleo que así anombré. Y en la dicha laguna Martín Alonso Pinçón, capitán de la Pinta, mató otra sierpe[2] tal como la otra de ayer de siete palmos. Y fize tomar aquí del liñáloe quanto se falló.

Martes 23 de otubre

Quisiera oy partir para la ysla de Cuba que creo que deve ser Çipango según las señas que dan esta gente de la grandeza della y riqueza, y no me detené más aquí ni [andaré] esta ysla alrededor para yr a la población como tenía determinado para aver lengua con este rey o señor. Que es por no me detener mucho, pues veo que aquí no ay mina de oro y al rodear destas yslas a menester muchas maneras de viento, y no vienta así como los hombres querrían. Y pues es de andar adonde aya trato grande, digo que no es razón de se detener, salvo yr a camino y calar mucha tierra fasta topar en tierra muy provechosa aunque mi entender es que ésta

---

[1] Margin: No[ta].
[2] Margin: Yuana es ésta.

are giving. They call it Colba,[64] and say that there are large ships and many seafarers there. And then from this island to another which they call Bofío[65] which they also say is very large. And the others in between I shall see in passing and depending on whether I find a quantity of gold or spices, I shall decide what to do. But I am still determined to go to the mainland and to the city of Quinsay[66] to give Your Highnesses' letters to the Great Khan and ask for a reply and return with it.

Monday 22 October

All last night and today I have been here waiting to see if the king of this place or anyone else would bring gold or anything else of substance, and there came many of these people, similar to those from the other islands, naked and painted like them, some white, some red, some black and so on in many different ways. They were carrying spears and some balls of cotton to barter which they exchanged with some sailors here for pieces of glass, broken cups, and pieces of earthenware dishes. Some of them wore pieces of gold hanging from the nose, which they gladly gave for one of those bells for a sparrow-hawk's foot, and for glass beads and, indeed, for whatsoever little thing they are given. But there is so little that it is nothing. They also marvelled at our arrival and believed that we were from heaven. We drew water for the ships from a lake near here at what I called the Cabo del Isleo. In the lake Martín Alonso Pinzón, captain of the Pinta, killed another snake like the one yesterday, seven palms long. And I had as much aloe collected as we found.

Tuesday 23 October

I would like to set out today for the island of Cuba which I believe must be Cipangu,[67] to judge from the signs which these people give of its size and riches; I will no longer delay here nor around this island in order to go to the village to speak with this king or lord as I had planned. This is to avoid a long delay, for I can see that there is no gold mine here and to circle these islands requires wind from several directions, and the wind does not blow as men would wish it to. The best thing is to go where there is most business to be done, and I hold that it is not right to delay, but better to be on our way and keep moving until we find a profitable land. My understanding is, however, that this island is rich in spices, but I have no

sea muy provechosa de espeçería mas que yo no la cognozco, que llevo la mayor pena del mundo; que veo mill maneras de árboles que tienen cada uno su manera de fruta y verde agora como en España en el mes de mayo y junio y mill maneras de yervas, asimesmo con flores, y de todo no se cognosció salvo este liñáloe de que oy mandé también traer a la nao mucho para levar a Vuestras Altezas. Y no e dado ni doy la vela para Cuba porque no ay viento salvo calma muerta y llueve mucho y llovió ayer mucho sin hazer ningún frío, antes el día haze calor y las noches temperadas como en mayo en España en el Andaluzía.

Miércoles 24 de otubre

Esta noche a medianoche levanté las anclas de la ysla Ysabela del Cabo del Ysleo que es de la parte del norte adonde y[o] estava posado para yr a la ysla de Cuba adonde oy desta gente que era muy grande y de gran trato y avía en ella oro y especerías y naos grandes y mercaderes y me amostró que al güessudueste yría a ella. Y yo así lo tengo, porque creo que si es así como por señas que me hizieron todos los yndios destas yslas y aquellos que llevo yo en los navíos, porque por lengua no los entiendo, es la ysla de Cipango de que se cuentan cosas maravillosas. Y en las esp[h]eras que yo vi y en las pinturas de mapamundos es ella en esta comarca. Y así navegué fasta el día al güesudueste y amaneçiendo calmó el viento y llovió y así cassi toda la noche. Y estuve así con poco viento fasta que passava de mediodía y estonçes tornó a ventar muy amoroso y llevava todas mis velas de la nao: maestra y dos bonetas y triquete y çevadera y mezana y vela de gavia, y el batel por popa. Así anduve al camino fasta que anocheçió; y estonçes me quedava el Cabo Verde de la ysla Fernandina, el qual es de la parte de sur a la parte de güeste, me quedava al norueste y hazía de mí a él siete leguas. Y porque ventava ya rezio y no sabía yo quánto camino oviese fasta la dicha ysla de Cuba y por no la yr a demandar de noche, porque todas estas islas son muy fondas a no hallar fondo todo enderredor salvo a tyro de dos lombardas, y esto es todo manchado, un pedaço de roquedo y otro de arena, y por esto no se puede seguramente surgir salvo a vista de ojo. Y por tanto acordé de amaynar las velas todas, salvo el triquete, y andar con él y de a un rato creçía mucho el viento y hazía mucho camino de que dudava, y hera muy gran çerrazón y llovía. Mandé amaynar el trinquete y no anduvimos esta noche dos leguas, etc.

Jueves 25 de otubre

Navegó después del sol salido al güeste sudueste hasta las nueve oras. Andarían 5 leguas. Después mudó el camino al güeste. Andavan 8 millas por ora hasta la

knowledge of them, which gives me the greatest sorrow in the world; for I can see a thousand different kinds of trees each with its own kind of fruit and as green now as in Spain in the month of May and June, and a thousand kinds of herbs, also in flower, and of them all the only one I can identify is this aloe, of which I have today ordered a large quantity to be brought to the ship to carry back to Your Highnesses. I have not set sail for Cuba, nor can I, because there is no wind, only a dead calm, and it is raining heavily. It rained heavily yesterday but was not cold; on the contrary, it is hot during the day and the nights are as mild as in May in Spain, in Andalusia.

Wednesday 24 October

Last night at midnight I weighed anchor from the island of Isabela, from the Cabo del Isleo, which is on the north side and where I was stationed, to go to the island of Cuba which I hear from these people is very large and very busy and where there is gold and spices and great ships and merchants, and they indicated that I would get to it by going WSW. This I am doing, because I think that if it is as all the Indians of these islands and those I have on the ships say it is, in sign language because I do not understand their tongue, then it is the island of Cipangu of which so many marvellous tales are told. On the globes which I have seen and on the world maps Cipangu is in this area. And so I steered WSW until daybreak and at sunrise the wind dropped and it rained as it did nearly all night. And there I was with little wind until after midday and then it gently began to blow again and I hoisted all sail: mainsail and two bonnets, foresail, spritsail, mizzen, main topsail and the boat's sail on the poop.[68] *Thus I followed the course until nightfall, when Cabo Verde, which is on the S side of the western part of the island of Fernandina, lay to the NW of me about 7 leagues distant. And because the wind was now strong and I did not know how far it was to the island of Cuba and so as not to go looking for it at night, because all these islands are in very deep water and there is no bottom around them beyond two lombard shots, and what there is is patchy, rocky here and sandy there, so that one cannot anchor safely except by eye, I therefore decided to take in all sail except the foresail and proceed just with that, and after a while the wind grew stronger and I made a lot of headway of which I was unsure, and it was very cloudy and raining. I ordered the foresail to be furled and we made less than 2 leagues during the night, etc.*

Thursday 25 October

After sunrise he steered WSW until nine o'clock. They made about 5 leagues. Then he changed course to the W. They were making 8 miles an hour until one

una después de mediodía y de allí hasta las tres [6], y andarían 44 millas. Entonçes vieron tierra y eran siete o ocho yslas en luengo todas de norte a sur; distavan dellas 5 leguas, etc.

Viernes 26 de otubre

Estuvo de las dichas yslas de la parte del sur. Era todo baxo çinco o seys leguas; surgió por allí. Dixeron los yndios que llevava que avía dellas a Cuba andadura de día y medio con sus almadías que son navetas de un madero adonde no llevan vela. Estas son las canoas. Partió de allí para Cuba, porque por las señas que los yndios le davan de la grandeza y del oro y perlas della, pensava que era ella, conviene a saber, Çipango.

Sábado 27 de otubre

Levantó las anclas, salido el sol, de aquellas yslas que llamó las Yslas de Arena por el poco fondo que tenían de la parte del sur hasta seys leguas. Anduvo ocho millas por ora hasta la una del día al sursudueste y avrían andado 40 millas, y hasta la noche andarían 28 millas al mesmo camino; y antes de noche vieron tierra. Estuvieron la noche al reparo con mucha lluvia que llovió. Anduvieron el sábado fasta el poner del sol 17 leguas al sursudueste.

Domingo 28 días de otubre

Fue de allí en demanda de la ysla de Cuba al sursudueste a la tierra della más çercana y entró en un río muy hermoso y muy sin peligro de baxas ni de otros inconvenientes y toda la costa que anduvo por allí era muy hondo y muy limpio fasta tierra. Tenía la boca del río doze braças y es bien ancha para barloventear. Surgió dentro dizque a tyro de lo[m]barda. Dize el Almirante que nunca tan hermosa cosa vido, lleno de árboles, todo çercado el río, fermosos y verdes y diversos de los nuestros con flores y con su fruto cada uno de su manera. Aves muchas y paxaritos que cantavan muy dulçemente. Avía gran cantidad de palmas de otra manera que las de Guinea y de las nuestras, de una estatura mediana y los pies sin aquella camisa y las hojas muy grandes, con las quales cobijan las casas; la tierra muy llana. Saltó el Almirante en la barca y fue a tierra y llegó a dos casas que creyó ser de pescadores y que con temor se huyeron en una de la[s] quales halló un perro que nunca ladró; y en ambas casas halló redes de hilo de palma y cordeles y anzuelo de cuerno y fisgas de güesso y otros aparejos de

o'clock in the afternoon, and from then till three o'clock [six miles an hour], and they made about 44 miles. Then they sighted land and it was seven or eight islands in a line from north to south,[69] about 5 leagues away, etc.

Friday 26 October

He was S of the islands mentioned. It was shallow for 5 or 6 leagues; he anchored thereabouts. The Indians he had with him said that from those islands to Cuba was a day and a half's journey in their 'almadías' which are boats made out of a single piece of wood and without a sail. These are the canoes.[70] He left there for Cuba because, from the signs that the Indians gave of its size and the gold and pearls there, he thought that that must be it, that is to say, Cipangu.

Saturday 27 October

At sunrise he weighed anchor from those islands which he called the Islas de Arena on account of the shallow bottom they found for up to six leagues to the S.[71] He made eight miles an hour SSW until one o'clock and they made about 40 miles, by the evening they made about 28 miles on the same course; and before nightfall they saw land. They were on watch all night in heavy rain. Up to sunset on Saturday they made 17 leagues SSW.

Sunday 28 October

From there he made for the nearest point of the island of Cuba to the SSW and entered a very beautiful river[72] with no danger of shoals or other obstacles, and all the water wherever he went along that coast was very deep and free of shoals right up to the shore. The mouth of the river was twelve fathoms deep and it is wide enough to tack. He says that they went in to the length of a lombard shot. The Admiral says that he never saw anything so beautiful. The river was completely surrounded by beautiful green trees, all different from ours, each one with its own kind of flowers and fruit and many birds singing sweetly. There was a large number of palms, different from those of Guinea and our own, of medium height and without bark on the stems, and very large leaves with which they thatch their houses. The land is very flat. The Admiral jumped into the boat and went ashore and came to two houses which he thought were those of fishermen who had fled in fear, and in one of them he found a dog which did not bark;[73] and in both houses he found nets and ropes made from twisted palms and fish-hooks made of horn and harpoons of bone, and other fishing tackle, and

pescar y muchos huegos dentro y creyó que en cada una casa se ayuntan muchas personas. Mandó que no se tocase en cosa de todo ello y así se hizo. La yerva era grande como en el Andaluzía por abril y mayo. Halló verdolagas muchas y bledos. Tornóse a la barca y anduvo por el río arriba un buen rato y era, dizque, gran plazer ver aquellas verduras y arboledas y de las aves que no podía dexallas para se bolver. Dize que es aquella ysla la más hermosa que ojos ayan visto, llena de muy buenos puertos y ríos hondos y la mar que parecía que nunca se devía de alçar, porque la yerva de la playa llegava hasta quasi el agua, lo qual no suele llegar donde la mar es brava. Hasta entonçes no avía experimentado en todas aquellas yslas que la mar fuese brava. La isla dize que es llena de montañas muy hermosas aunque no son muy grandes en longura, salvo altas, y toda la otra tierra es alta de la manera de Çeçilia. Llena es de muchas aguas, según pudo entender de los yndios que consigo lleva que tomó en la ysla de Guanahaní, los quales le dizen por señas que ay diez ríos grandes y que con sus canoas no la pueden çercar en XX días. Quando yva a tierra con los navíos salieron dos almadías, o canoas, y como vieron que los marineros entravan en la barca y remavan para yr a ver el fondo del río para saber dónde avían de surgir, huyeron las canoas. Dezían los yndios que en aquella ysla avía minas de oro y perlas,[1] y vido el Almirante lugar apto para ellas y almejas que es señal dellas, y entendía el Almirante que allí venían naos del Gran Can y grandes, y que de allí a tierra firme avía jornada de diez días. Llamó el Almirante aquel río y puerto de San Salvador.

Lunes 29 de otubre

Alçó las anclas de aquel puerto y navegó al poniente para yr dizque a la çiudad donde le pareçía que le dezían los yndios que estava aquel rey. Una punta de la ysla le salía al norueste seys leguas de allí. Otra punta le salía al leste diez leguas. Andada otra legua vido un río no tan grande [de] entrada, al qual puso nombre el Río de la Luna. Anduvo hasta ora de bísperas. Vido otro río muy más grande que los otros y así se lo dixeron por señas los yndios y açerca de él vido buenas poblaçiones de casas; llamó al río el Río de Mares. Enbió dos barcas a una poblaçión por aver lengua y a una dellas un yndio de los que traya, porque ya los entendían algo y mostravan estar contentos con los cristianos, de las quales todos los hombres y mugeres y criaturas huyeron, desamparando[2] las casas con todo

---

[1] Margin: No[ta].
[2] MS: desmanparando.

many fires inside. He thought that many people live together in one house. He
ordered none of these things to be touched and the order was observed. The
grass was as high as in Andalusia in April and May. He found much purslane
and wild amaranth. He returned to the boat and went a good way up-river and it
was, he says, a great joy to see that greenery and the groves of trees and the
birds and he could not bear to leave it and return. He says that that island is the
most beautiful that eyes have ever seen, full of good harbours and deep rivers
and the sea which never seemed to be rough, because the vegetation on the
shore reached almost to the water, which does not normally happen where the
sea is rough. Until then he had not experienced in all those islands a rough sea.
He says the island is full of very beautiful mountains although they are not in
long ranges, but high, and the rest of the land is high in the manner of Sicily.
There is plenty of water, as far as he could understand from the Indians he had
with him whom he took from the island of Guanahaní, who told him by signs
that there are ten large rivers and that in their canoes they cannot circle the
island in 20 days.[74] When he was going ashore with the boats two 'almadías', or
canoes, came out and when they saw that the sailors were getting in the boats
and rowing out to judge the depth of the river, so that they would know where to
anchor, the canoes fled. The Indians said that there were gold mines and pearls
on that island, and the Admiral saw suitable places for them and mussels, which
is a sign of them, and the Admiral understood that large ships came there from
the Great Khan, and that from there to the mainland was a journey of 10 days.
The Admiral called that river and harbour San Salvador.

Monday 29 October

He weighed anchor from that harbour and steered W to go, as he says, to the
city where he thought the Indians said that the king was. One point of the island
jutted out 6 leagues from there to the NW. Another point jutted out to the E 10
leagues away. After another league he saw a river with not such a wide entrance
which he named the Río de la Luna.[75] He sailed until the hour of vespers. He
saw another river, much larger than the others, as the Indians had told him by
signs, and nearby he saw fair-sized groups of houses. He called the river Río de
Mares.[76] He sent two boats to a village to talk to them, and in one of the boats
he sent one of the Indians he had with him, because they were beginning to
understand them and were showing signs of growing to like the Christians. All
the men, women and children of the village fled leaving their houses unguarded

lo que tenían, y mandó el Almirante que no se tocase en cosa. Las casas dizque eran ya más hermosas que las que avía visto y creya que quanto más se allegase a la tierra firme serían mejores. Eran hecha[s] a manera de alfaneques muy grandes y pareçían tiendas en real sin conçierto de calles sino una acá y otra acullá y de dentro muy barridas y limpias y sus adereços muy compuestos. Todas son de ramos de palma, muy hermosas. Hallaron muchas estatuas en figura de mugeres y muchas cabeças en manera de cara[n]tona, muy bien labradas.[1] No sé si esto tienen por hermosura o adoran en ellas. Avía perros que jamás ladraron; avía avezitas salvajes mansas por sus casas; avía maravillosos adereços de redes y anzuelos y artifiçios de pescar. No le tocaron en cosa dello. Creyó que todos los de la costa devían de ser pescadores, que llevan el pescado la tierra dentro porque aquella ysla es muy grande y tan hermosa que no se hartava dezir bien della. Dize que halló árboles y frutas de muy maravilloso sabor, y dize que deve aver vacas en ella y otros ganados, porque vido cabeças en güesso que le pareçieron de vaca.[2] Aves y paxaritos y el cantar de los grillos en toda la noche con que se holgava[n] todos. Los ayres[3] sabrosos y dulçes de toda la noche ni frío ni callente, mas por el camino de las otras yslas [a] aquella dizque hazía gran calor y allí no, salvo templado como en mayo. Atrivuye el calor de las otras yslas por ser muy llanas y por el viento que trayan hasta allí ser leva[n]te y por eso cálido. El agua de aquellos ríos era salada a la boca; no supieron de dónde bevían los yndios aunque tenían en sus casas agua dulçe. En este río podía[n] los navíos boltejar para entrar y para salir y tienen muy buenas señas o marcas; tienen siete o ocho braças de fondo a la boca y dentro çinco. Toda aquella mar dize que le pareçe que deve ser siempre mansa como el río de Sevilla, y el agua aparejada para criar perlas. Halló caracoles grandes sin sabor no como los de España. Señala la disposiçión del río y del puerto que arriba dixo y nombró San Salvador, que tiene sus montañas hermosas y altas como la Peña de los Enamorados y una dellas tiene ençima otro montezillo a manera de una hermosa mezquita.[4] Estotro río y puerto en que agora estava tiene de la parte del sueste dos montañas así redondas y de la parte del güestenorueste un hermoso cabo llano que sale fuera.[5]

---

[1] Margin: No[ta].
[2] Margin: Devía de ser de manatí.
[3] MS: aryes.
[4] Margin: El puerto de Barocoa.
[5] Margin: O es éste el de Barocoa por lo que dize del cabo llano.

with their belongings in them and the Admiral ordered nothing to be touched. He says that the houses were even more beautiful than those he had seen and he believed that the nearer he got to the mainland the better they would be. They were built like very large tents and gave the impression of tents in a camp, without a street plan, one here and one there. Inside they are swept very clean and neatly furnished. They are all made of palm branches, very beautiful. They found many statues in the form of a woman and heads like carnival masks, very well carved. I do not know if they are intended as ornaments or whether they worship them. There were dogs which did not bark;[77] there were tame wild birds in the houses; there were marvellous stocks of nets and hooks and fishing tackle. They touched none of it. He thought that all those on the coast must be fishermen who take their catch inland, because that island is very large and so beautiful that he did not tire of speaking well of it. He says that he found trees and fruits with a marvellous taste, and says that there must be cattle and other stock there because he saw skulls which looked like those of cows.[78] Birds, and the song of crickets which delighted them all night. The breezes were sweet and gentle all night, neither cold nor hot, whereas on the way from the other islands to that one he says that it had been very hot, but not there, rather it was mild as in May. He attributes the heat of the other islands to their being very flat and to the wind which had blown up till then from the E, and was therefore hot. The water of those rivers was salt to the taste; they did not know where the Indians drank from, although they had fresh water in their houses. In this river the ships could manoeuvre in and out and there are very good land marks; there is a depth of 7 or 8 fathoms at the mouth and 5 further in. He says that it seems to him that the whole of that sea must always be calm like the river at Seville, and the water ideal for cultivating pearls. He found large snails but they had no taste, unlike those in Spain. He gives details of the river and harbour which earlier he named San Salvador which has its beautiful high mountains like the Peña de los Enamorados, and one of them has on top another small hill like a beautiful mosque.[79] The river and harbour at which he is now stationed has to the SE two rounded mountains[80] and to the WNW a fine flat cape that juts out.

Martes 30 de otubre

Salió del Río de Mares al norueste y vido cabo lleno de palmas y púsole Cabo de Palmas después de aver andado quinze leguas. Los yndios que yvan en la caravela Pinta dixeron que detrás de aquel cabo avía un río y del río a Cuba avía quatro jornadas; y dixo el capitán de la Pinta que entendía que esta Cuba era çiudad y que aquella tierra era tierra firme muy grande, que va mucho al norte, y que el rey de aquella tierra tenía guerra con el Gran Can, al qual ellos llamavan Cami, y a su tierra, o çiudad, Faba, y otros muchos nombres.[1] Determinó el Almirante de llegar a aquel río y enbiar un presente al rey de la tierra, y enbiarle la carta de los Reyes. Y para ella tenía un marinero que avía andado en Guinea en lo mismo, y çiertos yndios de Guanahaní que querían yr con él con que después los tornasen a su tierra. Al pareçer del Almirante distava de la línea equinocial 42 grados hazia la vanda del norte, si no está corrupta la letra de donde trasladé esto, y dize que avía de trabajar de yr al Gran Can que pensava que estava por allí o a la çiudad de Cathay que es del Gran Can, que dizque es muy grande según le fue dicho antes que partiese de España. Toda aquesta tierra dize ser baxa y hermosa y fonda la mar.

Miércoles 31 de otubre[2]

Toda la noche martes anduvo barloventeando y vido un río donde no pudo entrar por ser baxa la entrada y pensaron los yndios que pudieran entrar los navíos como entrava[n] sus canoas. Y navegando adelante halló un cabo que salía muy fuera y çercado de baxos y vido una concha o baya donde podían estar navío[s] pequeños y no lo pudo encabalgar porque el viento se avía tyrado del todo al norte y toda la costa se corría al nornorueste y sueste, y otro cabo que vido adelante le salía más afuera. Por esto y porque el çielo mostrava de ventar rezio se ovo de tornar al Río de Mares.[3]

---

[1] Margin: Muy a [o]scuras andavan todos por no entender a los yndios. Yo creo que la Cuba que los yndios les dezían era la provincia de Cubanacan de aquella ysla de Cuba que tiene minas de oro, etc. Toda esta tierra es la ysla de Cuba y no tierra firme.

[2] MS: noviembre.

[3] Margin: Por esto que dize aquí del viento que llevava es çierto que era Cuba por la costa que andava.

Tuesday 30 October

He left Río de Mares to the NW and, after having made 15 leagues, he saw a cape full of palms which he named Cabo de Palmas.[81] The Indians in the caravel Pinta said that behind that cape there was a river and that from the river to Cuba was 4 days' journey. The captain of the Pinta said that he understood that this Cuba was a city and that that land was a very large stretch of mainland which extends a long way to the N, and that the king of that land was at war with the Great Khan, whom they call Cami, and his land, or city, Faba, and many other names.[82] The Admiral decided to go as far as that river and send a gift to the king of that land and send him the Monarchs' letter. And for this purpose he had a sailor who had been in Guinea on the same business, and some Indians from Guanahaní who were prepared to go with him provided that afterwards they were returned to their land. In the Admiral's opinion he was 42 degrees N of the equinoctial line, if the text from which I have transcribed this is not corrupt,[83] and he says that he will endeavour to go to the Great Khan who he thought was in that region or to the city of Cathay which is in the Great Khan's possession, which he says is very large according to what he was told before he left Spain. All this land he says is low-lying and beautiful and the sea is deep.

Wednesday 31 October

All Tuesday night he tacked back and forth and saw a river which he could not enter because the mouth was shallow, the Indians thinking that the ships could enter just as their canoes did. Sailing on he found a cape which jutted out a long way and was surrounded by shoals and he saw an inlet or bay where small boats could anchor, but he could not make it because the wind had shifted due N and the whole coast ran NNW and SE, and another cape he could see ahead jutted out still further. For this reason, and because the sky threatened a stronger wind, he had to return to Río de Mares.

Jueves 1º de noviembre

En saliendo el sol enbió el Almirante las barcas a tierra a las casas que allí estavan y hallaron que eran toda la gente huyda. Y desde a buen rato pareçió un hombre, y mandó el Almirante que lo dexasen asegurar y bolviéronse las barcas. Y después de comer tornó a enbiar a tierra uno de los yndios que llevava, el qual desde lexos le dio bozes[1] diziendo que no oviesen miedo porque eran buena gente y no hazían mal a nadie, ni eran del Gran Can, antes dava[n] de lo suyo en muchas yslas que avían estado. Y echóse a nadar el yndio y fue a tierra, y dos de los de allí lo tomaron de braços[2] y lleváronlo a una casa donde se informaron de él. Y como fueron çiertos que no se les avía de hazer mal, se aseguraron y vinieron luego a los navíos más de diez y seys almadías o canoas con algodón hylado y otras cosillas suyas de las quales mandó el Almirante que no se tomasse nada, porque supiesen que no buscava el Almirante salvo oro a que ellos llaman nucay.[3] Y así en todo el día anduvieron y vinieron de tierra a los navíos, y fueron de los cristianos a tierra muy seguramente. El Almirante no vido a alguno dellos oro, pero dize el Almirante que vido a uno dellos un pedaço de plata labrado colgado a la nariz, que tuvo por señal que en la tierra avía plata.[4] Dixeron por señas que antes de tres días vernían muchos mercaderes de la tierra dentro a comprar de las cosas que allí llevan los cristianos, y darían nuevas del rey de aquella tierra, el qual según se pudo entender por las señas que davan, que estava de allí quatro jornadas porque ellos avían enbiado muchos por toda la tierra a le hazer saber del Almirante. *Esta gente,* dize el Almirante, *es de la misma calidad y costumbre de los otros hallados, sin ninguna secta que yo cognozca; que fasta oy [a] aquestos que traygo no e visto hazer ninguna oraçión;*[5] *antes dizen la Salve y el Ave María con las manos al çielo como le amuestran, y hazen la señal de la cruz. Toda la lengua también es una y todos amigos, y creo que sean todas estas yslas, y que tengan guerra con el Gran Chan a que ellos llaman Cavila y a la provinçia Bafan. Y así andan también desnudos como los otros.* Esto dize el Almirante. El río dize que es muy hondo, y en la boca pueden llegar los navío[s] con el bordo hasta tierra; no llega el agua dulçe a la boca con una legua, y es muy dulçe. *Y es çierto,* dize el Almirante, *que esta es la tierra firme*

---

[1] Margin: No[ta].
[2] Margin: No[ta].
[3] Margin: No[ta].
[4] Margin: No[ta].
[5] Margin: No[ta].

Thursday 1 November

At sunrise the Admiral sent the boats ashore to the houses nearby and they found that all the people had fled. After some time a man appeared and the Admiral ordered them to reassure him and the boats returned. After eating he once more sent ashore one of the Indians he had with him who called out to them from a distance saying that they should have no fear because these were good people and did no-one any harm, and were not from the Great Khan, but, on the contrary, had made gifts of their property on the other islands they had visited. And the Indian dived in and swam ashore and two of the others grabbed him by the arms and took him to a house where they questioned him. And when they were assured that no harm was to come to them, they grew in confidence and more than 16 'almadías' or canoes came to the ships with spun cotton and other things which the Admiral ordered should not be taken, so that they would know that the Admiral was looking solely for gold, which they call nucay. And so all day they went back and forth from the shore to the ships and from the Christians to the shore in perfect safety. The Admiral saw gold on none of them, but he says that he saw one of them with a piece of worked silver hanging from the nose, which he took to be an indication that there was silver in that land. They made signs that within three days many merchants would come from the interior to buy the things the Christians had with them, and they would give news of the king of that land, who, as far as he could understand from the signs they made, was four days' journey away, because they had sent many men throughout the land to inform him about the Admiral. *These people*, says the Admiral, *are of the same type and customs as the others I have found, with no religion that I can see. So far I have not seen these Indians I have with me say any prayers, but they say the Salve and the Ave María with their hands to heaven as they are taught to do, and they make the sign of the cross. They all speak the same language and are friendly with each other, as I believe all these islands are, and that they are at war with the Great Khan whom they call Cavila and the province, Bafan. They all go naked like the others.* This is what the Admiral says. He says the river is very deep, and that at the mouth the ships can come alongside the shore. The fresh water does not come within a league of the mouth, and it is very sweet. *It is certain*, says the Admiral, *that this is the mainland and that I am*, he says, *near*

y que estoy, dize él, *ante Zaytó y Quinsay*[1] *cien leguas poco más o poco menos lexos de lo uno y de lo otro y bien se amuestra por la mar que viene de otra suerte que fasta aquí no a venido y ayer que yva al norueste fallé que hazía frío.*

Viernes 2 de noviembre

Acordó el Almirante enbiar dos hombres españoles, el uno se llamava Rodrigo de Xerez que bivía en Ayamonte, y el otro era un Luys de Torres que avía bivido con el Adelantado de Murçia y avía sido judío y sabía dizque ebrayco y caldeo y aun algo arávigo. Y con estos enbió dos yndios: uno de los que consigo traya de Guanahaní, y el otro de aquellas casas que en el río estavan poblados. Dioles sartas de cuentas para comprar de comer si les faltase, y seys días de término para que bolviesen. Dioles muestras de espeçería para ver si alguna della topasen. Dioles ynstruçión de cómo avían de preguntar por el rey de aquella tierra y lo que le avían de hablar de partes de los Reyes de Castilla, cómo enbiavan al Almirante para que les diese de su parte sus cartas y un presente, y para saber de su estado y cobrar amistad con él, y favoreçelle en lo que oviese dellos menester, etc., y que supiesen de çiertas provinçias y puertos y ríos de que el Almirante tenía notiçia y quánto distavan de allí, etc. Aquí tomó el Almirante el altura con un quadrante esta noche y halló que estava 42 grados de la línea equinoçial,[2] y dize que por su cuenta halló que avía andado desde la ysla del Hierro mill y çiento y quarenta y dos leguas, y todavía afirma que aquella es tierra firme.

Sábado 3 de noviembre

En la mañana entró en la barca el Almirante y porque haze el río en la boca un gran lago el qual haze un singularíssimo puerto muy hondo y limpio de piedras muy buena playa para poner navíos a monte y mucha leña, entró por el río arriba hasta llegar al agua dulçe que sería çerca de dos leguas, y subió en un montezillo por descubrir algo de la tierra y no pudo ver nada por las grandes arboledas las quales muy frescas, odoríferas. Por lo qual dize no tener duda que no aya yervas aromáticas. Dize que todo era tan hermoso lo que vía, que no podía cansar los ojos de ver tanta lindeza, y los cantos de las aves y paxaritos. Vinieron en aquel día muchas almadías o canoas a los navíos a resgatar cosas de algodón filado y redes en que dormían, que son hamacas.

---

[1] Margin: Esta algaravía no entiendo yo.
[2] Margin: Esto es falso porque no está Cuba sino en ** grados.

Zaytó[84] and Quinsay, within a hundred leagues more or less of one or the other, as is shown by the sea which has a different character from the way it has been hitherto, and yesterday when I was going NW I found that it was cold.

Friday 2 November

The Admiral decided to send out two Spaniards, one called Rodrigo de Jerez who lived in Ayamonte, and the other a certain Luis de Torres who had lived with the Adelantado de Murcia and had been a Jew and, he says, knew Hebrew and Chaldean and some Arabic, and with them two Indians, one from among those he had brought from Guanahaní, and the other from those houses grouped by the river. He gave them strings of beads to buy food if they needed to, and a time limit of six days to return. He gave them samples of spices to see if they could come across any of them. He gave them instructions on how they should ask for the king of that land and what they should tell him on behalf of the Monarchs of Castile and how they had sent the Admiral on their behalf to present him with letters and a gift and to find out about his state and to establish friendship with him and assist him in whatever he should require of them, etc., and that they should find out about certain provinces and harbours and rivers of which the Admiral had information and how far away they were, etc. Here the Admiral took the latitude with a quadrant this evening and found that he was 42 degrees from the equinoctial line, and he says that by his calculations he found that he had travelled 1142 leagues from the island of Ferro, and he still affirms that this is the mainland.

Saturday 3 November

In the morning the Admiral got into the boat and, because the mouth of the river forms a great lake which makes a most exceptional deep harbour, free of rocks and with a good beach on which to careen the ships and a good supply of wood, he went up-river until he reached the fresh water, which would be around 2 leagues, and climbed a hill to take a look at the land, and he could see nothing for the great groves of trees which were very fresh and scented. For this reason he says that there is no doubt that there are aromatic herbs here. He says that everything he saw was so lovely that his eyes did not tire of seeing such beauty, nor could he weary of the birds singing. Many 'almadías' or canoes came that day to the ships to barter with cotton thread and the nets in which they slept, which are hammocks.

Domingo 4 de noviembre

Luego en amaneçiendo entró el Almirante en la barca y salió a tierra a caçar de las aves que el día antes avía visto. Después de buelto vino a él Martín Alonso Pinçón con dos pedaços de canela, y dixo que un portugués que tenía en su navío avía visto a un yndio que traya dos manojos della grandes, pero que no se la osó resgatar por la pena que el Almirante tenía puesta que nadie resgatase. Dezía más: que aquel yndio traya unas cosas bermejas como nuezes. El contramaestre de la Pinta dixo que avía hallado árboles de canela; fue el Almirante luego allá y halló que no eran. Mostró el Almirante a unos yndios de allí canela y pimienta, parez que de la que llevava de Castilla para muestra, y cognosçiéronla, dizque, y dixeron por señas que çerca de allí avía mucho de aquello al camino del sueste.[1] Mostróles oro y perlas, y respondieron çiertos viejos que en un lugar que llamaron Bohío[2] avía infinito y que lo trayan al cuello y a las orejas y a los braços y a las piernas, y también perlas. Entendió más: que dezían que avía naos grandes y mercaderías y todo esto era al sueste. Entendió también que lexos de allí avía hombres de un ojo, y otros con hoçicos de perros que comían los hombres, y que en tomando uno lo degollavan y le bevían la sangre, y le cortavan su natura.[3] Determinó de bolver a la nao el Almirante a esperar los dos hombres que avía enbiado, para determinar de partirse a buscar aquellas tierras, si no truxesen aquellos alguna buena nueva de lo que deseavan. Dize más el Almirante: *Esta gente es muy mansa y muy temerosa,*[4] *desnuda como dicho tengo, sin armas y sin ley. Estas tierras son muy fértiles; ellos las tienen llenas de mames que son como çanahorias que tienen sabor de castañas,*[5] *y tienen faxones y favas muy diversas de las nuestras y mucho algodón el qual no siembran y nace por los montes árboles grandes, y creo que en todo tiempo lo aya para coger porque vi lo[s] cogujos abiertos y otros que se abrían y flores todo en un árbol y otras mill maneras de frutas que me no es possible escrevir, y todo deve ser cosa provechosa.* Todo esto dize el Almirante.

---

[1] Margin: Devían mentir los yndios.
[2] Margin: Bohío llamavan los yndios a aquellas yslas a las casas y por eso creo que no entendía bien el Almirante. Ante devía de dezir por la ysla Española que llamavan Haití.
[3] Margin: Todo esto devían de dezir de los caribes.
[4] Margin: No[ta].
[5] Margin: Los ajes, o batatas son éstas.

Sunday 4 November

Then at dawn the Admiral got into the boat and went ashore to catch some of the birds he had seen the previous day. On his return Martín Alonso Pinzón came to him with two pieces of cinnamon,[85] and said that a Portuguese he had on his ship had seen an Indian who was carrying two great bundles of it, but that he did not dare to barter for it because of the penalty the Admiral had imposed on bartering. He also said that that Indian had some red things like nuts. The boatswain of the Pinta said that he had found cinnamon trees; the Admiral immediately went and looked but found that they were not. The Admiral showed the Indians there some cinnamon and pepper, apparently some which he had brought from Castile as a sample, and they recognised it, he says, and made signs that there was a lot of it nearby to the SE. He showed them gold and pearls, and some old men replied that in a place which they called Bohío there was a great deal of it and that they wore it around the neck and ears and arms and legs, and pearls also. He further understood that they said that there were large ships and merchandise and that all this was to the SE. He also understood that far away there were men with one eye, and others with a dog's snout, who ate men, and on capturing one would cut his throat and drink the blood and cut off his genitals. The Admiral decided to return to the ship to wait for the two men he had sent out, and determined to set off in search of those lands if they did not bring back any good news of what they wanted. The Admiral further says: *These people are very gentle and timorous, naked as I have said, without weapons and without laws. These lands are very fertile; they are full of 'mames',*[86] *which are like carrots with the flavour of chestnuts, and they have kidney beans and broad beans of many different types from ours, and much cotton which they do not sow but which grows wild in great trees on the hillsides, and I think that it is ready to be gathered all year round, because I saw the bolls open and others about to open and flowers all on one tree, and a thousand other types of fruit which it is impossible to put in writing, and it all must be of great value.* All this the Admiral says.

Lunes 5 de noviembre

En amaneziendo mandó poner la nao a monte y los otros navíos, pero no todos juntos sino que quedasen siempre dos en el lugar donde estavan por la seguridad, aunque dize que aquella gente era muy segura y sin temor se pudieran poner todos los navíos junto en monte. Estando así vino el contramaestre de l[a] Niña a pedir albriçias al Almirante porque avía hallado almáçiga mas no traya la muestra porque se le avía caydo. Prometióselas el Almirante y enbió a Rodrigo Sánches y a maestre Diego a los árboles y truxeron un poco della, la qual guardó para llevar a los Reyes y también del árbol y dize que se cognosçió que era almáçiga aunque se a de coger a sus tiempos, y que avía en aquella comarca para sacar mill quintales cada año. Halló dizque allí mucho de aquel palo que le pareçió lignáloe. Dize más: que aquel puerto de Mares[1] es de los mejores del mundo y mejores ayres y más mansa gente, y porque tiene un cabo de peña altillo, se puede hazer una fortaleza para que, si aquello saliese rico y cosa grande, estaría[n] allí los mercaderes seguros de qualquiera otras naçiones. Y dize: *Nuestro Señor, en cuyas manos están todas las victorias, adereza todo lo que fuere su servicio.* Dizque dixo un yndio por señas que el almáçiga era buena para quando les dolía el estómago.

Martes 6 de noviembre

Ayer en la noche, dize el Almirante, vinieron los dos hombres que avía embiado a ver la tierra dentro y le dixeron cómo avían andado doze leguas que avía hasta una población de çinquenta casas, donde dizque avría mill vezinos porque biven muchos en una casa.[2] Estas casas son de manera de alfaneques grandíssimos. Dixeron que los avían resçebido con gran solenidad según su costumbre. Y todos así hombres como mugeres los venían a ver, y aposentáronlos en las mejores casas. Los quales los tocavan y les besavan las manos y los pies maravillándose y creyendo que venían del çielo y así se lo davan a entender. Dávanles de comer de lo que tenían. Dixeron que en llegando los llevaron de braços los más honrrados del pueblo a la casa principal, y diéronles dos sillas en que se assentaron, y ellos todos se assentaron en el suelo en derredor dellos.[3] El yndio que con ellos yva les notificó la manera de bivir de los cristianos y cómo eran buena gente.

---

[1] Margin: Éste deve ser Barocoa.
[2] Margin: Señal de ser pacíficos.
[3] Margin: No[ta].

Monday 5 November

At dawn he ordered the flagship and the other ships to be beached for careening, not all at the same time, but so as to leave two in position for reasons of security, although he says that those people were very trustworthy and all the ships could be beached at the same time without fear. Meanwhile the boatswain of the Niña came to claim a reward of the Admiral because he had found mastic, but he did not have the sample with him because he had dropped it. The Admiral promised him the reward and sent Rodrigo Sánchez and master Diego off to the trees and they brought a little of it back, which the Admiral kept, together with a portion of the tree, to take back to the Monarchs, and he says that it was recognisably mastic, although it has to be harvested in season, and that there was enough in that region to collect a thousand quintales every year.[87] He says he found there a great deal of that wood which seemed to him to be aloe. He further says that that harbour of Mares is one of the best in the world with the mildest breezes and gentlest people, and because it has a cape with a rocky high point on which could be built a fortress, so that, if that area turned out to be rich and important, the merchants would be safe from any other nations. He says, *Our Lord, in whose hands lie all victories, disposes everything for his service.* He says that an Indian made signs that mastic was good for when they suffered stomach-ache.

Tuesday 6 November

Yesterday evening, says the Admiral, the two men he had sent out to investigate the interior returned and described how they had walked the 12 leagues to a village of 50 houses, where he says that there must have been 1000 people because many live together in one house. These houses are like very large tents. They said that they had been received with great solemnity after their custom, and they all, men and women alike, came to see them and put them up in the best houses. The people touched them and kissed their hands and feet and marvelled at them, believing that they had come from heaven and that is what they gave them to understand. They gave them things to eat from what they had. They said that when they arrived, the most honourable men of the village carried them on their shoulders to the main house and gave them two seats on which to sit, and they all sat on the floor around them. The Indian who accompanied them told them how the Christians lived and how they were good people. Then the men

Después saliéronse los hombres, y entraron las mugeres y sentáronse de la misma manera en derredor dellos, besándoles las manos y los pies, palpándolos, atentándolos si eran de carne y de güesso como ellos. Rogávanles que se estuviesen allí con ellos al menos por çinco días. Mostraron la canela y pimienta y otras espeçias que el Almirante les avía dado, y dixéronles por señas que mucha della avía çerca de allí al sueste, pero que en allí no sabían si la avía. Visto cómo no tenían recaudo de çiudad se bolvieron, y que si quisieran dar lugar a los que con ellos se querían venir que más de quinientos hombres y mugeres vinieran con ellos, porque pensavan que se bolvían al çielo. Vino, enpero, con ellos un principal del pueblo y un su hijo y un hombre suyo. Habló con ellos el Almirante, hízoles mucha honrra, señalóle muchas tierras e yslas que avía en aquellas partes. Pensó de traerlo a los Reyes, y dizque no supo qué se le antojó parez que de miedo; y de noche escuro quísose yr a tierra. Y el Almirante dizque porque tenía la nao en seco en tierra, no le queriendo enojar le dexó yr, diziendo que en amaneçiendo tornaría; el qual nunca tornó.[1] Hallaron los dos cristianos por el camino mucha gente que atravesava a sus pueblos mugeres y hombres con un tizón en la mano, yervas para tomar sus sahumerios que acostumbravan. No hallaron poblaçión por el camino de más de çinco casas, y todos les hazían el mismo acatamiento. Vieron muchas maneras de árboles e yervas y flores odoríferas. Vieron aves de muchas maneras diversas de las de España, salvo perdizes y ruyseñores que cantavan y ánsares que destos ay allí hartos. Bestias de quatro pies no vieron, salvo perros que no ladravan. La tierra muy fértil y muy labrada de aquellos niames[2] y faxo[n]es y havas muy diversas de las nuestras; eso mismo panizo y mucha cantidad de algodón cogido y filado y obrado, y que en una sola casa avían visto más de quinientas arrovas, y que se pudiera aver allí cada año quatro mill quintales.[3] Dize el Almirante que le pareçía que no lo sembravan y que da fruto todo el año: es muy fino, tiene el capillo grande. Todo lo que aquella gente tenía dizque dava por muy vil preçio, y que una gran espuerta de algodón dava por cabo de agujeta o otra cosa que se le dé. *Son gente,* dize el Almirante, *muy sin mal ni de guerra,*[4] *desnudos todos, hombres y mugeres, como sus madres los parió. Verdad es que las mugeres traen una cosa de algodón solamente, tan grande que le cobija su natura y no más. Y son ellas de muy buen acatamiento ni muy negras*[5] *salvo menos que canarias.*

---

[1] Margin: ¡Cómo necio!
[2] MS: ?mames.
[3] Margin: Y aun diez mill.
[4] Margin: No[ta].
[5] MS: negros.

went out and the women came in and sat around them in the same way, kissing their hands and feet, touching them to see if they were of flesh and blood like them. They asked them to stay there with them for at least five days. They showed them the cinnamon and the pepper and other spices which the Admiral had given them, and the people said in sign language that there was a lot of it nearby to the SE, but that they did not know if there was any thereabouts. When they found no indication of any city, they returned, and if they had allowed all those who wanted to do so, more than 500 men and women would have come with them, because they thought that they were returning to heaven. However, one of the elders of the village came with them with his son and a manservant. The Admiral spoke with them, paid them many courtesies, and he pointed out many lands and islands which there were in that region. The Admiral thought about bringing him back to the Monarchs, and says that he did not know what came over him but apparently out of fear and the dark night he wanted to go ashore. And the Admiral says that because the flagship was on dry land, not wishing to upset him he let him go. The Indian said that in the morning he would return; but he never came back. The two Christians found many people, men and women, on their journey who were on their way to their villages carrying a smouldering brand of herbs which they are accustomed to smoke.[88] They found no village on the way with more than five houses, and all treated them with the same respect. They saw many kinds of trees and plants and fragrant flowers. They saw birds of many kinds, different from those of Spain, except partridges and nightingales which sang,[89] and geese, of which there are a great many. They saw no four-legged animals except dogs which did not bark. The land was very fertile and cultivated with those 'niames' and kidney beans and broad beans all very unlike our own; likewise, Indian corn and a great quantity of cotton, picked and spun and woven; in a single house they had seen more than 500 arrobas, and 4000 quintales a year could be obtained there. The Admiral says that it seemed to him that they did not cultivate it and that it fruits all year round. It is very fine and produces large bolls. He says that everything those people had they gave for a very low price, and that they would give a great basket of cotton for the end of a leather thong or whatever else they are given. *They are people,* says the Admiral, *completely without evil or aggression, naked every one of them, men and women, as the day they were born. It is true that the women wear only a cotton garment, large enough to cover their genitals, but no more. They are very good looking, not very black, rather less so than the Canary Islanders. Most Serene Princes* (says

Tengo por dicho, sereníssimos prínçipes (dize aquí el Almirante) *que sabiendo la lengua dispuesta suya personas devotas religiosas que luego todos se tornarían cristianos. Y así espero en Nuestro Señor que Vuestras Altezas se determinarán a ello con mucha diligençia*[1] *para tornar a la Iglesia tan grandes pueblos y las convertirán, así como an destruído aquellos que no quisieron confessar el Padre y el Hijo y el Espíritu Sancto; y después de sus días, que todos somos mortales, dexarán sus reynos en muy tranquilo estado, y limpios de heregía y maldad, y serán bien resçebidos delante el Eterno Criador al qual plega de les dar larga vida y acreçentamiento grande de mayores reynos y señoríos, y voluntad y disposiçión para acreçentar la sancta religión cristiana, así como hasta aquí tienen fecho. Amén. Oy tiré la nao de monte y me despacho para partir el jueves en nombre de Dios e yr al sueste a buscar del oro y espeçerías y descobrir tierra.* Estas todas son palabras del Almirante, el qual pensó partir el jueves, pero porque le hizo el viento contrario no pudo partir hasta doze días de noviembre.

Lunes 12 de noviembre

Partió del puerto y río de Mares al rendir del quarto de alva para yr a una ysla que mucho affirmavan los yndios que traya que se llamava Baveque, adonde según dizen por señas que la gente della coge el oro con candelas de noche en la playa, y después con martillo dizque hazían vergas dello. Y para yr a ella era menester poner la proa al leste quarta del sueste. Después de aver andado ocho leguas por la costa delante halló un río, y dende andadas otras quatro halló otro río que pareçía muy caudaloso y mayor que ninguno de los otros que avía hallado. No se quiso detener ni entrar en alguno dellos por dos respectos: el uno y principal porque el tiempo y viento era bueno para yr en demanda de la dicha ysla de Babeque; lo otro porque si en él oviera alguna populosa o famosa çiudad çerca de la mar se pareçiera, y para yr por el río arriba eran menester navíos pequeños, lo que no eran los que llevava; y así se perdiera también mucho tiempo, y los semejantes ríos son cosa para descobrirse por sí. Toda aquella costa era poblada mayormente çerca del río a quien puso por nombre el Río del Sol. Dixo que el domingo antes, onze de noviembre, le avía pareçido que fuera bien tomar algunas personas de las de aquel río para llevar a los Reyes porque aprendieran nuestra lengua, para saber lo que hay en la tierra y porque bolviendo sean lenguas de los cristianos[2] y tomen nuestras costumbres y las cosas

---

[1] Margin: No[ta].
[2] Margin: No[ta].

Journal of Columbus

the Admiral at this point), *I hold that once dedicated and religious people knew their language and put it to use, they would all become Christians. And so I hope in Our Lord that Your Highnesses will determine with all speed to bring such great peoples to the Church and convert them, just as you have destroyed those who refused to confess the Father and the Son and the Holy Ghost; and at the end of your days, for we are all mortal, you will leave your kingdoms in tranquillity, free from heresy and evil, and will be well received before the Eternal Creator, whom it may please to grant you long life and great increase of your many kingdoms and possessions, and the will and the inclination to spread the holy Christian religion as you have done hitherto. Amen. Today I will refloat the flagship and I am readying myself to set out on Thursday in the name of God to go SE and seek the gold and spices and discover land.*[90] These are all the words of the Admiral, who planned to leave on Thursday, but because the wind was against him, could not depart until the twelfth day of November.

Monday 12 November

He left the harbour and river of Mares at the end of the dawn watch to go to an island which the Indians he had with him insisted was called Babeque[91] where, according to the signs they made, the people collect gold by candlelight at night on the beach and afterwards beat it into bars with hammers. To go there it was necessary to steer E by S. After sailing 8 leagues along the coast he found a river,[92] and after a further 4 leagues he found another[93] which seemed to be of great volume and larger than any of the others he had found. He did not wish to delay and enter any of them for two reasons: the first and principal one was that the weather and wind were favourable for searching for the island of Babeque; the other was that if there were any large or important city by the shore it would be apparent, and that to go up-river needed small ships, which his were not, and so much time would be wasted, and such rivers need to be explored in their own right. All that coast was populated, particularly near the river to which he gave the name Río del Sol.[94] He said that the previous Sunday, 11 November, it had seemed to him a good idea to capture some of the people from that river and to take them to the monarchs to learn our language, so that we would know what there is in this land and so that on their return they could act as interpreters for

de la fe. *Porque yo vi e cognozco* (dize el Almirante) *que esta gente no tiene secta ninguna ni son ydólatras, salvo muy mansos y sin saber qué sea mal*[1] *ni matar a otros ni prender y sin armas, y tan temerosos que a una persona de los nuestros fuyen çiento dellos aunque burlen con ellos. Y crédulos y cognosçedores que ay Dios en el çielo, e firmes que nosotros avemos venido del çielo, y muy presto[s] a qualquiera oración que nos les digamos que digan y hazen el señal de la cruz +. Así que deven Vuestras Altezas determinarse a los hazer cristianos, que creo que si comiençan en poco tiempo acabará[n] de los aver convertido a nuestra sancta fe multidumbre de pueblos, y cobrando grandes señoríos y riquezas y todos sus pueblos de la España. Porque sin duda es en estas tierras grandíssima suma de oro que no sin causa dizen estos yndios que yo traygo que ha en estas yslas lugares adonde cavan el oro y lo traen al pescueço, a las orejas, y a los braços, e a las piernas, y son manillas muy gruessas. Y también ha piedras y ha perlas preciosas y infinita espeçería. Y en este Río de Mares de adonde partí esta noche sin duda ha grandíssima cantidad de almáçiga, y mayor si mayor se quisiere hazer.*[2] *Porque los mismos árboles plantándolos prenden de ligero y ha muchos y muy grandes y tienen la hoja como lentisco y el fructo, salvo que es mayor así los árboles como la hoja como dize Plinio e yo e visto en la ysla de Xío en el Arcipiélago. Y mandé sangrar muchos destos árboles para ver si echaría[n] resina para la traer y como aya siempre llovido el tiempo que yo e estado en el dicho río no e podido aver della, salvo muy poquita que traygo a Vuestras Altezas; y también puede ser que no es el tiempo para los sangrar, que esto creo que conviene al tiempo que los árboles comiençan a salir del invierno y quieren echar la flor, y acá ya tienen el fruto quasi maduro agora. Y también aquí se avría grande suma de algodón, y creo que se vendería muy bien acá sin le llevar a España,*[3] *salvo a las grandes çiudades del Gran Can que se descubrirán sin duda, y otras muchas de otros señores que avrán en dicha servir a Vuestras Altezas y adonde se les darán de otras cosas de España y de las tierras de oriente,*[4] *pues éstas son a nos en poniente. Y aquí ha también infinito lignáloe aunque no es cosa para hazer gran caudal, mas del almáçiga es de entender bien, porque no la ha salvo en la dicha ysla de Xío, y creo que sacan dello bien çinquenta mill ducados, si mal no me acuerdo. Y ha aquí en la boca del dicho río el mejor puerto que fasta oy vi, limpio e ancho e fondo y buen lugar y asiento para hazer una villa e fuerte e que qualesquier navíos se puedan llegar el*

---

[1] Margin: No[ta].
[2] Margin: No[ta].
[3] Margin: No[ta].
[4] Margin: No[ta].

the Christians and adopt our customs and faith. *For I have seen and recognise (says the Admiral) that these people have no religion, nor are they idolaters, rather they are very gentle and know nothing of evil, nor murder nor theft nor weapons, and so timorous that a hundred of them flee from one of our people even though they may only be teasing them. They are trusting and know that there is a God in heaven, and firmly believe that we have come from heaven, and they are ready to repeat any prayer that we say to them and they make the sign of the cross +. So Your Highnesses must resolve to make them Christians, for I believe that once you begin you will in a short space succeed in converting to our faith a multitude of peoples and acquiring great kingdoms and riches and all their peoples for Spain. Because without doubt there is in these lands a huge amount of gold and not without reason do these Indians I have with me say that there are in these islands places where they dig up gold and wear it at the neck and from the ears and in very thick bracelets on their arms and legs. And there are also pearls and precious stones and an infinity of spices. And by this river of Mares which I left last night there is without doubt a huge quantity of mastic, and more if one cared to produce more, for the trees take well when planted, they are large and there are many of them and they have leaves like a lentisk and fruit too, except that it is larger, both the trees and the leaves, as Pliny says*[95] *and I have seen on the island of Chios in the Archipelago. I ordered many of these trees to be tapped to see if they would produce resin to bring back, and as it has been raining throughout the time that I have been at the said river I have not been able to obtain any, except a very little which I am bringing back to Your Highnesses; it may also be that it is not the season for tapping them, which is best done I believe when the trees are emerging from winter and about to flower, and here the fruit is already nearly ripe. There would also be a great amount of cotton here and I believe it could very well be sold here without taking it back to Spain, but taking it instead to the great cities of the Great Khan which will doubtless be discovered, and to the many other cities of other lords who will have great pleasure in serving Your Highnesses and where they will be supplied with other goods from Spain and the lands of the Orient, since these are to the W of us. There is also an infinite amount of aloe here, although this is not something from which to make a great fortune, but from mastic much can be expected because it is only found on the island of Chios and I believe that they earn a good 50000 ducats from it, if I remember rightly. And there is here at the mouth of the said river the best harbour I have yet seen, clean and wide and deep and a*

bordo a los muros, e tierra muy temperada y alta y muy buenas aguas. Así que ayer vino a bordo de la nao una almadía con seys mancebos y los çinco entraron en la nao; estos mandé detener e los traygo.[1] Y después enbié a una casa que es de la parte del río del poniente y truxeron siete cabeças de mugeres entre chicas e grandes y tres niños. Esto hize porque mejor se comportan los hombres en España aviendo mugeres de su tierra que sin ellas, porque ya otras muchas vezes se acaeçió traer hombres de Guinea para que deprendiesen la lengua en Portugal y después que bolvían y pensavan de se aprovechar dellos en su tierra por la buena compañía que le avían hecho y dádivas que se les avían dado, en llegando en tierra jamás parecía[n].[2] Otros no lo hazían así. Así que teniendo sus mugeres ternán gana de negociar lo que se les encargare y también estas mugeres mucho enseñarán a los nuestros su lengua. La qual es toda una en todas estas yslas de Yndia, y todos se entienden y todas las andan con sus almadías, lo que no han en Guinea adonde es mill maneras de lenguas que la una no entiende la otra. Esta noche vino a bordo en una almadía el marido de una destas mugeres y padre de tres fijos, un macho y dos fembras, y dixo que yo le dexase venir con ellos y a mí me aplogó mucho,[3] y quedan agora todos consolados con él, que deven todos ser parientes y él es ya hombre de 45 años. Todas estas palabras son formales del Almirante. Dize también arriba que hazía algún frío, y por esto que no le fuera buen consejo en invierno navegar al norte para descubrir.[4] Navegó este lunes hasta el sol puesto 18 leguas al leste quarta del sueste hasta un cabo a que puso por nombre el Cabo de Cuba.

Martes 13 de noviembre

Esta noche toda estuvo a la corda como dizen los marineros que es andar barloventeando y no andar nada, por ver un abra que es un abertura de sierras como entre sierra y sierra que le començó a ver al poner del sol, adonde se mostravan dos grandíssimas montañas,[5] y pareçía que se apartava la tierra de Cuba con aquella de Bofío, y esto dezían los yndios que consigo llevavan por señas. Venido el día claro dio las velas sobre la tierra y passó una punta que le

---

[1] Margin: No fue lo mejor del mundo esto.

[2] Margin: ¡Mira qué maravilla!

[3] Margin: ¿Por qué no le distes sus hijos?

[4] Margin: De esto que aquí dize pareçe que si navegara hazia el norte en dos días sin duda descubriera la Florida.

[5] Margin: Estas montañas eran la una el Cabo de Cuba que se llama la Punta de Mahiçi...

good place and situation for a town and fortress alongside whose walls any ship would be able to draw, and a very temperate land and high and with very good water. Yesterday a canoe drew up beside the ship with six youths, of whom five came aboard; these I ordered to be detained and I am bringing them with me. Later I sent to a house on the western side of the river and they brought back seven women, young and adult, and three children. I did this so that the men would behave themselves better in Spain with their own women than without them, because on many other occasions men have been brought from Guinea to learn the language in Portugal and when they were returned and it was thought that some advantage might be gained from them in their own land in consideration for the good treatment they had received and the presents they had been given, once they arrived on land, they never appeared again. Others did not do this. If they have their women they will be more willing to provide the cooperation expected of them and these women will also do much to teach our men their language. The language is one and the same in all these islands of India; they all understand each other and go about the islands in their canoes, which is not the case in Guinea where there are a thousand different languages, with one not understanding the other. This evening there came alongside in a canoe the husband of one of these women and father of three children, one male and two female, and asked me to let him come with them, which I was very pleased to do, and they are all now consoled with him. They must all be related and he is already a man of 45 years. These are all the Admiral's own words. He also says earlier that it was somewhat cold and that for this reason it would not be a good idea to sail north to discover in winter. This Monday he sailed till sunset 18 leagues E by S as far as a cape to which he gave the name Cabo de Cuba.[96]

Tuesday 13 November

All this night he stood 'a la corda' as the mariners say, which is beating back and forth without going anywhere, in order to take a look at a gap or opening between the mountains, as between one range and another, which he first saw at sunset. The appearance of two very large mountains gave the impression that the land of Cuba was separated from that of Bofío, and this is what the Indians they had with them said in sign language. When daylight came he sailed towards the

pareçió anoche obra de dos leguas, y entró en un grande golpho çinco leguas al sursudueste y le quedavan otras çinco para llegar al cabo adonde en medio de dos grandes montes hazía un degollado el qual no pudo determinar si era entrada de mar. Y porque deseava yr a la ysla que llamavan Baneque, adonde tenía nueva según él entendía que avía mucho oro la qual isla le salía al leste, como no vido alguna grande población para ponerse al rigor del viento que le creçía más que nunca hasta allí, acordó de hazerse a la mar y andar al leste con el viento que era norte, y andava 8 millas cada ora y desde las diez del día que tomó aquella derrota hasta el poner del sol anduvo 56[1] millas que son 14 leguas al leste desde el Cabo de Cuba. Y de la otra tierra de Bohío que le quedava a sotavento comença ndo del cabo del sobredicho golpho descubrió a su pareçer 80 millas que son XX leguas, y corríase toda aquella costa lesueste y güesnorueste.

Miércoles 14 de noviembre

Toda la noche de ayer anduvo al reparo y barloventeando (porque dezía que no era razón de navegar entre aquellas yslas de noche hasta que las oviese descubierto) porque los yndios que traya le dixeron ayer martes que avría tres jornadas desde el Río de Mares hasta la ysla de Baneque que se deve entender jornadas de sus almadías que pueden andar 7 leguas, y el viento también le escaseava y aviendo de yr al leste, no podía sino a la quarta del sueste y por otros ynconvinientes que allí refiere, se ovo de tener hasta la mañana. Al salir del sol determinó de yr a buscar puerto, porque de norte se avía mudado el viento al nordeste, y si puerto no hallara, fuérale neçessario bolver atrás a los puertos que dexava en la ysla de Cuba. Llegó a tierra aviendo andado aquella noche 24 millas al leste quarta del sueste. Anduvo al sur *** millas hasta tierra, adonde vio muchas entradas y muchas ysletas y puertos, y porque el viento era mucho y la mar muy alterada, no osó acometer a entrar; antes corrió por la costa al norueste quarta del güeste mirando si avía puerto y vido que avía muchos pero no muy claros. Después de aver andado así 64 millas, halló una entrada muy honda, ancha un quarto de milla y buen puerto y río, donde entró y puso la proa al sursudueste y después al sur hasta llegar al sueste, todo de buena anchura y muy fondo, donde vido tantas yslas que no las pudo contar todas, de buena grandeza y muy altas tierras llenas de diversos árboles de mill maneras e infinitas palmas. Maravillóse en gran manera ver tantas yslas y tan altas y

---

[1] MS + seys.

land and passed a point which last night seemed to him a matter of 2 leagues away, and entered a great gulf 5 leagues SSW and there were still another 5 leagues before he reached the cape where there was a defile between two large hills; he could not tell if it was a sea inlet. And because he wished to go to the island they called Baneque where, as far as he understood from information he had, there was a lot of gold, which island was to the E of him, as he saw no large settlement where he could shelter from the wind which was growing stronger than it ever had been until then, he decided to go out to sea and sail E before the wind, which was N, and he made 8 miles an hour and from ten o'clock when he took that course until sunset he made 56 miles which is 14 leagues E from Cabo de Cuba. And of the other land of Bohío to leeward, starting from the cape of the aforementioned gulf, he discovered what he reckoned to be 80 miles, which is 20 leagues, and all that coast ran ESE and WNW.[97]

Wednesday 14 November

All last night he cautiously beat about (because he said that it was not a good idea to sail among those islands at night until he had investigated them), and because the Indians he had with him told him yesterday, Tuesday, that it would be three days' journey from Río de Mares to the island of Baneque, that is days in their canoes which can make 7 leagues; and because the wind was also against them, and having to go E he could only manage E by S, and for other hindrances to which he there refers, he had to hold fast until the morning. At sunrise he decided to seek a harbour, because the wind had veered from N to NE and if he did not find a harbour he would need to turn back to those harbours he left on the island of Cuba. He reached land that night having made 24 miles E by S. He went S *** miles to the shore where he saw many inlets and many islets and harbours, and because the wind was strong and the sea very rough he did not dare to attempt to enter; instead he ran NW by W along the coast looking for a harbour and he saw that there were several, but not very clear. After making 64 miles in this way he found a very deep inlet, a quarter of a mile wide, and a good harbour and river which he entered and turn the prow SSW and then S until he reached the SE, all of it of good width and very deep, where he saw so many islands that he could not count all of them, all of good size and very high lands, full of a thousand different kinds of trees and an infinite number of palms. He marvelled greatly at seeing so many islands and so high, and assures the

çertifica a los Reyes que las montañas que desde antier a visto por estas costas y las destas yslas, que le pareçe que no las ay más altas en el mundo, ni tan hermosas y claras, sin niebla ni nieve, y al pie dellas grandíssimo fondo. Y dize que cree que estas yslas son aquellas innumerables que en los mapamundos en fin de oriente se ponen. Y dixo que creía que avía grandíssimas riquezas y piedras preçiosas y espeçería en ellas, y que duran muy mucho al sur y se ensanchan a toda parte. Púsoles nombre la Mar de Nuestra Señora, y al puerto que está çerca de la boca de la entrada de las dichas yslas, puso Puerto del Príncipe, en el qual no entró más de velle desde fuera hasta otra buelta que dio el sábado de la semana venidera, como allí pareçerá. Dize tantas y tales cosas de la fertilidad y hermosura y altura destas yslas que halló en este puerto, que dize a los Reyes que no se maravillen de encareçellas tanto, porque los çertifica que cree que no dize la çentéssima parte. Algunas dellas que pareçía que llegan al çielo y hechas como puntas de diamantes; otras que sobre su gran altura tienen ençima como una mesa, y al pie dellas fondo grandíssimo que podrá llegar a ellas una grandíssima carraca, todas llenas de arboledas y sin peñas.

Jueves 15 de noviembre

Acordó de andallas estas yslas con las barcas de los navíos y dize maravillas dellas, y que halló almáçiga e infinito lignáloe, y algunas dellas eran labradas de las rayzes de que hazen su pan los yndios, y halló aver ençendido huego en algunos lugares. Agua dulçe no vido. Gente avía alguna y huyeron. En todo lo que anduvo halló hondo de quinze y diez [y] seys braças y todo basa que quiere dezir que el suelo de abaxo es arena y no peñas, lo que mucho desean los marineros, porque las peñas cortan los cables de las anclas de las naos.

Viernes 16 de noviembre

Porque en todas las partes, yslas y tierras donde entrava dexava siempre puesta una cruz, entró en la barca y fue a la boca de aquellos puertos y en una punta de la tierra halló dos maderos muy grandes, uno más largo que el otro y el uno sobre el otro hechos cruz, que dizque un carpintero no los pudiera poner más proporcionados. Y adorada aquella cruz, mandó hazer de los mismos maderos una muy grande y alta cruz. Halló cañas por aquella playa que no sabía dónde

Monarchs that it seems to him that there can be no higher mountains in the world than those which since the day before yesterday he has seen on this coast and those of these islands, nor any so beautiful and clear without cloud or snow, and with such deep waters at their foot. And he says that he believes that these are the innumerable islands which appear on world maps at the eastern edge. He says that he thought that there were great riches and precious stones and spices on them, and that they extend a long way to the S and spread out in every direction. He called them the Mar de Nuestra Señora,[98] and the harbour near the mouth of the entrance to those islands he called Puerto del Príncipe. He looked at it from outside but did not enter it until he returned on the Saturday of the following week, as will be apparent.[99] He says so many and such things about the fertility and beauty and loftiness of these islands which he found in this harbour, that he tells the Monarchs that they should not wonder that he enthuses so much, because he assures them that he believes that he is not telling the hundredth part. Some of them seem to reach the sky and are pointed like diamonds; others form a sort of table at their highest peak and at their foot the water is very deep so that a huge carrack could reach them, and they are all full of groves and not rocky.

Thursday 15 November

He decided to sail among these islands in the ships' boats and says marvellous things about them, and that he found mastic and an infinite amount of aloe; some of them were cultivated for the roots from which the Indians make their bread,[100] and he found that fires had been started in some places. He saw no fresh water. There were some people who fled. Wherever he sailed he found a depth of 15 or 16 fathoms and it was all 'basa', which means that the bottom is sandy and not rocky, which sailors very much hope for, because rocks cut the ships' anchor cables.

Friday 16 November

Because wherever he went, whatever island or land he entered, he always left a cross in position, he got into the boat and went to the mouth of one of those harbours and on a point of land he found two large pieces of timber, one larger than the other and one on top of the other like a cross, so that he says that a carpenter could not have made them better proportioned. When they had worshipped that cross he ordered a very large high cross to be made out of the same pieces of wood. He found canes on that beach but did not know where they came

naçían, y creya que las traería algún río y las echava a la playa y tenía en esto razón. Fue a una cala dentro de la entrada del puerto de la parte del sueste (cala es una entrada angosta que entra el agua del mar en la tierra); allí hazía un alto de piedra y peña como cabo, y al pie de él era muy fondo, que la mayor carraca del mundo pudiera poner el bordo en tierra y avía un lugar o rincón donde podían estar seys navíos sin anclas como en una sala. Pareçióle que se podía hazer allí una fortaleza a poca costa, si en algún tiempo en aquella mar de yslas resultase algún resgate famoso. Bolviéndose a la nao halló los yndios que consigo traya que pescavan caracoles muy grandes que en aquellas mares ay, y hizo entrar la gente allí e buscar si avía nácaras que son las hostias donde crían las perlas, y hallaron muchas, pero no perlas, y atribuyólo a que no devía de ser el tiempo dellas que creya él que era por mayo y junio. Hallaron los marineros un animal que parecía taso o taxo. Pescaron también con redes y hallaron un pece entre otros muchos que pareçía proprio puerco, no como tonina; el qual dizque era todo concha muy tiesta, y no tenía cosa blanda sino la cola y los ojos y un agujero debaxo della para expeler sus superfluydades. Mandólo salar para llevar que lo viesen los Reyes.

Sábado 17 de noviembre

Entró en la barca por la mañana y fue a ver las yslas que no avía visto por la vanda del sudueste. Vido muchas otras y muy fértiles y muy graçiosas y entremedio dellas muy gran fondo. Algunas dellas dividían arroyos de agua dulçe, y creya que aquella agua y arroyos salían de algunas fuentes que manavan en los altos de las sierras de las yslas. De aquí yendo adelante halló una ribera de agua muy hermosa y dulçe y salía muy fría por lo enxuto della. Avía un prado muy lindo y palmas muchas y altíssimas más que las que avía visto. Halló nuezes grandes de las de Yndia,[1] creo que dize, y ratones grandes de los de Yndia también, y cangrejos grandíssimos. Aves vido muchas y olor vehemente de almizque y creyó que lo devía de aver allí. Este día, de seys mancevos que tomó en el Río de Mares que mandó que fuesen en la caravela Niña, se huyeron los dos más viejos.

Domingo 18 de noviembre

Salió en las barcas otra vez con mucha gente de los navíos y fue a poner la gran cruz que avía mandado hazer de los dichos dos maderos a la boca de la entrada

---

[1] Margin: Hutías devían de ser.

from, and believed that some river must have brought them and left them on the shore, and in this he was correct. He went to a creek within the entrance to the harbour to the SE (a creek is a narrow entrance where the sea enters the land); there was there a high rocky point like a cape, and at the foot of it the water was very deep, so that the largest carrack in the world could come alongside and there was a place or corner where six ships could lie without anchors, as if in a drawing room. It seemed to him that a fortress could be made there at little cost, if a large volume of trade should ever develop in that sea of islands. Returning to the ship he found that the Indians he had with him were fishing for the large snails which there are in those waters, and he had the men dive in and see if there were oysters, which are the hosts in which pearls grow, and they found many, but no pearls, and he attributed this to the fact that it could not be the season for them which he believed was in May and June. The sailors found an animal which looked like a badger.[101] They also fished with nets and found a fish among many others which looked just like a pig, not like a tunny, which he says was all very hard shell, and had nothing soft except the tail and the eyes and a hole underneath to expel waste.[102] He ordered it to be salted in order to take it back for the Monarchs to see.

Saturday 17 November

He entered the boat in the morning and went to see the islands which he had not seen, to the SW. He saw many others, very fertile and delightful with deep water between them. Some of them were crossed by streams of fresh water and he believed that water and those streams came from some springs which flowed from the high peaks of the islands. Proceeding from here he found a stream of beautiful sweet water which flowed very cold through a narrow crack. There was a very pretty meadow and many palms which were much taller than those he had seen. He found large nuts like those of India,[103] I think he says, and large mice like those of India too, and very large crabs. He saw many birds, and there was a strong smell of musk and he thought there must be some there. On this day, of the six youths he took from the Río de Mares and ordered to travel in the caravel Niña, the two eldest ran away.

Sunday 18 November

He went out in the boats again with many of the ships' crew and went to set up the great cross which he had ordered made from the two timbers mentioned

del dicho Puerto del Prínçipe en un lugar vistoso y descubierto de árboles; ella muy alta y muy hermosa vista. Dize que la mar creçe y descreçe allí mucho más que en otro puerto de lo que por aquella tierra aya visto y que no es más maravilla por las muchas yslas, y que la marea es al revés de las nuestras, porque allí la luna al sudueste quarta del sur es baxamar en aquel puesto. No partió de aquí por ser domingo.

Lunes 19 de noviembre

Partió antes que el sol saliese y con calma, y después al mediodía ventó algo al leste y navegó al nornordeste. Al poner del sol le quedava el Puerto del Prínçipe al sursudueste y estaría de él siete leguas. Vido la ysla de Baneque al leste justo de la qual estaría 60 millas. Navegó toda esta noche al nordeste; escasso andaría 60 millas y hasta las diez del día martes otras doze que son por todas 18 leguas, y al nordeste quarta del norte.

Martes 20 de noviembre

Quedávanle el Baneque o las yslas del Baneque al lesueste de donde salía el viento que llevava contrario. Y viendo que no se mudava y la mar se alterava, determinó de dar la buelta al Puerto del Prínçipe, de donde avía salido, que le quedava XXV leguas. No quiso yr a la ysleta que llamó Ysabela que le estava 12 leguas[1] que pudiera yr a surgir aquel día, por dos razones: la una porque vido dos yslas al sur, las quería ver; la otra porque los yndios que traya que avía tomado en Guanahaní que llamó San Salvador que estava ocho leguas de aquella Ysabela, no se le fuesen, de los quales dizque tiene necessidad y por traellos a Castilla etc. Tenían dizque entendido que en hallando oro los avía el Almirante de dexar tornar a su tierra. Llegó en paraje del Puerto del Prínçipe, pero no lo pudo tomar porque era de noche y porque lo decayeron las corrientes al norueste. Tornó a dar la buelta y puso la proa al nordeste con viento rezio; amansó y mudóse el viento al terçero quarto de la noche, puso la proa en el leste quarta del nordeste; el viento era susueste; y mudóse al alva de todo en sur y tocava en el sueste. Salido el [sol] marcó el Puerto del Prínçipe y quedávale al sudueste y quasi a la quarta del güeste y estaría de él 48 millas que son 12 leguas.

---

[1] MS: lueguas.

earlier, at the mouth of the entrance to the Puerto del Príncipe, in a clearing in a prominent position. It was very high and made a fine sight. He says that the sea rises and falls there much more than in any other harbour he has seen in that land and that it is not surprising considering the many islands, and that the tide is the opposite of our own, because there when the moon is SW by S it is low tide in that position.[104] He did not set out from here because it was Sunday.

Monday 19 November

He left before sunrise in a calm, and later, at midday, there was a light E wind and he steered NNE. At sunset Puerto del Príncipe lay SSW, at about 7 leagues. He saw the island of Baneque due E at a distance of about 60 miles. All that night he steered NE; he would have made scarcely 60 miles and by 10 o'clock on Tuesday another 12 NE by N, which is 18 leagues in all.

Tuesday 20 November

Baneque, or the islands of Baneque, lay ESE, the direction from which the wind which was against him was blowing. And seeing that there was no change and that the sea was getting up, he decided to return to Puerto del Príncipe, from where he had set out, which lay 25 leagues away. He did not wish to go to the island he called Isabela which was 12 leagues away, and off which he could have anchored that day, for two reasons: one because he saw two islands to the S and wished to examine them; the other so that the Indians he had with him and had taken in Guanahaní, which he called San Salvador and which was 8 leagues from Isabela, would not escape. He says that he needs them and wishes to take them to Castile, etc. He says that they had understood that once he had found gold the Admiral would allow them to return to their own land. He reached the area of Puerto del Príncipe but could not make it because it was night and because the currents were taking him NW. He turned round again and steered NE with a strong wind; the wind fell and changed at the third watch of the night, he steered E by N; the wind was SSE; at dawn it shifted due S and slightly SE. After sunrise he took a bearing on Puerto del Príncipe which lay SW and almost SW by W and he must have been 48 miles distant, which is 12 leagues.

Miércoles 21 de noviembre

Al sol salido navegó al leste con viento sur. Anduvo poco por la mar contraria. Hasta oras de bísperas ovo andado 24 millas. Después se mudó el viento al leste y anduvo al sur quarta del sueste y al poner del sol avía andado 12 millas. Aquí se halló el Almirante en 42 grados de la línea equinoçial a la parte del norte como en el Puerto de Mares. Pero aquí dize que tiene suspenso el quadrante hasta llegar a tierra que lo adobe. Por manera que le pareçía que no devía distar tanto y tenía razón porque no era possible como no estén estas yslas sino en *** grados. Para creer que el quadrante andava bueno le movía ver dizque el norte tan alto como en Castilla. Y si esto es verdad mucho allegado y alto andava con la Florida. Pero ¿dónde están luego agora estas yslas que entre manos traya? Ayudava a esto que hazía dizque gran calor, pero claro es que si estuviera en la costa de la Florida que no oviera calor, sino frío. Y es también manifiesto que en quarenta y dos grados en ninguna parte de la tierra se cree hazer calor, si no fuese por alguna causa de per açidens, lo que hasta oy no creo yo que se sabe. Por este calor que allí el Almirante dize que padecía, arguye que en estas yndias y por allí donde andava devía de aver mucho oro. Este día se apartó Martín Alonso Pinçón[1] con la caravela Pinta, sin obediençia y voluntad del Almirante, por cudiçia, dizque pensando que un yndio que el Almirante avía mandado poner en aquella caravela le avía de dar mucho oro. Y así se fue sin esperar, sin causa de mal tiempo, sino porque quiso. Y dize aquí el Almirante: *Otras muchas me tiene hecho y dicho.*

Jueves 22

Miércoles en la noche navegó al sur quarta del sueste con el viento leste y era quasi calma. Al terçero quarto ventó nornordeste. Todavía yva al sur por ver aquella tierra que por allí le quedava y quando salió el sol se halló tan lexos como el día passado por las corrientes contrarias y quedávale la tierra quarenta millas. Esta noche Martín Alonso siguió el camino del leste para yr a la ysla de Baneque donde dizen los yndios que ay mucho oro, el qual yva a vista del Almirante y avría hasta él 16 millas. Anduvo el Almirante toda la noche la buelta de tierra y hizo tomar algunas de las velas y tener farol toda la noche porque le pareçió que venía hazia él, y la noche hizo muy clara y el ventizillo bueno para venir a él si quisiera.

---

[1] Margin: No[ta].

Wednesday 21 November

At sunrise he steered E with a S wind. He made little headway as the sea was against him. By vespers he had made 24 miles. Then the wind changed to the E and he went S by E and by sunset he had made 12 miles. At this point the Admiral found himself 42 degrees N of the equinoctial line, as at the Puerto de Mares. But here he says that he has suspended the use of the quadrant until he reaches land and can repair it. For it seemed to him that he could not be so far N, and he was right because it was not possible since these islands are only *** degrees N. He says that seeing the North Star as high as in Castile led him to believe that the quadrant was accurate. If that were so he was very nearly as far N as Florida.[105] But in that case where were these islands which he had close at hand? This was confirmed by the fact that it was, he says, very hot, but it is clear that if he had been off the Florida coast it would not have been hot but cold. And it is also obvious that nowhere on earth is it thought to be hot at 42 degrees, unless it were an exceptional case, which I do not believe has been known to this day. From this heat which the Admiral says he was suffering, he argues that in these Indies and in the area in which he was sailing there must be much gold. On this day Martín Alonso Pinzón went off with the caravel Pinta, without leave and against the Admiral's will, out of cupidity, he says, thinking that an Indian the Admiral had ordered to be placed on that caravel would give him much gold. And so he went off without waiting, and not because the weather was bad but because he wanted to. And the Admiral says here, *He has done me many other wrongs in deed and word.*

Thursday 22

On Wednesday night he steered S by E with the wind E and almost calm. At the third watch it blew NNE. He was still sailing S to examine that land which lay in that direction and when the sun rose he found himself as far away as the day before because of the contrary currents, and the shore was 40 miles away. During this night Martín Alonso followed the route to the E to go to the island of Baneque where the Indians say there is much gold; he was within sight of the Admiral at a distance of about 16 miles. All night the Admiral ran parallel to the land and had some sail taken in and a lantern lit all night because it seemed to him that Martín Alonso was approaching, and the night was very clear and the breeze favourable for returning to him if he wished.

Viernes 23 de noviembre

Navegó el Almirante todo el día hazia la tierra al sur, siempre con poco viento y la corriente nunca le dexó llegar a ella, antes estava oy tan lexos della al poner del sol como en la mañana. El viento era lesnordeste y razonable para yr al sur sino que era poco. Y sobre este cabo encavalga otra tierra o cabo que va también al leste a quien aquellos yndios que llevava llamava[n] Bohío, la qual dezían que era muy grande y que avía en ella gente que tenía un ojo en la frente,[1] y otros que se llamavan caníbales, a quien mostravan tener gran miedo. Y desque vieron que lleva este camino dizque no podían hablar porque los comían, y que son gente muy armada. El Almirante dize que bien cree que avía algo dello, mas que pues eran armados sería gente de razón, y creya que avrían captivado algunos y que porque no bolvían a sus tierras, dirían que los comían. Lo mismo creyan de los cristianos y del Almirante al prinçipio que algunos los vieron.

Sábado 24 de noviembre

Navegó aquella noche toda y a la ora de tercia del día tomó la tierra sobre la Ysla Llana en aquel mismo lugar donde avía arribado la semana passada quando yva a la ysla de Baneque. Al principio no osó llegar a la tierra porque le pareçió que aquella abra de sierras rompía la mar mucho en ella. Y en fin llegó a la Mar de Nuestra Señora donde avía las muchas yslas, y entró en el puerto que está junto a la boca de la entrada de las yslas, y dize que si él antes supiera este puerto y no se ocupara en ver las yslas de la Mar de Nuestra Señora, no le fuera neçessario bolver atrás, aunque dize que lo da por bien empleado por aver visto las dichas yslas. Así que llegando a tierra enbió la barca y tentó el puerto y halló muy buena barra, honda de seys braços y hasta veynte, y limpio, todo basa. Entró en él poniendo la proa al sudueste y después, bolviendo al güeste, quedando la Ysla Llana de la parte del norte, la qual con otra su vezina hazen una laguna de mar en que cabrían todas las naos de España y podían estar seguras sin amarras de todos los vientos.[2] Y esta entrada de la parte del sueste que se entra poniendo la proa al susudueste tiene la salida al güeste muy honda y muy ancha. Así que se puede passar entremedio de la dichas yslas, y por cognoscimiento dellas, a quien viniese de la mar de la parte del norte que es su travesía desta costa, están las dichas yslas al pie de una grande montaña que es su longura de

---

[1] Margin: Por aquí pareçe quán poco los entendía.
[2] Margin: Este deve ser el puerto que llamó Sancta Cathalina porque llegó a él su bíspera.

Friday 23 November

All day the Admiral headed for the land to the S, always with a light wind, and the current never allowed him to reach it; on the contrary, he was as far away today at sunset as he had been in the morning. The wind was ENE and reasonable for going S but it was very light. And beyond this cape he saw another higher land or cape which also extends to the E, which those Indians he had with him called Bohío and which they said was very large, and there were people there with one eye in their forehead, and others called cannibals[106] of whom they appeared to be in great fear. And when they saw that he was taking this course he said that they were speechless with fear that they would be eaten, and because they are heavily armed. The Admiral says that he well believes that there was something in what they said but that since they were armed they must be an intelligent people[107] and he believed that they must have captured some of them and that because they did not return home the others would say that they had been eaten. They believed the same thing of the Christians and the Admiral when some first saw them.

Saturday 24 November

He sailed the whole of that night and at the hour of terce he made land at Isla Llana at the same spot at which he had landed last week when he was going to the island of Baneque.[108] At first he dared not approach the shore because it seemed that the sea was breaking very violently in the opening between the mountains. He eventually reached the Mar de Nuestra Señora where the many islands were, and he entered the harbour near the mouth of the entrance to the islands and says that if he had known about this harbour then and had not spent time looking at the islands in the Mar de Nuestra Señora he would not have needed to turn back, although he says that he considers it worthwhile to have seen those islands.[109] On making land he sent out the boat and sounded the harbour and found a very good entrance, from 6 to 20 fathoms deep and with a clean, sandy bottom. He entered, steering SW and then, turning to the W, with the Isla Llana lying to the N. This island and a neighbouring one form a sea lagoon in which all the ships of Spain could lie safe from all winds without anchors. And this entrance on the SE side, which is entered by steering SSW, has a very deep and wide exit to the W. So that it is possible to pass between the two islands, and so that anyone coming from the sea to the N along this coast would recognise these islands, they are at the foot of a large mountain running E-W

leste güeste y es harto luenga y más alta y luenga que ninguna de todas las otras que están en esta costa adonde ay infinitas. Y haze fuera una restinga al luengo de la dicha montaña como un banco que llega hasta la entrada. Todo esto de la parte del sueste, y también de la parte de la Ysla Llana haze otra restinga, aunque ésta es pequeña. Y así entremedias de ambas ay grande anchura y fondo grande como dicho es. Luego a la entrada a la parte del sueste dentro en el mismo puerto vieron un río grande y muy hermoso y de más agua que hasta entonçes avían visto, y que benía[1] el agua dulçe hasta la mar. A la entrada tiene un banco, mas después de entra[r] es muy hondo de ocho y nueve braças. Está todo lleno de palmas y de muchas arboledas como los otros.

Domingo 25 de noviembre

Antes del sol salido entró en la barca y fue a ver un cabo o punta de tierra al sueste de la ysleta Llana obra de una legua y media, porque le pareçía que devía de aver algún río bueno. Luego a la entrada del cabo de la parte del sueste, andando dos tiros de ballesta, vio venir un grande arroyo de muy linda agua que deçendía de una montaña abaxo y hazía gran ruydo. Fue al río y vio en él unas piedras reluzir con unas manchas en ellas de color de oro,[2] y acordóse que en el río Tejo que al pie de él junto a la mar se halla oro, y pareçióle que çierto devía de tener oro.[3] Y mandó coger çiertas de aquellas piedras para llevar a los Reyes. Estando así, dan bozes los moços grumetes diziendo que vían pinales. Miró por la sierra y vídolos tan grandes y tan maravillosos que no podía encareçer su altura y derechura como husos gordos y delgado[s],[4] donde cognosçió que se podían hazer navíos e infinita tablazón y másteles para las mayores naos de España. Vido robles y madroños y un buen río y aparejo para hazer sierras de agua. La tierra y los ayres más templados que hasta allí, por la altura y hermosura de las sierras. Vido por la playa muchas otras piedras de color de hierro, y otras que dezían algunos que eran de minas de plata, todas las quales trae el río. Allí cojó una entena y mástel para la mezana de la caravela Niña. Llegó a la boca del río y entró en una cala al pie de aquel cabo de la parte del sueste muy honda y grande, en que cabrían çient naos sin alguna amarra ni anclas. Y el puerto que los ojos otro tal nunca vieron. La[s] sierras altíssimas de

---

[1] MS: bevía.
[2] Margin: Estas devía[n ser] piedras de mangasita.
[3] Margin: No ay duda sino que allí lo avía.
[4] Margin: Ay los pinos almirables.

which is very long and higher and longer than any other along this coast, where there are an infinite number. To seaward there is a reef which runs the length of the mountain like a bar and extends as far as the entrance. This is all to the SE, but there is also another reef on the side of the Isla Llana, but this is a small one. Between them both it is very wide and deep, as has been said. Then at the SE entrance of the same harbour they saw a large and very beautiful river with more water than they had seen until then, and the fresh water reached right up to the sea. There is a bank at the entrance, but once entered it is eight or nine fathoms deep. It is surrounded by palms and many groves like the others.

Sunday 25 November

Before sunrise he entered the boat and went to look at a cape or point of land to the SE of the Llana islet, about a league and a half away,[110] because it seemed to him that there must be a good river there. Then at the beginning of the cape on the SE side, two crossbow shots further on, he saw a great stream of lovely water rushing down from a mountain and making a great noise. He went to the river and saw shining in it some stones with streaks on them the colour of gold, and he remembered that there is gold at the mouth of the river Tagus, and it seemed to him that there must be gold here. He ordered some of those stones to be collected to take them to the Monarchs. Meanwhile the ships' boys shouted that they could see pine forests. He looked across the sierra and saw them, so large and so marvellous that he could not do justice to their height and straightness, like thick and slender spindles. He recognised that ships could be built, and quantities of decking and masts, for the largest ships in Spain. He saw oaks and strawberry trees[111] and a good river and location for a sawmill. The land and the breezes were milder than before because of the altitude and beauty of the mountains. He saw on the beach many other stones the colour of iron, and others which some said were from silver mines, all of which the river brings down. He cut a yard and a mizzen mast for the caravel Niña. He reached the mouth of the river and entered a very large, deep creek at the foot of that cape on the SE side, in which a hundred ships could lie without cables or anchors.[112] And the harbour was such that eyes never saw another like it. The mountains

las quales descendían muchas aguas lindíssimas; todas las sierras llenas de pinos y por todo aquello diversíssimas y hermosíssimas florestas de árboles. Otros dos o tres ríos le quedavan atrás. Encareçe todo esto en gran manera a los Reyes y muestra aver resçebido de verlo y mayormente los pinos inextimable alegría y gozo,[1] porque se podían hazer allí quantos navíos desearen, trayendo los adereços si no fuere madera y pez que allí se ha[lla] harta. Y afirma no encareçello la çentíssima parte de lo que es, y que plugo a Nuestro Señor de le mostrar siempre una cosa mejor que otra, y siempre en lo que hasta allí avía descubierto yva de bien en mejor, ansí en las tierras y arboledas y yervas y frutos y flores como en las gentes, y siempre de diversa manera, y así en un lugar como en otro; lo mismo en los puertos y en las aguas. Y finalmente dize que quando el que lo vee le es tan grande admiraçión, quánto más será a quien lo oyere, y que nadie lo podrá creer si no lo viere.

Lunes 26 de noviembre

Al salir del sol levantó las anclas del Puerto de Sancta Cathalina adonde estava dentro de la Ysla Llana, y navegó de luengo de la costa con poco viento sudueste al camino del Cabo del Pico que era al sueste. Llegó al cabo tarde porque le calmó el viento, y llegado vido al sueste quarta del leste otro cabo que estaría de él 60 millas. Y de allí vido otro cabo que estaría hazia el navío al sueste quarta del sur, y parecióle que estaría de él 20 millas, al qual puso nombre el Cabo de Campana, al qual no pudo llegar de día porque le tornó a calmar del todo el viento. Andaría en todo aquel día 32 millas que son 8 leguas, dentro de las quales notó y marcó nueve puertos muy señalados los quales todos los marineros hazían maravillas, y çinco ríos grandes porque yva siempre junto con tierra, para verlo bien todo. Toda aquella tierra es montañas altíssimas muy hermosas y no secas ni de peñas, sino todas andables y valles hermosíssimos. Y así los valles como las montañas eran llenos de árboles altos y frescos que era gloria mirarlos y pareçía que eran muchos pinales. Y también detrás del dicho Cabo del Pico de la parte del sueste, están dos ysletas que terná cada una en çerco dos leguas, y dentro dellas tres maravillosos puertos y dos grandes ríos. En toda esta costa no vido poblado ninguno desde la mar; podría ser averlo y ay señales dello, porque dondequiera que saltavan en tierra hallavan señales de aver gente y huegos muchos. Estimava que la tierra que oy vido de la parte del sueste del Cabo de Canpana era la ysla que llamavan los yndios Bohío y paréçelo porque el dicho

---

[1] Margin: Todo çierto con gran razón.

were very high and many beautiful streams flowed down from them; all the mountains were full of pines and all over there were many varied and most beautiful groves of trees. Two or three more rivers lay behind. He praises all this very highly to the Monarchs and evidently received great pleasure and joy at seeing it, particularly the pines, because as many ships as were desired could be built there, if all the materials were brought out, except the wood, of which there was plenty. And he assures them that he is not praising the hundredth part of what it is, and that it pleased Our Lord always to show him something better and in what he had discovered until then he always went from good to better, both in the lands and groves and herbs and fruits and flowers and in the people, always different wherever he went; the same thing was true of the harbours and the waters. And finally he says that when it is such a wonder to one who has seen it, how much more wonderful must it be to one who hears about it, and that no one will be able to believe it unless he sees it.

Monday 26 November

At sunrise he weighed anchor from the Puerto de Santa Catalina[113] where he lay inside Isla Llana, and steered along the coast with a light SW wind in the direction of the Cabo del Pico which was to the SE.[114] He reached the cape late because the wind dropped and when he arrived he saw to the SE by E another cape about 60 miles away.[115] And from there he saw another cape which must have been SE by S of the ship and it seemed to him to be about 20 miles away. He named it Cabo de Campana;[116] he could not reach it that day because the wind again dropped completely. He made about 32 miles, which is 8 leagues, in the whole of that day, during which he observed and recorded nine outstanding harbours which all the sailors marvelled at, and five large rivers, because he hugged the shore so as to see everything properly. All that land has very high and very beautiful mountains, not bare and rugged, but all accessible and with very beautiful valleys. The valleys and the mountains alike are full of tall leafy trees which were glorious to see and there seemed to be many pine groves. Beyond the Cabo del Pico on the SE side there are also two islets, each about 2 leagues round, with three marvellous harbours and two large rivers beyond them.[117] On all this coast he saw no village at all from the sea; there may be some and there are signs that this is so because wherever they went ashore they found signs of people and many fires. He judged that the land which he saw today to the SE of Cabo de Campana was the island which the Indians called Bohío and it seems to

cabo está apartado de aquella tierra. Toda la gente que hasta oy a hallado dizque tiene grandíssimo temor de los de Caniba o Canima, y dizen que biven en esta ysla de Bohío, la qual debe de ser muy grande según le pareçe,[1] y cree que van a tomar a aquellos a sus tierras y casas como sean muy cobardes, y no saber de armas. Y a esta causa le parece que aquellos yndios que traya no suelen poblarse a la costa de la mar por ser vezinos a esta tierra, los quales dizque después que le vieron tomar la buelta desta tierra no podían hablar, temiendo que los avían de comer, y no les podía quitar el temor, y dezían que no tenían sino un ojo y la cara de perro,[2] y creya el Almirante que mentían, y sentía el Almirante que devían de ser del señorío del Gran Can que los captivavan.

Martes 27 de noviembre

Ayer al poner del sol llegó çerca de un cabo que llamó Campana y porque el çielo claro y el viento poco, no quiso yr a tierra a surgir aunque tenía de sotavento çinco o seys puertos maravillosos, porque se detenía más de lo que quería por el apetito y delectaçión que tenía y resçevía de ver y mirar la hermosura y frescura de aquellas tierras dondequiera que entrava, y por no se tardar en proseguir lo que pretendía. Por estas razones se tuvo aquella noche a la corda y temporejar hasta el día. Y porque los aguajes y corrientes lo avían echado aquella noche más de çinco o seys leguas al sueste adelante de donde avía anocheçido, y le avía pareçido la tierra de Campana, y allende aquel cabo pareçía una grande entrada que mostrava dividir una tierra de otra y hazía como ysla en medio, acordó bolver atrás con viento sudueste y vino adonde le avía pareçido el abertura, y halló que no era sino una grande baya y al cabo della de la parte del sueste un cabo en el qual ay una montaña alta y quadrada que pareçía ysla. Saltó el viento en el norte y tornó a tomar la buelta del sueste por correr la costa y descubrir todo lo que por allí oviese. Y vido luego al pie de aquel Cabo de Campana un puerto maravilloso y un gran río, y de [allí] a un quarto de legua otro río, y de allí a media legua otro río, y dende a otra media legua otro río, y dende a una legua otro río, y dende a otra otro río, y dende a otro quarto otro río, y dende a otra legua otro río grande, desde el qual hasta el Cabo de Campana avría 20 millas y le quedan al sueste. Y los más destos ríos tenían grandes entradas y anchas y limpias con sus puertos maravillosos para naos grandíssimas, sin bancos de arena ni de piedras ni restringas. Viniendo así

---

[1] Margin: Este Bohío devía ser la ysla Española.
[2] Margin: No los entendían.

be so because the said cape is separate from that land. All the people he has come across before today live, he says, in great fear of those from Caniba or Canima, and they say that they live on this island of Bohío, which must be very large, as it seems to him, and he believes that the Caniba go and capture the lands and houses of these people as they are so cowardly and know nothing of arms. For this reason he believes that those Indians he had with him tend not to inhabit the seashore because they are close to this land. He says that when they saw him head for this land, they were speechless for fear that they would be eaten, and he could not reassure them, and they said that they had only one eye and the face of a dog, and the Admiral believed that they were lying, and felt that those who captured them must be subjects of the Great Khan.

Tuesday 27 November

Yesterday at sunset he arrived near a cape which he called Campana and because the sky was clear and there was little wind he did not wish to go and anchor offshore, although he had five or six marvellous harbours to leeward, because he was being delayed more than he wished by the desire he had for and the delight he gained from seeing the beauty and freshness of those lands, wherever he went ashore, and so as not to delay the pursuit of his objective. For these reasons he spent that night beating about and standing off till daybreak. Because the tides and currents had carried him during the night more than 5 or 6 leagues SE beyond where he had been at nightfall, and the land of Campana had come into view, and beyond that cape there appeared a huge inlet which seemed to divide one land from another and formed a sort of island in between, he decided to return with the wind SW to where the opening had appeared, and he found that it was merely a large bay with a cape at the SE end on which there is a high, square mountain which seemed to be an island.[118] The wind veered N and he resumed the course SE to skirt the coast and discover all that there might be there. He then saw at the foot of that Cabo de Campana a marvellous harbour and a great river, and a quarter of a league from there another river, and half a league further another river, and from there another half league on another river, and after a further league another river, and after a further league another river, and after a further quarter another river, and then another league further on another large river at about 20 miles SE of the Cabo de Campana. Most of these rivers had large, clear, wide entrances with marvellous harbours for the largest ships, without sandbanks or rocks or reefs. Continuing in this way

por la costa a la parte del sueste del dicho postrero río, halló una grande población la mayor que hasta oy aya hallado, y vido venir infinita gente a la ribera de la mar dando grandes bozes todos desnudos con sus azagayas en la mano. Deseó de hablar con ellos y amaynó las velas y surgió, y enbió las barcas de la nao y de la caravela, por manera ordenados que no hiziesen daño alguno a los yndios ni lo resçibiesen, mandando que les diesen algunas cosillas de aquellos resgates. Los yndios hizieron adamanes de no los dexar saltar en tierra y resistillos. Y viendo que las barcas se allegavan más a tierra y que no les avían miedo, se apartaron de la mar. Y creyendo que saliendo dos o tres hombres de las barcas no temieran, salieron tres cristianos diziendo que no oviesen miedo en su lengua porque sabían algo della por la conversaçión de los que traen consigo. En fin, dieron todos a huyr. Ni grande ni chico quedó. Fueron los tres cristianos a las casas que son de paja y de la hechura de las otras que avían visto, y no hallaron a nadie ni cosa en alguna dellas. Bolviéronse a los navíos y alçaron velas a mediodía para yr a un cabo hermoso que quedava al leste que avría hasta él ocho leguas. Aviendo andado media legua por la misma baya vido el Almirante a la parte del sur un singularíssimo puerto y de la parte del sueste unas tierras hermosas a maravilla así como una vega montuosa dentro en estas montañas, y parecían grandes humos y grandes poblaçiones en ellas y las tierras muy labradas, por lo qual determinó de se baxar a este puerto y provar si podía aver lengua o prática con ellos; el qual era tal que si a los otros puertos avía alabado, éste dize que alabava más con las tierras y templança y comarca dellas y poblaçión. Dize maravillas de la lindeza de la tierra y de los árboles donde ay pinos y palmas[1] y de la grande vega que aunque no es llana de llano[2] que va al sursueste, pero es llana de montes llanos y baxos, la más hermosa cosa del mundo, y salen por ella muchas riberas de aguas que desçienden destas montañas. Después de surgida la nao saltó el Almirante en la barca para soldar el puerto que es como una escodilla y quando fue frontero de la boca al sur halló una entrada de un río que tenía de anchura que podía entrar una galera por ella y de tal manera que no se vía hasta que se llegase a ella, y entrando por ella tanto como longura de la barca tenía çinco braças[3] y de ocho de hondo. Andando por ella fue cosa maravillosa y las arboledas y frescuras y el agua claríssima y las aves y amenidad, que dize que le parecía que no quisiera salir de allí. Yva diziendo a los hombres que llevava en su compañía que para hazer

---

[1] Margin: Siempre donde ay palmas de las muy altas es fertilíssima tierra.
[2] Margin: Dize que no es llana de llano. Quiere dezir que no es rasa.
[3] MS: braços.

Journal of Columbus

along the coast SE from this last river mentioned he found a large village, the largest he has found until today, and he saw countless people on the seashore, shouting loudly, all naked, and with spears in their hands. He wanted to speak to them and he lowered the sails and anchored and sent the boats from the flagship and the caravel in an orderly manner so as not to harm any of the Indians nor to suffer any themselves, and with instructions to give them some trinkets from the objects for barter. The Indians made gestures that they would not allow them to land and would resist them, and when they saw that the boats were drawing nearer the land and that they were not afraid, they all left the shore. Believing that if two or three men landed from the boats they would not be afraid, three Christians disembarked telling them not to be afraid, in their own language because they knew a little of it from their contact with the Indians they have with them. Eventually they all fled. No one remained, large or small. The three Christians went to the houses, which are made of straw and of the same design as the others they had seen, and they did not find anyone or anything in any of them. They returned to the ships and set sail at midday to go to a beautiful cape which lay to the E about 8 leagues away. Having sailed half a league across the same bay, the Admiral saw a very remarkable harbour[119] to the S and to the SE some marvellously beautiful lands, like a hilly stretch of land within these mountains, and smoke appeared from many fires and there were many villages and the land was highly cultivated, for which reasons he decided to run down to this harbour and see if he could talk to them. The harbour was such that, if he had praised other harbours, he says that this one he praised more for the countryside, the temperate climate and the neighbouring villages. He speaks wonders about the beauty of the land and of the trees which include pines and palms, and of the great plain which stretches to the SSE; although it is not flat in the usual sense of the word, it has low, gently rounded hills, and is the most beautiful thing in the world, and many streams of water which come down from these mountains flow across it. After anchoring the flagship, the Admiral got into the boat to take soundings in the harbour, which is like a basin, and when he was opposite the entrance to the S he found the mouth of a river which was wide enough for a galley to enter yet it could not be seen until it was reached, and on entering a boat's length within it was found to be from 5 to 8 fathoms deep. It was marvellous to sail along it with the fresh green woods and the clear water and the birds and the beautiful surroundings and he says that he felt that he did not want to leave. He told the men in his company that a thousand tongues

relación a los Reyes de las cosas que vían, no bastaran mill lenguas a referillo, y su mano para lo escrevir, que le pareçía que estava encantado. Deseava que aquello vieran muchas otras personas prudentes y de crédito, de las quales dize ser çierto que no encareçieran estas cosas menos que él. Dize más el Almirante aquí estas palabras: *Quánto será el benefiçio que de aquí se puede aver yo no lo escrivo. Es çierto, Señores Prínçipes, que donde ay tales tierras que deve de aver infinitas cosas de provecho, mas yo no me detengo en ningúnd puerto porque querría ver todas las más tierras que yo pudiese para hazer relación dellas a Vuestras Altezas; y también no sé la lengua y la gente destas tierras no me entienden ni yo ni otro que yo tenga a ellos. Y estos yndios que yo traygo muchas vezes le entiendo una cosa por otra al contrario;*[1] *ni fío mucho dellos, porque muchas vezes an provado a fugir. Mas agora, plaziendo a Nuestro Señor, veré lo más que yo pudiere y poco a poco andaré entendiendo y cognosçiendo y faré enseñar esta lengua a personas de mi casa porque veo que es toda la lengua una fasta aquí. Y después se sabrán los benefiçios y se trabajará de hazer todos estos pueblos cristianos, porque de ligero se hará, porque ellos no tienen secta ninguna ni son ydólatras. Y Vuestras Altezas mandarán hazer en estas partes çiudad e fortaleza, y se convertirán estas tierras. Y çertifico a Vuestras Altezas que debaxo del sol no me pareçe que las pueda aver mejores, en fertilidad, en temperançia de frío y calor, en abundançia de aguas buenas y sanas y no como los ríos de Guinea que son todos pestilençia. Porque, loado Nuestro Señor, hasta oy de toda mi gente no a avido persona que le aya mal la cabeça ni estado en cama por dolençia, salvo un viejo de dolor de piedra de que él estava toda su vida apassionado, y luego sanó al cabo de dos días. Esto que digo es en todos tres los navíos. Así que plazerá a Dios que Vuestras Altezas enbiarán acá o vernán hombres doctos y verán después la verdad de todo. Y porque atrás tengo hablado del sitio de villa e fortaleza en el Río de Mares por el buen puerto y por la comarca es çierto que todo es verdad lo que yo dixe mas no a ninguna conparaçión de allá aquí ni de la Mar de Nuestra Señora. Porque aquí deve aver infra la tierra grandes poblaçiones y gente ynumerable y cosas de grande provecho. Porque aquí y en todo lo otro descubierto y tengo esperança de descubrir antes que yo vaya a Castilla, digo que terná toda la cristiandad negoçiaçión en ellas, quanto más la España a quien deve estar subjecto todo. Y digo que Vuestras Altezas no deven consentir que aquí trate ni faga pie ningúnd estrangero, salvo cathólicos cristianos, pues esto fue el fin y el comienço del propósito que fuese por acreçentamiento y gloria de la religión cristiana,*[2] *ni*

---

[1] Margin: No[ta].
[2] Margin: No[ta].

would not suffice to give the monarchs an account of what they had seen, and his hand could not write it for he seemed to be enchanted. He wished that many other cautious and trustworthy people could see it all, and he says that he is certain that they would not praise these things less than he did. The Admiral goes on in these words: *Of how great will be the benefits which could be had from this land, I write nothing. It is certain, Sovereign Princes, that where there are such lands there must be innumerable things of value, but I am not delaying in any harbour because I would like to see as many lands as possible in order to give Your Highnesses an account of them. Furthermore, I do not know the language and the people of these lands do not understand me, nor do I nor anyone I have with me understand them. With the Indians I have with me I often understand one thing for another, the wrong way round, and I have no great confidence in them because they have often tried to run away. But now, may it please Our Lord, I shall see as much as I can and gradually I shall understand and know more and shall teach this language to people of my household, because I can see that the language is all one until now. And later the benefits will become known and an effort will be made to make all these people Christians, for it will easily be done because they have no religion and are not idolaters. And Your Highnesses will order a city and fortress to be built in these parts, and these lands will be converted. I assure Your Highnesses that there cannot be better lands under the sun for fertility, temperate climate, neither cold nor hot, abundance of good, healthy water and not as in the rivers in Guinea which are all pestilential. Because, Our Lord be praised, to this day there has been not one person from all my crew who has had a headache nor been in bed with illness, except an old man with a kidney stone from which he had suffered all his life and then after two days was cured. What I say goes for all three ships. And so it will please God if Your Highnesses send here or if there come here learned men who will then see the truth of everything. And since I spoke earlier about a site for a town and fortress at Río de Mares on account of the harbour and the surroundings, it is certainly true what I said, but there is no comparison between there and here, or with the Mar de Nuestra Señora. For here there must be great townships and innumerable people inland and things of great benefit. For here and in all the other places I have discovered and hope to discover before I return to Castile I say that the whole of Christendom will come to do business, and above all Spain, to which all must be subject. And I say that Your Highnesses must not allow any foreigner to trade or set foot here unless he be a Catholic Christian, for this was the end and the beginning of the enterprise, that it should be for the promotion and glory of the Christian religion, and no one who is not a good Christian should come*

venir a estas partes ninguno que no sea buen cristiano. Todas son sus palabras. Subió allí por el río arriba y halló unos braços del río y rodeando el puerto halló a la boca del río estava[n] unas arboledas muy graciosas como una muy deleytable güerta y allí halló una almadía o canoa hecha de un madero tan grande como una fusta de doze bancos, muy hermosa, varada debaxo de una ataraçana o ramada hecha de madera y cubierta de grandes hojas de palma, por manera que ni el sol ni el agua le podían hazer daño. Y dize que allí era el proprio lugar para hazer una villa o çiudad y fortaleza por el buen puerto, buenas aguas, buenas tierras, buenas comarcas y mucha leña.

Miércoles 28 de noviembre

Estúvose en aquel puerto aquel día porque llovía y hazía gran çerrazón aunque podía correr toda la costa con el viento que era sudueste y fuera a popa; pero porque no pudiera ver bien la tierra y no sabiéndola es peligroso a los navíos no se partió. Salieron a tierra la gente de los navíos a lavar su ropa; entraron algunos dellos un rato por la tierra adentro. Hallaron grandes poblaciones y las casas vazías porque se avían huydo todos. Tornáro[n]se por otro río abaxo, mayor que aquel donde estavan en el puerto.

Jueves 29 de noviembre

Porque llovía y el çielo estava de la manera çerrado que ayer no se partió. Llegaron algunos de los cristianos a otra poblaçión çerca de la parte de norueste y no hallaron en las casas a nadie ni nada. Y en el camino toparon con un viejo que no les pudo huyr; tomáronle y dixéronle que no le querían hazer mal, y diéronle algunas cosillas del resgate y dexáronlo. El Almirante quisiera vello para vestillo y tomar lengua de él, porque le contentava mucho la felicidad de aquella tierra y disposición que para poblar en ella avía y juzgava que devía de aver grandes poblaçiones. Hallaron en una casa un pan de çera que truxo a los Reyes, y dize que donde çera ay también deve aver otras mil cosas buenas.[1] Hallaron también los marineros en casa una cabeça de hombre dentro en un çestillo cubierto con otro cestillo y colgado de un poste de la casa, y de la misma manera hallaron otra en otra poblaçión. Creyó el Almirante que devía[n] ser de algunos prinçipales del linaje, porque aquellas casas era[n] de manera que se acojen en ellas mucha gente en una sola, y deven ser parientes desçendientes de uno solo.

---

[1] Margin: Esta çera vino allí de Yucatán y por esto creo que esta tierra es Cuba.

to these parts. These are all his words. He went up-river and found some tributaries and in sailing around the harbour found at the mouth of the river some very lovely groves like a delightful orchard, and there he found an 'almadía' or canoe made of a piece of wood as large as a 12-seat galley, very beautiful, beached under a roof or canopy made of wood and covered in large palm leaves so that neither sun nor rain could damage it. And he says that there was a suitable place to build a town or city and fortress, on account of the good harbour, good water, good land, good surroundings and plentiful timber.

Wednesday 28 November

He spent that day in that harbour because it was raining and the clouds were very dark although he could have skirted the coast with the wind SW which would have been astern; but he did not set out because he could not have seen the land properly and it is dangerous for the ships if one does not know the land. The crews went ashore to wash their clothes; some went inland for a while. They found large villages and the houses empty because everyone had fled. They returned downstream along another river, larger than that in which they were stationed.

Thursday 29 November

Because it was raining and the sky was overcast in the same way as yesterday he did not set out. Some of the Christians went to another village nearby to the NW and found no one there and nothing in the houses. On the way they came across an old man who could not run away; they captured him and told him that they meant him no harm, and gave him some trinkets from the barter and let him go. The Admiral would have liked to see him to give him some clothes and speak with him, because he was very pleased by the congenial nature of that land and its suitability for settlement, and judged that there must be large towns there. In one house they found a cake of wax which he brought back for the Monarchs, and he says that where there is wax there must also be a thousand other good things.[120] The sailors also found in one house a man's head in a small basket covered with another basket and hanging from a post of the house, and they found another similar in another village. The Admiral believed that they must be the heads of some prominent ancestors,[121] because those houses were such that many people live together in one house, and they must be related to each other and descended from one man.

Viernes 30 de noviembre

No se pudo partir porque el viento era levante, muy contrario a su camino. Envió ocho hombres bien armados y con ellos dos yndios de los que traya para que viesen aquellos pueblos de la tierra dentro y por aver lengua. Llegaron a muchas casas y no hallaron a nadie ni nada, que todos se avían huydo. Vieron quatro mançebos que estavan cavando en su[s] heredades. Así como vieron los cristianos dieron a huyr; no los pudieron alcançar. Anduvieron dizque mucho camino. Vieron muchas poblaçiones y tierra fertilíssima y toda labrada y grandes riberas de agua, y çerca de una vieron una almadía o canoa de noventa y çinco palmos de longura de un solo madero, muy hermosa, y que en ella cabrían y navegarían çiento y çinquenta personas.

Sábado 1º día de diziembre

No se partió, por la misma causa del viento contrario y porque llovía mucho. Asentó una cruz grande a la entrada de aquel puerto que creo llamó el Puerto Sancto sobre unas peñas bivas. La punta es aquella que está de la parte del sueste a la entrada del puerto. Y quien oviere de entrar en este puerto se deve llegar más sobre la parte del norueste de aquella punta que sobre la otra del sueste, puesto que al pie de ambas, junto con la peña, ay doze braços de hondo y muy limpio. Más a la entrada del puerto, sobre la punta del sueste, ay una baxa que sobreagua, la qual dista de la punta tanto que se podría passar entremedias, aviendo neçessidad, porque al pie de la baxa y del cabo todo es fondo de doze y de quinze braças[1] y a la entrada se a de poner la proa al sudueste.

Domingo 2 de diziembre

Todavía fue contrario el viento y no pudo partir. Dize que todas las noches del mundo vienta terral, y que todas las naos que allí estuvieren non ayan miedo de toda la tormenta del mundo porque no puede recalar dentro por una baxa que está al principio del puerto, etc. En la boca de aquel río dizque halló un grumete çiertas piedras que pareçen tener oro; trúxolas para mostrar a los Reyes. Dize que ay por allí a tyro de lombarda grandes ríos.

---

[1] MS: braços.

Friday 30 November

He could not set out because the wind was E, completely wrong for his course. He sent out 8 well-armed men and two of the Indians he had with him to investigate the peoples inland and speak to them. They came across many houses but did not find anyone or anything because they had all fled. They saw four youths who were digging their fields. As soon as they saw the Christians they fled; they could not catch up with them. They walked, he says, a great distance. They saw many villages and very fertile land, all cultivated, and great streams of water, and near one of them they saw an 'almadía' or canoe 95 palms long and of a single piece of wood, very beautiful, in which 150 people could sit and travel.

Saturday 1 December

He did not set out for the same reason, a head wind, and because it was raining heavily. He set up a great cross on a craggy point at the mouth of that harbour which I believe he called Puerto Santo. The point is the one on the SE side of the harbour entrance. Anyone needing to enter this harbour must keep more to the NW side of that point than to the SE, because at the foot of both points, next to the crag, it is 12 fathoms deep and very clear. Nearer the harbour entrance, on the SE side, there is a reef which breaks the surface, far enough from the point to allow passage between them, if necessary, because at the foot of the reef and the cape it is fully 12 to 15 fathoms deep and at the entrance one needs to steer SW.

Sunday 2 December

The wind was still against him and he could not set out. He says that every night without exception there is a land breeze and that however many ships there might be there, they need have no fear of any storm whatsoever because it could not penetrate the harbour because of a sandbank at the entrance, etc. He says that at the mouth of that river a ship's boy found some stones which appear to contain gold; he brought them back to show the Monarchs. He says that a lombard shot away there are large rivers.

Lunes 3 de diziembre

Por causa de que hazía siempre tiempo contrario no partía de aquel puerto, y acordó de yr a ver un cabo muy hermoso un quarto de legua del puerto de la parte del sueste. Fue con las barcas y alguna gente armada. Al pie del cabo avía una boca de un buen río, puesta la proa al sueste para entrar, y tenía çient passos de anchura; tenía una braça de fondo a la entrada o en la boca, pero dentro avía doze braças y çinco y quatro y dos, y cabrían en él quantos navíos ay en España. Dexando un braço de aquel río fue al sueste y halló una caleta en que vido çinco muy grandes almadías que los yndios llaman canoas como fustas muy hermosas y labradas que era, dizque, era plazer vellas, y al pie del monte vido todo labrado. Estavan debaxo de árboles muy espessos, y yendo por un camino que salía a ellas fueron a dar a una ataraçana muy bien ordenada y cubierta que ni sol ni agua no les podía hazer daño, y debaxo della avía otra canoa hecha de un madero como las otras como una fusta de diez y siete bancos que era plazer ver las labores que tenía y su hermosura. Subió una montaña arriba y después hallóla toda llana y senbrada de muchas cosas de la tierra y calabaças que era gloria vella, y en medio della estava una gran población. Dio de súbito sobre la gente del pueblo y como los vieron hombres y mugeres dan de huyr. Asegurólos el yndio que llevava consigo de los que traya, diziendo que no oviesen miedo, que gente buena era. Hízolos dar el Almirante cascaveles y sortijas de latón, y contezuelas de vidro verdes y amarillas, con que fueron muy contentos. Visto que no tenían oro ni otra cosa preçiosa y que bastava dexallos seguros, y que toda la comarca era poblada y huydos los demás de miedo,[1] y çertifica el Almirante a los Reyes que diez hombres hagan huyr a diez mill, tan cobardes y medrosos son que ni traen armas salvo una[s] varas y en el cabo dellas un palillo agudo tostado, acordó bolverse. Dize que las varas se las quitó todas con buena manera, resgatándoselas de manera que todas las dieron. Tornados adonde avían dexado las barcas, enbió çiertos cristianos al lugar por donde subieron, porque le avía pareçido que avía visto un gran colmenar. Antes que viniesen los que avía enbiado, ayuntáronse muchos yndios y vinieron a las barcas donde ya se avía el Almirante recogido con su gente toda. Uno dellos se adelantó en el río junto con la popa de la barca, y hizo una grande plática que el Almirante no entendía, salvo que los otros yndios de quando en quando alçavan las manos al çielo y davan una grande boz. Pensava el Almirante que lo aseguravan y que les plazía de su venida, pero vido al yndio que consigo traya demudarse la cara y

---

[1] Margin: No[ta].

Monday 3 December

Because the weather was still against him he did not set out from that harbour and decided to go and see a very beautiful cape[122] a quarter of a league SE of the harbour. He went with the boats and some armed men. At the foot of the cape there was the mouth of a good river;[123] he steered SE to enter, and it was 100 paces wide; it was a fathom deep at the entrance or mouth, but within it was 12, 5, 4 and 2 fathoms, and it would take all the ships of Spain. Taking a branch of that river he went SE and found a creek where he saw 5 very large 'almadías' which the Indians call canoes, like galleys, very beautifully made and, he says, a pleasure to look at, and at the foot of the hill he saw that everywhere was cultivated. The canoes were beneath very thick trees, and taking a path which led to them, they came upon a very well contrived canopy, providing cover against damage from sun and rain, and underneath there was another canoe made from a single piece of wood like the others and like a 17-seat galley, and it was a great pleasure to see the beauty of it and the workmanship. He climbed up a mountain and then found the terrain flat and planted with gourds and many of the local crops that were a joy to behold, and in the centre was a large village. He came suddenly upon the village folk and as soon as they saw them all the men and women took flight. The Indian he had there, one of those he had brought with him, reassured them, saying that they should not be afraid for these were good people. The Admiral ordered them to be given hawks' bells and brass rings, and green and yellow glass beads, with which they were very pleased. Seeing that they had no gold or any other thing of value, and that it was enough to leave them in peace, and that the whole surrounding area was inhabited and the majority had fled in fear - and the Admiral assures the Monarchs that ten men are enough to frighten off ten thousand of them; they are so cowardly and timorous that they carry no arms except some spears with sharp fire-hardened tips - he decided to return. He says that he took away all the spears from them in a proper manner, bartering for them so that they gave all of them. Returning to where they had left the boats, he sent some Christians to the place to which they had climbed, because he thought that he had seen a great beehive. Before those he had sent could return, many Indians gathered together and came to the boats where the Admiral and all his men had reassembled. One of the Indians waded into the river up to the stern of the boat and made a great speech which the Admiral did not understand, except that the other Indians from time to time raised their hands to the sky and gave a great shout. The Admiral thought that they were reassuring him and were pleased at his coming, but he saw the face of the Indian

amarillo como la çera y temblava mucho, diziendo por señas que el Almirante se fuese fuera del río, que los querían matar. Y llegóse a un cristiano que tenía una ballesta armada y mostróla a los yndios y entendió el Almirante que les dezía que los matarían todos, porque aquella ballesta tyrava lexos y matava. También tomó una espada y la sacó de la vayna, mostrándosela, diziendo lo mismo. Lo qual oydo por ellos dieron todos a huyr, quedando todavía temblando el dicho yndio de cobardía y poco coraçón y era hombre de buena estatura y rezio. No quiso el Almirante salir del río, antes hizo remar en tierra hazia donde ellos estavan, que eran muy muchos, todos tyñidos de colorado y desnudos como sus madres los parió y algunos dellos con penachos en la cabeça y otras plumas, todos con sus manojos de azagayas. *Lleguéme a ellos y diles algunos bocados de pan y demandéles las azagayas,*[1] *y dávales por ellas a unos un cascavelito, a otros una sortizuela de latón, a otros unas contezuelas, por manera que todos se apaziguaron y vinieron todos a las barcas y davan quanto tenían por quequiera*[2] *que les davan. Los marineros avían muerto una tortuga y la cáscara estava en la barca en pedaços, y los grumetes dávanles della como la uña, y los yndios les davan un manojo de azagayas. Ellos son gente como los otros que e hallado* (dize el Almirante) *y de la misma creençia y creyan que veníamos del çielo y de lo que tienen luego lo dan por qualquiera cosa que les den sin dezir que es poco y creo que así harían de espeçería y de oro si lo tuviesen. Vide una casa hermosa no muy grande y de dos puertas, porque así son todas, y entré en ella y vide una obra maravillosa como cámaras hechas por una çierta manera que no lo sabría dezir, y colgado al çielo della caracoles y otras cosas. Yo pensé que era templo y los llamé y dixe por señas si hazían en ella oración; dixeron que no, y subió uno dellos arriba y me dava todo quanto allí avía y dello tomé algo.*

Martes 4 de diziembre

Hízose a la vela con poco viento y salió de aquel puerto que nombró Puerto Santo. A las dos leguas vido un buen río de que ayer habló. Fue de luengo de costa y corríase toda la tierra, passado el dicho cabo, lessueste y güesnorueste hasta el Cabo Lindo que está al Cabo del Monte al leste quarta del sueste y ay de uno a otro çinco leguas. Del Cabo del Monte a legua y media ay un gran río algo angosto; pareçió que tenía buena entrada y era muy hondo, y de allí a tres quartos de legua vido otro grandíssimo río y deve venir de muy lexos. En la boca

---

[1] Margin: No[ta].
[2] MS: por que quequiera.

he had with him change colour and go as yellow as wax, and he trembled greatly, saying in sign language that the Admiral should leave the river because they intended to kill them. He went up to a Christian who had a loaded crossbow and showed it to the Indians and the Admiral understood that he was telling them that they would kill them all, because that crossbow fired and killed at a distance. He also took a sword and drew it out of its sheath, showing it to them and saying the same thing. When they heard this they all started to run off, and the Indian remained still trembling with cowardice and faint-heartedness although he was well built and a strong man. The Admiral refused to leave the river and instead ordered the men to row towards the shore where they were, and there were many of them, all painted red and as naked as their mothers bore them and some with feathers on their head and other plumes, and all with their bundles of spears. *I went up to them and gave them pieces of bread and asked for their spears, and in exchange gave some of them a hawk's bell and others a brass ring and others some beads, so that they all were all pacified and came down to the boats and gave everything they had for whatever was given to them. The sailors had killed a turtle and the shell was in pieces in the boat, and the boys gave them a piece of it about the size of a fingernail, and the Indians gave them a bundle of spears. They are like the other peoples I have found* (says the Admiral) *with the same beliefs, and they believed that we had come from heaven and they give what they have for whatever they are given without saying that it is too little and I believe that they would do the same with spices and gold if they had any. I saw a beautiful house, not very large and with two doors, for that is how they all are, and I went in and saw a marvellous arrangement of rooms which I could not describe, and hanging from the roof were shells and other things. I thought that it was a temple and I called them and asked in sign language if they said prayers there; they said not, and one of them climbed up and gave me everything that was there and I took some of it.*

Tuesday 4 December

He set sail with a light wind and left that harbour which he called Puerto Santo. After two leagues he saw a good river of which he spoke yesterday. He went along the coast and skirted the land which ran ESE and WNW, past the cape already mentioned as far as Cabo Lindo,[124] which is 5 leagues from the Cabo del Monte[125] E by S. A league and a half from Cabo del Monte there is a large, somewhat narrow river; it seemed to have a good entrance and was very deep, and three quarters of a league from there he saw another very large river which must have come from very far off. It was a good 100 paces at the mouth, without

tenía bien çien passos y en ella ningún banco y en la boca ocho braças y buena entrada *porque lo enbié a ver y sondar con la barca y viene el agua dulce hasta dentro en la mar*, y es de los caudalosos que avía hallado, y deve aver grandes poblaciones. Después del Cabo Lindo ay una grande baya que sería buen pozo por lesnordeste y suest[e] y sursudueste.

Miércoles 5 de diziembre

Toda esta noche anduvo a la corda sobre el Cabo Lindo, adonde anocheció, por ver la tierra que yva al leste, y al salir del sol vido otro cabo al leste a dos leguas y media. Passado aquél vido que la costa bolvía al sur y tomava del sudueste y vido luego un cabo muy hermoso y alto a la dicha derrota,[1] y distava desotro siete leguas. Quisiera yr allá, pero por el deseo que tenía de yr a la ysla de Baneque que le quedava según dezían los yndios que llevava al nordeste, lo dexó. Tanpoco pudo yr al Baneque porque el viento que llevava era nordeste. Yendo así miró al sueste y vido tierra y era una ysla muy grande,[2] de la qual ya tenía dizque informaçión de los yndios, a que llamavan ellos Bohío, poblada de gente. Desta gente dizque los de Cuba o Juana[3] y de todas estotras yslas tienen gran miedo porque dizque comían los hombres. Otras cosas le contavan los dichos yndios por señas muy maravillosas, mas el Almirante no dizque las creya, sino que devían tener más astuçia y mejor yngenio los de aquella ysla Bohío para los captivar que ellos, porque eran muy flacos de coraçón. Así que porque el tiempo era nordeste y tomava del norte, determinó de dexar a Cuba o Juana,[4] que hasta entonçes avía tenido por tierra firme por su grandeza porque bien avría andado en un paraje çiento y veynte leguas, y partió al sueste quarta del leste, puesto que la tierra que él avía visto se hazía al sueste, dava este reguardo porque siempre el viento rodea del norte para el nordeste y de allí al leste y sueste. Cargó mucho el viento y llevava todas sus velas, la mar llana y la corriente que le ayudava por manera que hasta la una después de mediodía desde la mañana hazía de camino 8 millas por ora y eran seys oras aún no complidas porque dize que allí eran las noches çerca de quinze oras. Después anduvo diez millas por ora, y así andaría hasta el poner del sol 88 millas que son 22 leguas todo al sueste. Y porque se hazía noche mandó a la caravela Niña que se adelantasse

---

[1] Margin: Este deve ser la punta de Maysí que es la postrera de Cuba.
[2] Margin: Esta es la Española según pareçe,
[3] Margin: Aquí pareçe que devía de aver puesto nombre el Almirante a Cuba Juana.
[4] Margin: No[ta].

a sandbar, and 8 fathoms deep and with a good entrance *because I sent men to inspect it and take soundings in the boat and the fresh water flows right into the sea*, and it has one of the greatest volumes of water he had seen, and there must be large villages along it. Beyond Cabo Lindo there is a great bay which would be a good shelter from the ENE, SE and SSE.

Wednesday 5 December

All this night he beat about off Cabo Lindo, where he was at nightfall, in order to see the land which ran to the E, and at sunrise he saw another cape 2 and a half leagues to the E. Beyond that he saw that the coast turned S again and trended SW and he saw a very beautiful high cape[126] in that direction, seven leagues distant from the previous one. He would have liked to go there, but he left it out of the desire he had to go to the island of Baneque which, according to the Indians he had with him, lay to the NE. Yet he could not go to Baneque either because the wind he had was NE. Proceeding in this way, he looked to the SE and saw land and it was a very large island,[127] about which he says he already had information from the Indians, who called it Bohío, that it was inhabited. He says that the inhabitants of Cuba or Juana[128] and of all the other islands are afraid of these people because it is said that they eat men. The Indians told him many other marvellous things by signs, but the Admiral says that he did not believe them, thinking rather that the Indians from the island of Bohío must be more astute and skilled at capture than the others, who are very faint-hearted. Because the weather was from the NE veering N, he decided to leave Cuba or Juana which up to that point he had regarded as the mainland on account of its size, because he must have sailed a good 120 leagues in one direction, and he set out on a course SE by E since the land which he had seen lay to the SE and he took this precaution because the wind veers from the N to NE and from there to E and SE. The wind got up and he set all sail, the sea was flat and the current was favourable so that from morning to 1 o'clock in the afternoon he made 8 miles an hour, not quite for 6 hours because he says that the nights there were nearly 15 hours long. Afterwards he made 10 miles an hour and so made about 88 miles or 22 leagues by sunset, all to the SE. And because night was falling he

para ver con día el puerto porque era velera; y llegando a la boca del puerto que era como la baya de Cáliz y porque era ya de noche, enbió a su barca que sondase el puerto. La qual llevó lumbre de candela, y antes que el Almirante llegasse adonde la caravela estava barloventeando y esperando que la barca le hiziese señas para entrar en el puerto, apagósele la lumbre a la barca. La caravela, como no vido lumbre, corrió de largo y hizo lumbre al Almirante y llegado a ella contaron lo que avía acaeçido. Estando en esto los de la barca hizieron otra lumbre: la caravela fue a ella y el Almirante no pudo y estuvo toda aquella noche barloventeando.

Jueves 6 de diziembre

Quando amaneció se halló quatro leguas del puerto; púsole nombre Puerto María y vido un cabo hermoso al sur quarta del sudueste al qual puso nombre Cabo de l[a] Estrella y pareçióle que era la postrera tierra de aquella ysla hazia el sur y estaría el Almirante de él xxviii millas. Pareçíale otra tierra como ysla no grande al leste y estaría de él 40 millas. Quedávale otro cabo muy hermoso y bien hecho a quien puso nombre Cabo del Elefante al leste quarta del sueste y distávale ya 54 millas. Quedávale otro cabo al lessueste al que puso nombre el Cabo de Çinquin; estaría de él 28 millas. Quedávale una gran scisura o abertura o abra a la mar que le pareçió ser río al sueste y tomava de la quarta del leste, avría de él a la abra 20 millas. Pareçíale que entre el Cabo del Elifante del de Çinquin avía una grandíssima entrada y algunos de los marineros dezían que era apartamiento de ysla; aquella puso por nombre la Ysla de la Tortuga. Aquella ysla grande pareçía altíssima tierra no çerrada con montes, sino rasa como hermosas campiñas y pareçe toda labrada o grande parte della y parecían las sementeras como trigo en el mes de mayo en la campiña de Córdova. Viéronse muchos huegos aquella noche, y de día muchos humos como atalayas que pareçía estar sobre aviso de alguna gente con quien tuviesen guerra. Toda la costa desta tierra va al leste. A oras de bísperas entró en el puerto dicho y púsole nombre Puerto de San Nicolao porque era día de Sant Nicolás por honrra suya, y a la entrada de él se maravilló de su hermosura y bondad.[1] Y aunque tiene mucho alabados los puertos de Cuba, pero sin duda dize él que no es menos éste, antes los sobrepuja y ninguno le es semejante. En boca y entrada tiene legua y media de ancho y se pone la proa al sursueste puesto que por la grande anchura se puede poner la proa adonde quisieren. Va desta manera al sursueste

---

[1] Margin: No entiendo cómo este puerto puso arriba Puerto María y agora de San Nicolás.

ordered the caravel Niña, because she was faster, to go ahead to inspect the harbour while it was still daylight. Arriving at the mouth of the harbour, which was like the bay of Cádiz, and because it was already night, she sent her boat to take soundings in the harbour. It carried a light, and before the Admiral could reach the point where the caravel was lying to and waiting for the boat to give a sign to enter the harbour, the boat's light went out. The caravel, because she could not see any light, ran out and lit the way for the Admiral and when he caught up with her they told him what had happened. Meanwhile the men in the boat lit another torch: the caravel followed it but the Admiral could not and lay to all that night.

Thursday 6 December

At dawn he found himself 4 leagues from the harbour. He named it Puerto María[129] and saw a beautiful cape S by W which he called Cabo de la Estrella[130] and it seemed to be the southernmost point of that island and he would have been 28 miles from it. Another piece of land like a small island appeared to the E about 40 miles away. Another very beautiful and well formed cape lay 54 miles E by S and he called it Cabo del Elefante.[131] Another cape lay to the ESE, to which he gave the name Cabo de Cinquin;[132] it must have been 28 miles from him. To the SE and SE by E lay a large ravine or breach or opening into the sea which seemed to be a river, and there would have been 20 miles between him and the opening. It seemed to him that between the Cabo del Elefante and Cabo Cinquin there was a wide inlet and some of the sailors said that it divided the island in two. He gave it the name of Isla de la Tortuga.[133] That large island seemed to be a very high land, not capped with mountains, but flat like beautiful meadows and it appears to be completely cultivated or nearly so and the crops looked like wheat in the month of May in the fields around Córdoba. They saw many fires that night, and during the day many smoke signals from lookouts as if the island were on guard against people with whom they were at war. All the coast of this land stretches E. At the hour of vespers he entered the harbour mentioned and gave it the name Puerto de San Nicolás in honour of St Nicholas because it was his feast day, and at the entrance he marvelled at its beauty and pleasant aspect.[134] Although he has praised the harbours of Cuba very highly, without doubt he says that this one is no less to be praised, but rather it excels them and none are its equal. At the mouth and entrance it is a league and a half wide and one should steer SSE, although because of the great width one could

dos leguas, y a la entrada de él por la parte del sur se haze como una angla y de allí se sigue así ygual hasta el cabo adonde está una playa muy hermosa y un campo de árboles de mill maneras y todos cargados de frutas que creya el Almirante ser de especerías y nuezes moscadas sino que no estava[n] maduras y no se cognoscían, y un río en medio de la playa. El hondo deste puerto es maravilloso, que hasta llegar a la tierra en longura de una [nao] no llegó la sondaresa o plomada al fondo con quarenta braças, y ay hasta esta longura el hondo de xv braças y muy limpio. Y así es todo el dicho puerto de cada cabo hondo dentro a una passada de tierra de 15 braças y limpio; y desta manera es toda la costa muy hondable y limpia que no pareçe una sola baxa. Y al pie della tanto como longura de un remo de barca de tierra tiene çinco braças. Y después de la longura del dicho puerto yendo al sursueste, en la qual longura pueden barloventear mill carracas, bojó un braço del puerto al nordeste por la tierra dentro una grande media legua y siempre en una misma anchura como que lo hizieran por un cordel; el qual queda de manera que estando en aquel braço que será de anchura de veynte y çinco passos no se puede ver la boca de la entrada grande de manera que queda puerto çerrado, y el fondo deste braço es así en el comienço hasta la fin de onze braças y todo basa o arena limpia, y hasta tierra y poner los bordos en las yervas tiene ocho braças. Es todo el puerto muy ayroso y desabahado de árboles, rasó. Toda esta ysla le pareçió de más peñas que ninguna otra que aya hallado. Los árboles más pequeños y muchos dellos de la naturaleza de España como carrascos y madroños y otros y lo mismo de las yervas. Es tierra muy alta y toda campiña o rasa y de muy buenos ayres y no se a visto tanto frío como allí aunque no es de contar por frío mas díxolo al respecto de las otras tierras. Hazia enfrente de aquel puerto una hermosa vega y en medio della el río susodicho; y en aquella comarca (dize) deve aver grandes poblaçiones según se vían las almadías con que navegan, tantas y tan grandes dellas como una fusta de 15 bancos. Todos los yndios huyeron y huyan como vían los navíos. Los que consiguo de las ysletas traya tenían tanta gana de yr a su tierra,[1] que pensava (dize el Almirante) que después que se partiese de allí los tenía de llevar a sus casas, y que ya lo tenían por sospechoso porque no lleva el camino de su casa, por lo qual dize que ni les creya lo que le dezían, ni los entendía bien ni ellos a él, y dizque avían el mayor miedo del mundo de la gente de aquella ysla. Así que por querer aver lengua con la gente de aquella ysla, le fuera neçessario detenerse algunos días en aquel puerto, pero no lo hazía por ver mucha tierra y por dudar que el tiempo le duraría. Esperava en Nuestro Seño

---

[1] Margin: Con razón.

steer any course one wished. It stretches 2 leagues SSE in this way and on the southern side of the entrance there is a sort of promontory and from there it continues in the same way up to the cape where there is a very beautiful beach and a field with a thousand different kinds of trees all laden with fruit which the Admiral believed to be spices and nutmegs except that they were not ripe and could not be identified, and there is a river in the centre of the beach. This harbour is marvellously deep, for a 40-fathom plumb line did not strike bottom until a [ship's] length from the shore, and within that distance it is 15 fathoms and very clear. The whole harbour from one cape to the other is like that, 15 fathoms and clear up to a pace from the shore; and the whole coast is just as deep and clear and there is not a single shoal, and at the foot it is 5 fathoms deep an oar's length from the shore. Having gone the length of this harbour to the SSE, in which space a thousand carracks could lay to, an arm of the harbour extended inland to the NE for a good half league and at a width as constant as if it had been measured out by a rope. This arm lies in such a way that within its width, of some 25 paces, the mouth of the main entrance cannot be seen, so that it becomes a closed harbour, and the depth of this arm is a constant 11 fathoms from beginning to end with a clean sandy bottom throughout, and 8 fathoms deep right up to the edge, with the ship's gunwales against the grass. The whole harbour is very breezy, flat and unsheltered by trees. All this island seemed to him rockier than any other he had found. The trees are smaller and many of them of the same type as in Spain, such as oaks and strawberry trees and others, and the same is true of the plants. It is a very high land and all fields and open country and with very good breezes and it has not been so cold as there, although it cannot really be counted cold, except in comparison with the other lands. Opposite that harbour is a beautiful plain and in the middle, the river mentioned earlier. In that area, he says, there must be large villages to judge from the look of the canoes in which they travel, of which there are so many and as large as a galley of 15 benches. All the Indians fled and ran off as soon as they saw the ships. Those he had with him from the islands wanted so much to return to their land that he thought (says the Admiral) that after he left there he had to take them home, and they were already suspicious of him because he did not steer a course for their home. For this reason he says that he did not believe what they told him, and he could not understand them properly, nor they him, and he says that they were as afraid as anything of the people of that island. He would have to delay several days in that harbour if he wanted to speak to the people of that island, but did not do so because there was much land to see and he doubted if he would have time. He hoped in Our Lord that the Indians he had

que los yndios que traya sabrían su lengua y él la suya y después tornaría y hablara con aquella gente, *y plazerá a Su Magestad* (dize él) *que hallara algún buen resgate de oro antes que buelva.*

Viernes 7 de diziembre

Al rendir del quarto del alva dio las velas y salió de aquel Puerto de Sant Nicolás y navegó con el viento sudueste al nordeste dos leguas hasta un cabo que haze el Cheranero, y quedávale al sueste un angla, y el Cabo de la Estrella al sudueste y distava del Almirante 24 millas. De allí navegó al leste luengo de costa hasta el Cabo Çinquin que sería 48 millas; verdad es que las veynte fueron al leste quarta del nordeste. Y aquella costa es tierra toda muy alta y muy grande fondo; hasta dar en tierra es de veynte y treynta braças y fuera tanto como un tiro de lombarda no se halla fondo lo qual todo lo provó el Almirante aquel [día] por la costa, mucho a su plazer con el viento sudueste. El angla que arriba dixo llega dizque al Puerto de San Nicolás tanto como tyro de una lombarda que si aquel espacio se atajase o cortase quedaría hecha ysla lo demás. Bojaría en el çerco 3 [o] 4 millas. Toda aquella tierra era muy alta y no de árboles grandes, sino como carrascos y madroños propria dizque tierra de Castilla. Antes que llegase al dicho Cabo Çinquin con dos leguas halló un a[n]grezuela como la abertura de una montaña por la qual descubrió un valle grandíssimo y vídolo todo senbrado como çevadas y sintió que devía de aver en aquel valle grandes poblaçiones y a las espaldas de él avía grandes montañas y muy altas. Y quando llegó al Cabo de Çinquin le demorava el cabo de la Ysla Tortuga[1] al nordeste, y avría treynta y dos millas. Y sobre este Cabo Çinquin a tyro de una lombarda está una peña en la mar que sale en alto que se puede ver bien. Y estando el Almirante sobre el dicho cabo le demorava el Cabo del Elifante al leste quarta del sueste y avría hasta él 70 millas y toda tierra muy alta. Y a cabo de seys leguas halló una grande angla, y vido por la tierra dentro muy grandes valles y campiñas y montañas altíssimas todo a semejança de Castilla. Y dende a ocho millas halló un río muy hondo, sino que era angosto, aunque bien pudiera entrar en él una carraca y la boca toda limpia, sin banco ni baxas. Y dende a diez y seys millas halló un puerto muy ancho y muy hondo[2] hasta no hallar fondo en la entrada ni a las bordas a tres passos salvo 15 braças[3] y va dentro un quarto de legua. Y

---

[1] Margin: La Tortuga.
[2] Margin: Otro puerto de la Consolación.
[3] MS: braços.

Journal of Columbus

with him would learn his language and he theirs and later he would return and talk to those people, *and please God* (he says) *that I find a good supply of gold to barter before I return.*

Friday 7 December

At the end of the dawn watch he set sail and left that Puerto de San Nicolás and steered NE with a SW wind for 2 leagues as far as a cape he calls Cheranero,[135] and to his SE lay a promontory and the Cabo de la Estrella was 24 miles SW of the Admiral. From there he steered E along the coast to Cabo Cinquin, which would be 48 miles; in truth, 20 of them were E by N. All the land on that coast is very high and the water is very deep; it is 20 to 30 fathoms right up to the shore and at the distance of a lombard shot there is no bottom, all of which the Admiral proved that day to his great delight, sailing along the coast with the wind SW. The promontory he mentioned above extends he says as far as a lombard shot from the Puerto de San Nicolás so that if that space were cut through the rest would be an island of 3 or 4 miles round. All that land was very high and with no large trees, but only oaks and strawberry trees just like, he says, the land in Castile. Two leagues before he reached Cabo Cinquin he found an opening like a ravine in a mountain through which he discovered a very large valley which he could see was sown with what looked like barley and he felt that there must be large villages in that valley and behind it there were large high mountains. When he reached Cabo Cinquin the cape of the island of Tortuga lay about 32 miles NE. At about a lombard shot off Cabo Cinquin there is a rock which rises up out of the sea and is easily seen. When the Admiral stood off that cape the Cabo del Elefante lay E by S about 70 miles away and all the land was very high. Six leagues further on he found a large bay[136] and saw inland very large valleys and fields and very high mountains, all just like Castile. Eight miles further on he found a very deep river, but very narrow, although a carrack could perfectly well enter, and the entrance was completely clear with no sandbanks or shoals. Sixteen miles further on he found a very wide harbour,[137] so deep that there was no bottom at the entrance and 15 fathoms three paces from the shore, and it stretches a quarter of a league inland. Although it was still very early, about 1

puesto que fuese aun muy temprano como la una después de mediodía y el viento era a popa y rezio, pero porque el çielo mostrava querer llover mucho y avía gran çerrazón que es peligrosa aun para la tierra que se sabe quanto más en la que no se sabe, acordó de entrar en el puerto al qual llamó Puerto de la Conçepçión, y salió a tierra en un río no muy grande que está al cabo del puerto que viene por unas vegas y campiñas que era maravilla ver su hermosura. Llevó redes para pescar y antes que llegase a tierra saltó una liça como las de España propria en la barca que hasta entonces no avía visto peçe que pareçiese a los de Castilla. Los marineros pescaron y mataron otras, y lenguados y otros peçes como los de Castilla. Anduvo un poco por aquella tierra que es toda labrada y oyó cantar el ruyseñor[1] y otros paxaritos como los de Castilla. Vieron çinco hombres, mas no les quisieron aguardar sino huyr. Halló arraynán y otros árboles y yervas como las de Castilla y así es la tierra y las montañas.

Sábado 8 de diziembre

Allí en aquel puerto les llovió mucho con viento norte muy rezio. El puerto es seguro de todos los vientos excepto norte puesto que no le puede hazer daño alguno porque la resaca es grande que no da lugar a que la nao labore sobre las amarras ni el agua del río. Después de medianoche se tornó el viento al nordeste y después al leste, de los quales vientos es aquel puerto bien abrigado por la Ysla de la Tortuga,[2] que está frontera a 36 millas.

Domingo 9 de diziembre

Este día llovió y hizo tiempo de invierno como en Castilla por otubre. No avía visto poblaçión sino una casa muy hermosa en el Puerto de Sant Nicolás, y mejor hecha que en otras partes de las que avía visto. *La ysla es muy grande y,* dize el Almirante, *no será mucho que boje dozientas leguas.* A visto que es toda muy labrada; creya que devían ser las poblaçiones lexos de la mar de donde veen quándo llegava, y así huyan todos y llevavan consigo todo lo que tenían y hazían ahumadas como gente de guerra. *Este puerto tiene en la boca mill passos que es un quarto de legua; en ella ni ay banco ni baxa, antes no se halla quasi fondo hasta en tierra a la orilla de la mar, y hazia dentro en luengo va tres mill passos todo limpio y basa que qualquiera nao puede surgir en él sin miedo y entrar sin*

---

[1] Margin: El ruyseñor oyeron cantar.
[2] Margin: La ysla Tortuga.

o'clock in the afternoon, and the wind was astern and strong, because the sky looked like heavy rain and the sky was very overcast, which is dangerous even when one knows the shore and even more so when one does not, he decided to enter the harbour which he called Puerto de la Concepción, and he went ashore in a small river at the end of the harbour which flows across some meadows and fields whose beauty was marvellous to see. He took nets to fish and before he reached the shore a mullet just like those in Spain jumped into the boat. Until then he had not seen a fish which looked like those of Castile. The sailors fished and killed others, and soles and other fish like those of Castile. He took a short walk across that land which is all cultivated and heard a nightingale sing and other birds like those of Castile. They saw 5 men, but they would not wait and took flight. He found myrtle and other trees and plants like those of Castile, for that is what the land and the mountains are like.

Saturday 8 December

There in that harbour it rained very heavily with a very strong N wind. The harbour is protected from all winds except the N, and that cannot do any harm because the undertow is strong and prevents the ship from dragging at its moorings in the river. After midnight the wind turned NE and then E, from which winds that harbour is well protected by the island of Tortuga which lies facing it, 36 miles away.

Sunday 9 December

This day it rained and the weather was wintry like October in Castile. He had not seen any village except a very beautiful house in Puerto de San Nicolás, and better made than those in other parts he had seen. *The island is very large and*, says the Admiral, *it would not be surprising if it were not 200 leagues in circumference.* He has seen that it is all very cultivated; he thought the villages must be a distance from the sea, and from there they can see when he arrived and so they all fled and took with them what they had and lit beacons like soldiers. *This harbour is a thousand paces wide at the entrance, which is a quarter of a league. There are no sandbanks or shoals in it; on the contrary, there is hardly any bottom until the shoreline, and inside, it extends for three thousand paces, all clean and sandy so that any ship can anchor there without fear and enter without precaution.* At the

reguardo. *Al cabo de él tiene dos bocas de ríos que traen poca agua. Enfrente de él ay unas vegas las más hermosas del mundo y quasi semejables a las tierras de Castilla, antes éstas tienen ventaja* por lo qual puso nombre a la dicha ysla la Ysla Española.[1]

Lunes 10 de diziembre

Ventó mucho el nordeste y hízole garrar las anclas medio cable de que se maravilló el Almirante y echólo a que las anclas estavan mucho a tierra y venía sobre ella el viento. Y visto que era contrario para yr donde pretendía, embió seys hombres bien adereçados de armas a tierra que fuesen dos o tres leguas dentro en la tierra para ver si pudieran aver lengua. Fueron y bo[l]vieron no aviendo hallado gente ni casas: hallaron enpero unas cabañas y caminos muy anchos y lugares donde avían hecho lumbre muchos. Vieron las mejores tierras del mundo y hallaron árboles de almáciga muchos y truxeron della y dixeron que avía mucha salvo que no es agora el tiempo para cogella porque no quaja.

Martes 11 de diziembre

No partió por el viento que todavía era leste y nordeste. Frontero de aquel puerto como está dicho está la Ysla de la Tortuga y pareçe grande ysla, y va la costa della quasi como la Española y puede aver de la una a la otra a lo más diez leguas, conviene a saber, desde el cabo de Çinquin a la cabeça de la Tortuga, la qual está al norte de la Española. Después la costa della se corre al sur. Dize que quería ver aquel entremedio destas dos yslas por ver la ysla Española que es la más hermosa cosa del mundo y porque según le dezían los yndios que traya por allí se avía de yr a la ysla de Baneque. Los quales le dezía[n] que era ysla muy grande y de muy grandes montañas y ríos y valles, y dizían que la ysla de Bohío era mayor que la Juana a que llaman Cuba y que no está çercada de agua y pareçe dar a entender ser tierra firme que es aquí detrás desta Española a que ellos llaman Caritaba y que es cosa ynfinita. Y quasi traen razón que ellos sean trabajados de gente astuta, porque todas estas yslas biven con gran miedo de los de Caniba. *Y así torno a dezir como otras vezes dixe,* dize él, *que Caniba no es otra cosa sino la gente del Gran Can que deve ser aquí muy vezino, y terná navíos y vernán a captivarlos y como no buelven creen que se los [han] comido. Cada día entendemos más a estos yndios y ellos a nosotros puesto que muchas vezes ayan*

---

[1] Margin: Aquí puso el Almirante nombre a la Española.

end there are two rivers which bring down very little water. Opposite there are some meadows, the most beautiful in the world and almost comparable with the lands of Castile, although these have the advantage, for which reason he gave the island the name the Isla Española.[138]

Monday 10 December

The wind blew hard from the NE and made him pull in the anchors by half a cable, which surprised the Admiral who put it down to the fact that the anchors were near the land and the wind was blowing towards the land. Seeing that it was against the direction in which he wanted to go, he sent six well-armed men ashore to go two or three leagues inland to see if they could make contact. They went and returned, having found no people or houses, although they found some huts and wide paths and places where many people had made fires. They saw the best lands in the world and found many mastic trees and brought some of it and said that there was a great deal, except that now is not the time to collect it because it is not forming a gum.

Tuesday 11 December

He did not set out because of the wind which was still E and NE. Opposite that harbour, as has been said, is the island of Tortuga and it appears to be a large island, and its coast runs almost exactly as does that of Española, and there can be at the most 10 leagues from one to the other, that is to say from Cabo Cinquin to the head of Tortuga which is to the N of Española. Afterwards the coast of Española runs S. He says that he would like to explore the strait between these two islands to see the island of Española which is the most beautiful thing in the world and because, according to what the Indians he had with him were saying, they had to go that way to get to the island of Baneque.[139] They told him that it was a very large island and with very large mountains and rivers and valleys, and they said that the island of Bohío was larger than that of Juana, which they call Cuba,[140] and that it is not surrounded by water, by which they seem to mean that it is mainland which is here behind this island of Española, which they call Caritaba,[141] and is endless. It seems that they are harassed by people of intelligence because all these islands live in great fear of those from Caniba. *I repeat what I have said before*, he says, *that Caniba is quite simply the people of the Great Khan who must be very close by, and must have ships in which they come and capture them, and because they do not return they believe that they have been eaten.*[142] *Day by day we understand these Indians more and they us, even*

entendido uno por otro, dize el Almirante. Enbió gente a tierra. Hallaron mucha almáçiga sin quajarse; dize que las aguas lo deven hazer, y que en Xío la cogen por março y que en enero la cogerían en aquellas[1] tierras por ser tan templadas. Pescaron muchos pescados como los de Castilla: alvures, salmones, pijotas, gallos, pámpanos, liças, corvinas, camarones, y vieron sardinas. Hallaron mucho lignáloe.

Miércoles 12 de diziembre

No partió aqueste día por la misma causa del viento contrario dicha. Puso un[a] gran cruz a la entrada del puerto de la parte del hueste en un alto muy vistoso *en señal* (dize él) *que Vuestras Altezas tienen la tierra por suya y principalmente por señal de Jesucristo Nuestro Señor, y honrra de la cristiandad.* La qual puesta, tres marineros metiéronse por el monte a ver los árboles y yervas, y oyeron un gran golpe de gente todos desnudos como los de atrás a los quales llamaron e fueron tras ellos, pero dieron los yndios a huyr. Y finalmente tomaron una muger que no pudieron más *porque yo,* él dize, *les avía mandado que tomasen algunos para honrallos y hazelles perder el miedo y se oviese alguna cosa de provecho, como no pareçe poder ser otra cosa segúnd la fermosura de la tierra;* y así truxeron la muger muy moça y hermosa a la nao y habló con aquellos yndios porque todos tenían una lengua. Hyzola el Almirante vestir y diole cuentas de vidro y cascaveles y sortijas de latón, y tornóla enbiar a tierra muy honrradamente según su costumbre, y enbió algunas personas de la nao con ella, y tres de los yndios que llevava consigo porque hablasen con aquella gente. Los marineros que yvan en la barca quando la llevavan a tierra dixeron al Almirante que ya no quisiera salir de la nao sino quedarse con las otras mugeres yndias[2] que avía hecho tomar en el Puerto de Mares de la ysla Juana de Cuba. Todos estos yndios que venían con aquella yndia dizque venían en una canoa, que es su caravela en que navegan de alguna parte, y quando asomaron a la entrada del puerto y vieron los navíos bolviéronse atrás y dexaron la canoa por allí en algún lugar y fuéronse camino de su poblaçión. Ella mostrava el paraje de la poblaçión. Traya esta muger un pedaçito de oro en la nariz, que era señal que avía en aquella ysla oro.

---

[1] MS: ?aquestas corrected to ?aquellas.
[2] Margin: No[ta].

though they have often understood one thing for another, says the Admiral. He some sent men ashore. They found much mastic which had not yet formed gum; he says that the rain must cause this and that on Chios they gather it in March and that in these lands they would gather it in January because they are so temperate. They caught many fish like those of Castile: dace, salmon, hake, dory, pampano, mullet, congers, shrimps, and they saw sardines. They found much aloe.

Wednesday 12 December

He did not set out on this day for the same reason stated, a head wind. He set up a great cross at the entrance of the harbour on the western side on a very conspicuous high point, *as a sign* (he says) *that Your Highnesses hold the land as your own and principally as a sign of Our Lord Jesus Christ, and in honour of Christendom*. When it had been set up, three sailors set off across the scrub to look at the trees and plants, and they heard a great crowd of people, all naked like those they had seen previously. They called and ran after them, but the Indians took flight. Eventually they caught a woman but they could not manage more, *for I*, he says, *had ordered them to capture some Indians to do them honour and reassure them so that we could gain something of value, for to judge from the beauty of the land it seems that there could not but be gains to be made*. So they brought the woman, very young and beautiful, to the ship and she spoke with those Indians because they all have the same language. The Admiral had her dressed and gave her some glass beads and hawks' bells and brass rings, and sent her back again to the land with great ceremony as was his custom, and sent some men from the ship with her, and three of the Indians he had with him to speak to the people there. The sailors who went in the boat told the Admiral that when they were taking her to land she did not want to leave the ship but preferred to remain with the other Indian women whom he had had captured at Puerto de Mares on the island of Juana or Cuba. All the Indians who came with that Indian woman, he says, came in a canoe, which is their form of caravel in which they travel about, and when they rounded the harbour entrance and saw the ships they turned back and left the canoe there and went off in the direction of their village. She showed them the whereabouts of the village. This woman was wearing a small piece of gold in her nose, which was a sign that there was gold on that island.

Jueves 13 de diziembre

Volviero[n] los tres hombres que avía enbiado el Almirante con la muger a tres oras de la noche, y no fueron con ella hasta la població[n] porque les pareçió lexos o porque tuvieron miedo. Dixeron que otro día vernían mucha gente a los navíos, porque ya devían de estar asegurados por las nuevas que daría la muger. El Almirante con deseo de saber si avía alguna cosa de provecho en aquella tierra, y por aver alguna lengua con aquella gente por ser la tierra tan hermosa y fértil y tomasen gana de servir a los Reyes, determinó de tornar a enbiar a la població[n] confiando en las nuevas que la yndia avría dado de los cristianos ser buena gente, para lo qual escogió nueve hombres bien adereçados de armas y aptos para semejante negocio, con los quales fue un yndio de los que traya. Estos fueron a la població[n] que estava quatro leguas y media al sueste, la qual hallaron en un grandíssimo valle, y vazía porque como sintieron yr los cristianos todos huyeron dexando quanto tenían la tierra dentro. La població[n] era de mil casas y de más de tres mill hombres. El yndio que llevavan los cristianos corrió tras ellos dando bozes diziendo que no oviesen miedo, que los cristianos no eran de Caniba mas antes eran del çielo y que davan muchas cosas hermosas a todos los que hallavan. Tanto les imprimió lo que dezía que se aseguraron[1] y vinieron juntos dellos más de dos mill, y todos venían a los cristianos y les ponían las manos sobre cabeça que era señal de gran reverençia y amistad, los quales estavan todos temblando hasta que mucho los aseguraro[n]. Dixeron los cristianos que después que ya estavan sin temor, yvan todos a sus casas y cada uno les traya de lo que tenía de comer que es pan de niamas que son unas rayzes como rávanos grandes que naçen, que siembra[n] y naçen y plantan en todas estas tierras, y es su vida, y hazen dellas pan y cuezen y asan y tienen sabor proprio de castañas y no ay quien no crea comiéndolas que no sean castañas. Y dávanles pan y pescados y de lo que tenían. Y porque los yndios que traya en el navío tenían entendido que el Almirante deseava tener algún papagayo, parez que aquel yndio que yva con los cristianos díxoles algo desto, y así les truxeron papagayos y les davan quanto les pedían sin querer nada por ello.[2] Rogávanles que no se viniesen aquella noche y que les darían otras muchas cosas que tenían en la sierra. Al tiempo que toda aquella gente estava junta con los cristianos, vieron venir una gran batalla o multitud de gente con el marido de la muger que

---

[1] Margin: No[ta].
[2] Margin: No[ta].

Thursday 13 December

The three men whom the Admiral had sent with the woman returned at three o'clock in the morning, and they did not go with her to the village because it seemed to them a long way off, or because they were afraid. They said that the next day many people would come to the ships, because they must by then have been reassured by the news which the woman would give. Out of a desire to see if there were anything of value in that land and to speak to the people, and because the land was very beautiful and fertile, and so that they should wish to serve the Monarchs, the Admiral decided to send again to the village, trusting in the report which the Indian woman would have given that the Christians were good people. For this purpose he chose nine men, well-supplied with arms and suitable for the task, and with them went one of the Indians he had with him. They went to the village which was 4 and a half leagues to the SE and which they found in a huge valley,[143] but it was empty because as soon as they heard the Christians coming they all fled into the interior leaving whatever they had behind. The village consisted of a thousand houses and more than three thousand people. The Indian whom the Christians had with them ran after them shouting that they should have no fear because the Christians were not from Caniba but from heaven and that they gave many beautiful things to all those they encountered. What he said so impressed them that they were reassured and more than two thousand of them all gathered together, and came to the Christians and put their hands on their heads, which was a sign of great reverence and friendship, and they were all trembling until they were greatly reassured. The Christians said that when they were no longer afraid, they all went to their houses and each person brought something of what he had to eat, which is bread made of 'niamas' which are roots like large radishes which grow there and which they sow and plant and cultivate in all these lands, and which is their staple food.[144] They make bread from them and boil and roast them and they taste just like chestnuts and no one would think, eating them, that they were not chestnuts. And they gave them bread and fish and whatever they had. And because the Indians he had with him in the ship had understood that the Admiral wanted to have a parrot, it seems that the Indian who had gone with the Christians said something about this and so they brought them parrots and gave them as much as they asked without wanting anything in exchange. They asked them not to return that night[145] and said that they would give them many other things which they had in the mountains. While all those people were together with the Christians they saw a great army or crowd of people coming towards them with the husband of the woman whom the Admiral had treated with

avía el Almirante honrrado y enbiado la qual trayan cavallera sobre sus hombros y venían a dar graçias a los cristianos por la honrra que el Almirante le avía hecho y dádivas que le avía dado. Dixeron los cristianos al Almirante que era toda gente más hermosa y de mejor condiçión que ninguna otra de las que avían hasta allí hallado; pero dize el Almirante que no sabe cómo puedan ser de mejor condiçión que las otras, dando a entender que todas las que avían en las otras yslas hallado era[n] de muy buena condiçión. Quanto a la hermosura dezían los cristianos que no avía comparaçión, así en los hombres como en las mugeres, y que son blancos más que los otros, y que entre los otros vieron dos mugeres moças tan blancas como podían ser en España.[1] Dixeron también de la hermosura de las tierras que vieron, que ninguna comparaçión tienen las de Castilla las mejores en hermosura y en bondad. Y el Almirante así lo vía por las que a visto y por las que tenía presentes. Y dizíanle que las que vía ninguna comparación tenían con aquellas de aquel valle, ni la campiña de Córdova llegava [a] aquella con tanta differençia como tiene el día de la noche. Dezían que todas aquellas tierras estavan labradas y que por medio de aquel valle passava un río muy ancho y grande que podía regar todas las tierras. Estavan todos los árboles verdes y llenos de fruta, y las yervas todas floridas y muy altas; los caminos muy anchos y buenos; los ayres eran como en abril en Castilla; cantava el ruyseñor y otros paxaritos como en el dicho mes en España, que dizen que era la mayor dulçura del mundo. Las noches cantavan algunos paxaritos suavemente; los grillos y ranas se oyan muchas; los pescados, como en España; vieron muchos almáçigos y lignáloe y algodonales. Oro no hallaron, y no es maravilla en tan poco tiempo no se halle.[2] Tomó aquí el Almirante experiençia de qué oras era el día y la noche y de sol a sol halló que passaron veynte ampolletas que son de a media ora, aunque dize que allí puede aver defecto porque o no la buelven tan presto o dexa de passar algo. Dize también que halló por el quadrante que estava de la línea equinoçial 34 grados.[3]

Viernes 14 de diziembre

Salió de aquel Puerto de la Conçepción con terral y luego desde a poco calmó y así lo experimentó cada día de los que por allí estuvo. Después vino viento levante; navegó con él al nornordeste, llegó a la Ysla de la Tortuga, vido una

---

[1] Margin: No[ta].
[2] Margin: No[ta].
[3] Margin: Esto es impossible.

respect and sent back and whom they were carrying on their shoulders. They came to thank the Christians for the honour which the Admiral had done her and the gifts which he had given her. The Christians told the Admiral that all these people were more handsome and of better character than any they had found until then; but the Admiral says that he does not know how they can be of better character than the others, by which he means that all those he had found on the other islands were of very good character. As far as beauty was concerned, the Christians said that there was no comparison, in both men and women, and that they are whiter than the others and that among them[146] they saw two young women who were as white as they would be in Spain. They also said, speaking of the beauty of the lands they saw, that there was no comparison with the most beautiful and fertile lands of Castile. And the Admiral saw that it was so of both the lands he has seen and those he had before him. And they told him that there was no comparison between those which he could see and those in that valley, and that even the fields around Córdoba could not match them, being as different from them as the day is from the night. They said that all those lands were cultivated and that through the middle of that valley flowed a very wide river, large enough to irrigate all the fields. All the trees were green and full of fruit and all the plants in flower and very tall; the tracks were very wide and good; the air was like April in Castile; the nightingale and other small birds sang as they do in that month in Spain, and they say that it is the most delightful thing in the world. At night some birds sang sweetly; many crickets and frogs could be heard; the fish were like those in Spain; they saw many mastic and aloe and cotton trees. They found no gold, and it is not surprising that it could not be found in so short a time. Here the Admiral measured the length of the day and night in hours, and found that from sunrise to sunset was 20 half-hour glasses, although he says that there could be an error either because they do not turn the glass soon enough or because some of the sand has not passed through. He also says that he found from his quadrant that he was 34 degrees from the equinoctial line.[147]

Friday 14 December

He set out from Puerto de la Concepción with a land breeze which then dropped and he found this to be the case every day he was there. Then an E wind blew and he steered NNE, reached the island of Tortuga, spotted a point which he

punta della que llamó la Punta Pierna que estava al lesnordeste de la cabeça de la ysla y avría 12 millas, y de allí descubrió otra punta que llamó la Punta Lançada en la misma derrota del nordeste que avría diez y seys millas. Y así desde la cabeça de la Tortuga hasta la Punta Aguda avría 44 millas que son onze leguas al lesnordeste. En aquel camino avía algunos pedaços de playa grandes. Esta isla de la Tortuga es tierra muy alta pero no montañosa y es muy hermosa y muy poblada de gente como la de la ysla Española, y la tierra así toda labrada que pareçía ver la campiña de Córdova. Visto que el viento le era contrario y no podía yr a la ysla Baneque, acordó tornarse al Puerto de la Conçepçión de donde avía salido, y no pudo cobrar un río que está de la parte del leste del dicho puerto dos leguas.

Sábado 15 de diziembre

Salió del Puerto de la Conçepçión otra vez para su camino pero en saliendo del puerto ventó leste rezio su contrario y tomó la buelta de la Tortuga hasta ella y de allí dio buelta para ver aquel río que ayer quisiera ver y tomar y no pudo, y desta buelta tampoco lo pudo tomar, aunque surgió media legua de sotavento en una playa, buen surgidero y limpio. Amarrados sus navíos, fue con las barcas a ver el río y entró por un braço de mar que está antes de media legua y no era la boca. Bolvió y halló la boca que no tenía aún una braça, y venía muy rezio. Entró con las barcas por él para llegar a las poblaçiones que los que antier avía enbiado avían visto, y mandó echar la sirga en tierra y tyrando los marineros della subieron las barcas dos tiros de lombarda y no pudo andar más por la reziura de la corriente del río. Vido algunas casas y el valle grande donde están las poblaçiones y dixo que otra cosa más hermosa no avía visto por medio del qual valle viene aquel río. Vido también gente a la entrada del río, mas todos dieron a huyr. Dize más: que aquella gente deve ser muy caçada, pues bive con tanto temor, porque en llegando que llegan a qualquiera parte luego hazen ahumadas de las atalayas por toda la tierra y esto más en esta ysla Española y en la Tortuga que también es grande ysla, que en las otras que atrás dexava. Puso nombre al valle Valle del Parayso,[1] y al río Guadalquivir, porque dizque así viene tan grande como Guadalquivir por Córdova y a las veras o riberas de él, playa de piedras muy hermosas, y todo andable.

---

[1] Margin: Valle del Parayso.

called Punta Pierna which was to the ENE of the head of the island at a distance of about 12 miles, and from there he sighted another point which he called Punta Lanzada in the same direction to the NE about 16 miles away. So from the head of Tortuga to the Punta Aguda would be about 44 miles which is 11 leagues ENE.[148] On that course there were some large stretches of beach. This island of Tortuga is very high but not mountainous, and it is very beautiful and inhabited like the island of Española, and all the land is cultivated so that he thought he was looking at the fields around Córdoba. Seeing that the wind was against him and that he could not go to the island of Baneque, he decided to turn back to Puerto de la Concepción from where he had set out, and he could not gain entrance to a river which was 2 leagues E of that harbour.

Saturday 15 December

He again set out from Puerto de la Concepción to go on his way, but as he left the harbour the wind blew strongly from the E against him and he steered towards Tortuga and on reaching it turned back in order to take a look at that river which he wanted to see yesterday but which he could not reach. He could not make it from this pass either, although he anchored off a beach half a league downwind, a good clean anchorage. Having moored the ships he went with the boats to explore the river and entered via an arm of the sea half a league short of it and which is not the river mouth. He turned back and found the mouth which was not even a fathom deep and the current was very fierce. He went in with the boats to make for the villages which those whom he had sent out the day before yesterday had seen, and ordered the line to be thrown ashore, and with the sailors pulling on it they dragged the boats along two lombard shots but could not go further because of the strength of the current in the river. He saw some houses and the great valley where the villages lay and said that he had seen nothing more beautiful than that valley with the river flowing through it. He also saw people at the mouth of the river, but they all fled. He says further that those people must be hunted a great deal because they live in such fear, for wherever they land they make smoke signals from beacons right across the land and more so in the island of Española and in Tortuga, which is also a large island, than in the other islands he had left behind him. He named the valley Valle del Paraíso, and the river, Guadalquivir,[149] because he says that it flows as strongly as the Guadalquivir at Córdoba, and on the edges or banks there is a beach with beautiful stones, and it is all suitable for walking.

Domingo 16 de diziembre

A la medianoche con el ventezuelo de tierra dio las velas por salir de aquel golpho, y viniendo del bordo de la ysla Española yendo a la bolina porque luego a ora de terçia ventó leste, a medio golpho halló una canoa con un yndio solo en ella, de que se maravillava el Almirante cómo se podía tener sobre el agua siendo el viento grande. Hyzolo meter en la nao a él y a su canoa, y halagado diole cuentas de vidro, cascaveles y sortijas de latón, y llevólo en la nao hasta tierra a una población que estava de allí diez y seys millas junto a la mar, donde surgió el Almirante y halló buen surgidero en la playa junto a la población que pareçía ser de nuevo hecha porque todas las casas eran nuevas. El yndio fuése luego con su canoa a tierra, y da nuevas del Almirante y de los cristianos ser buena gente, puesto que ya las tenían por lo passado de las otras donde avían ydo los seys cristianos, y luego vinieron más de quinientos hombres, y desde a poco vino el rey dellos, todos en la playa juntos a los navíos porque estavan surgidos muy çerca de tierra. Luego uno a uno y muchos a muchos venían a la nao, sin traer consigo cosa alguna, puesto que algunos trayan algunos granos de oro finíssimo a las orejas o en la nariz el qual luego davan de buena gana. Mandó hazer honrra a todos el Almirante, y dize él: *porque son la mejor gente del mundo y más mansa.*[1] *Y sobre todo* (dize) *que tengo mucha esperança en Nuestro Señor que Vuestras Altezas los harán todos cristianos y serán todos suyos que por suyos los tengo.* Vido también que el dicho rey estava en la playa y que todos le hazían acatamiento. Embióle un presente el Almirante, el qual dizque rescibió con mucho estado, y que sería moço de hasta veynte y un años, y que tenía un ayo viejo y otros consejeros que le consejavan y respondían, y que él hablava muy pocas palabras. Uno de los yndios que traya el Almirante habló con él, y le dixo cómo venían los cristianos del çielo, y que andava en busca de oro,[2] y que quería yr a la ysla de Baneque; y él respondió que bien era y que en la dicha ysla avía mucho oro. El qual amostró al alguazil del Almirante que le llevó el presente, el camino que avía de llevar y que en dos días yría de allí a ella, y que si de su tierra avían menester algo lo daría de muy buena voluntad. Este rey y todos los otros andavan desnudos como sus madres los parieron, y así las mugeres sin algún empacho. Y son los más hermosos hombres y mugeres que hasta allí ovieron hallado, harto blancos, que si vestidos anduviesen y se guardasen del sol y del ayre serían quasi tan blancos como en España. Porque esta

---

[1] Margin: No[ta].
[2] Margin: Satis improportionabi[li]ter hac habent.

Sunday 16 December

He set sail at midnight to leave that gulf with the land breeze, and coming from the coast of the island of Española and sailing close-hauled because presently at the hour of terce the wind blew from the E, in mid gulf he encountered a canoe with a solitary Indian in it, and the Admiral was amazed at the way he managed to remain afloat with the strong wind. He had him and his canoe brought aboard the ship and flattered him with some presents of glass beads, hawks' bells and brass rings, and took him by ship to land at a coastal village which was 16 miles away, where the Admiral anchored and found a good anchorage off the beach next to the village, which seemed to be newly built as all the houses were new. The Indian then went ashore with his canoe and gave news that the Admiral and the Christians were good people, although they already knew about what had happened to the others whom the six Christians had visited, and then more than 500 men came, and shortly afterwards their king, and they all gathered together on the beach near the ships, for they had anchored very close to the shore. Then one by one and many at a time they came to the flagship, bringing nothing with them, although some wore in their ears or noses some grains of the finest gold which they gave very willingly. The Admiral ordered that all should be treated with respect, and he says: *because they are the finest people in the world and the most gentle. And above all* (he says) *I dearly hope in Our Lord that Your Highnesses will make them all Christians and they will all be your subjects, for I already consider them to be so.* He also saw that the said king was on the beach and that they all showed him respect. The Admiral sent him a present, which he says he received with great ceremony, and that he must have been a youth of about 21 years, and that he had an elderly tutor and other counsellors who advised him and answered for him, and that he himself spoke very few words. One of the Indians the Admiral had with him spoke to him and told him how the Christians came from heaven, and that the Admiral was searching for gold and that he wished to go to the island of Baneque. He replied that this was good and that there was much gold on that island. He pointed out to the Admiral's bailiff[150] who had taken the present to him the route which he should take and said that in two days he would get from there to the island, and that if they needed anything from his land he would give it very willingly. This king and all the others were as naked as their mothers bore them, and the women too, without a hint of shame. They are the most handsome men and women they had seen up to that time, so fair that if they wore clothes and kept out of the sun and the wind they would almost be as white as those in Spain. Because this land is quite cold and the

tierra es harto fría y la mejor que lengua puede dezir. Es muy alta y sobre el mayor monte podrían arar bueyes, y hecha toda de campiñas y valles. En toda Castilla no ay tierra que se pueda comparar a ella en hermosura y bondad. Toda esta ysla y la de la Tortuga son todas labradas como la campiña de Córdova. Tienen sembrado en ellas ajes, que son unos ramillos que plantan, y al pie dellos naçen unas rayzes como çanahorias, que sirven por pan y rallan y amassan y hazen pan dellas, y después tornan a plantar el mismo ramillo en otra parte y torna a dar quatro y çinco de aquellas rayzes que son muy sabrosas proprio gusto de castañas. Aquí las ay las más gordas y buenas que avía visto en ninguna [parte] porque también dizque de aquellas avía en Guinea. Las de aquel lugar eran tan gordas como la pierna, y aquella gente todos dizque eran gordos y valientes y no flacos, como los otros que antes avía hallado, y de muy dulçe conversaçión sin secta. Y los árboles de allí dizque eran tan viçiosos, que las hojas dexavan de ser verdes y eran prietas de verdura.[1] Era cosa de maravilla ver aquellos valles y los ríos y buenas aguas, y las tierras para pan, para ganado de toda suerte, de que ellos no tienen alguna; para güertas y para todas las cosas del mundo que el hombre sepa pedir. Después a la tarde vino el rey a la nao. El Almirante le hizo la honrra que devía[2] y le hizo dezir cómo era de los Reyes de Castilla, los quales eran los mayores príncipes del mundo. Mas ni los yndios que el Almirante traya que eran los intérpretes creyan nada,[3] ni el rey tampoco, sino creyan que venían del çielo, y que los reynos de los Reyes de Castilla eran en el çielo y no en este mundo. Pusiéronle de comer al rey de las cosas de Castilla, y él comía un bocado y después dávalo todo a sus consejeros y al ayo y a los demás que metió consigo. *Crean Vuestras Altezas que estas tierras son en tanta cantidad buenas y fértiles y en especial estas desta ysla Española, que no ay persona que lo sepa dezir, y nadie lo puede creer si no lo viese. Y crean que esta ysla y todas las otras son así suyas como Castilla,[4] que aquí no falta salvo assiento y mandarles hazer lo que quisieren, porque yo con esta gente que traygo que no son muchos, correría todas estas yslas sin afrenta, que ya e visto solos tres destos marineros desçender en tierra y aver multitud destos yndios y todos huyr sin que les quisiesen hazer mal. Ellos no tienen armas y son todos desnudos y de ningún ingenio en las armas y muy cobardes que mill no aguardarían tres, y así son buenos para les mandar y les hazer trabajar sembrar y hazer todo lo otro que fuere*

---

[1] Margin: No[ta].
[2] Margin: No[ta].
[3] Margin: No[ta].
[4] Margin: No[ta].

finest that words can describe. It is very high and on the highest hill oxen could plough, and it consisted entirely of fields and valleys. In all Castile there is no land which could be compared with it in beauty or bounteousness. All this island and that of Tortuga are cultivated like the fields around Córdoba. They have it all sown with 'ajes'[151] which are small branches which they plant and at the foot of which grow roots like carrots, which are used for bread and which they grate and knead and make into bread, and then they plant the same branch in another spot and it again produces four or five of those roots which are very tasty, with the exact flavour of chestnuts. Those here are the fattest and best he had seen anywhere for he also says that there were some of them in Guinea. Those here were as thick as a leg and he says that all these people were well-built and strapping and not thin like the others he had found previously, and very quietly spoken and without religion. And he says that the trees there are so lush that the leaves were so green that they were not green but black. It was a marvellous thing to see those valleys and rivers and fine waters, and the fields suitable for producing bread and for raising stock of all kinds, of which they have none, and for orchards and for everything on earth that man could ask for. Later that afternoon the king came to the flagship. The Admiral treated him with due respect and had him told how he came from the King and Queen of Castile who were the greatest princes in the world. But the Indians the Admiral had with him who were the interpreters believed none of it, nor did the king; they believed instead that they came from heaven, and that the kingdoms of the King and Queen of Castile were in the sky and not on this earth. They gave the king some things to eat from Castile, and he ate a mouthful and then gave it all to his counsellors and to the tutor and to the others he had with him. *Believe me, Your Highnesses, that these lands are so good and so fertile, particularly these of this island of Española, that there is no one who could describe them, and no one could believe it unless they saw it. Be sure that this island and all the others are as much your own as is Castile, for all that is needed here is a seat of government and to command them to do what you wish, for I with these people I have with me, who are not many, could travel throughout these islands unopposed, and I have seen three of these sailors go ashore alone where there was a crowd of these Indians, and they have all run off, without anyone wishing them any harm. They have no weapons and are all naked and with no experience of arms and very timid, so that a thousand of them would not stand up to three of us, and so they are suitable to*

menester,[1] *y que hagan villas y se enseñen a andar vestidos y a nuestras costumbres.*

Lunes 17 de diziembre

Ventó aquella noche reziamente viento lesnordeste; no se alteró mucho la mar porque lo estorva y escuda la Ysla de la Tortuga que está frontera y haze abrigo. Así estuvo allí aqueste día. Embió a pescar los marineros con redes. Holgáronse mucho con los cristianos los yndios, y truxéronles ciertas flechas de los de Caniba o de los caníbales, y son de las espigas de cañas y enxiérenles uno[s] palillos tostados y agudos y son muy largos. Mostráronles dos hombres que les faltavan algunos pedaços de carne de su cuerpo y hiziéronles entender que los caníbales los avían comido a bocados; el Almirante no lo creyó. Tornó a embiar çiertos cristianos a la población, y a trueque de contezuelas de vidro rescataron algunos pedaços de oro labrado en hoja delgada.[2] Vieron a uno que tuvo el Almirante por governador de aquella provinçia, que llamavan caçique, un pedaço tan grande como la mano de aquella hoja de oro y pareçía que lo quería resgatar; el qual se fue a su casa, y los otros quedaron en la plaça. Y él hazía hazer pedaçuelos de aquella pieça, y trayendo cada vez un pedaçuelo resgatávalo. Después que no ovo más dixo por señas que él avía enbiado por más y que otro día lo traerían. Estas cosas todas y la manera dellos y sus costumbres y mansedumbre y consejo muestra de ser gente más despierta y entendida que otros que hasta allí oviese hallado, dize el Almirante. *En la tarde vino allí una canoa de la Ysla de la Tortuga con bien quarenta hombres,*[3] *y en llegando a la playa toda la gente del pueblo que estava junta se assentaron todos en señal de paz, y algunos de la canoa y quasi todos desçendieron en tierra. El caçique se levantó solo y con palabras que parecían de amenazas*[4] *los hizo bolver a la canoa y les echava agua, y tomava piedras de la playa y las echava en el agua y después que ya todos con mucha obediencia se pusieron y enbarcaron en la canoa, él tomó una piedra y la puso en la mano a mi alguazil para que la tyrase, al qual yo avía enbiado a tierra y al escrivano y a otros para ver si trayan algo que aprovechase, y el alguazil no les quiso tyrar. Allí mostró mucho aquel caçique que se favoreçía con el Almirante. La canoa se fue luego, y dixeron al Almirante*

---

[1] Margin: No[ta]. Algo más parece aquí estenderse el Almirante de lo que devría.
[2] Margin: Resgataron oro.
[3] Margin: No[ta].
[4] Margin: No[ta].

take orders and be made to work, sow and do anything else that may be needed, and build towns and be taught to wear clothes and adopt our customs.

Monday 17 December

The wind blew strongly that night from the ENE; the sea did not get up very much, as it is protected and shielded by the island of Tortuga which is opposite and forms a shelter. So he remained there throughout this day. He sent the men to fish with nets. The Indians enjoyed themselves very much with the Christians and brought them certain arrows belonging to the Caniba or Cannibals, and they are made from the stem of a reed with fire-hardened points inserted at the tip and are very long. They showed them two men with pieces of flesh missing from their bodies and gave them to understand that the cannibals had eaten mouthfuls of them. The Admiral did not believe it. He sent some Christians to the village again and in exchange for small glass beads they bartered some pieces of gold beaten into thin leaves. They saw one man, whom the Admiral took to be the governor of that province and whom they called 'cacique',[152] with a piece of that gold leaf as large as a hand and he seemed to want to trade it. He went home and the others remained in the square, and he was cutting that piece into smaller pieces and bringing one at a time to barter with them. When there was none left he made signs that he had sent for more and that they would bring it the following day. All these things, their behaviour and customs, their gentleness and good sense shows them to be a more alert and understanding people than the others he had found until now, says the Admiral. *In the afternoon there came a canoe from the island of Tortuga with a good 40 men, and when they arrived on the beach all the people of the nearby town sat down as a sign of peace and some if not most of those in the canoe came ashore. The cacique stood up alone and in words which seemed to be threatening made them go back to the canoe and threw water at them and took stones from the beach and threw them in the water and when they had all very obediently got into the canoe, he took a stone and put it in the hand of my bailiff, whom I had sent ashore with the secretary and others to see if they could bring back anything of value, for him to throw it, and the bailiff declined to throw it at them. In this way that cacique made it very clear that he favoured the Admiral. The canoe then went off and when it had left they told the Admiral that there*

después de yda que en la Tortuga avía más oro que en la ysla Española porque es más çerca de Baneque. Dixo el Almirante que no creya que en aquella ysla Española ni en la Tortuga oviese minas de oro, sino que lo trayan de Baneque, y que traen poco, porque no tiene[n] aquellos qué dar por ello. Y aquella tierra es tan gruessa que no a menester que trabajen mucho para sustentarse ni para vestirse, como anden desnudos. Y creya el Almirante que estava muy çerca de la fuente y que Nuestro Señor le avía de mostrar dónde nasçe el oro. Tenía nueva que de allí al Baneque avía quatro jornadas que podrían ser xxx o xl leguas que en un día de buen tiempo se podían andar.[1]

Martes 18 de diziembre

Estovo en aquella playa surto este día porque no avía viento y también porque avía dicho el caçique que avía de traer oro, no porque tuviese en mucho el Almirante el oro (dizque) que podía traer pues allí no avía minas sino por saber mejor de dónde lo trayan. Luego en amaneçiendo mandó ataviar la nao y la caravela de armas y vanderas por la fiesta que era este día de Sancta María de la O o comemoraçión de la Anunçiaçión. Tyráronse muchos tyros de lombardas y el rey de aquella ysla Española (dize el Almirante) avía madrugado de su casa que devía de distar çinco leguas de allí según pudo juzgar y llegó a ora de terçia a aquella poblaçión, donde ya estavan algunos de la nao que el Almirante avía enbiado para ver si venía oro, los quales dixeron que venían con el rey más de dozientos hombres, y que lo trayan en unas andas quatro hombres y era moço como arriba se dixo. Oy estando el Almirante comiendo debaxo del castillo llegó a la nao con toda su gente.[2] Y dize el Almirante a los Reyes: *Sin duda pareçiera bien a Vuestras Altezas su estado y acatamiento que todos le tienen puesto que todos andan desnudos. El así como entró en la nao halló que estava comiendo a la mesa debaxo del castillo de popa y él a buen andar se vino a sentar a par de mí, y no me quiso dar lugar que yo me saliese a él ni me levantase de la mesa, salvo que yo comiese. Yo pensé que él ternía a bien de comer de nuestras viandas; mandé luego traerle cosas que él comiesse. Y quando entró debaxo del castillo hizo señas con la mano que todos los suyos quedasen fuera y así lo hizieron con la mayor priesa y acatamiento del mundo y se assentaron todos en la cubierta, salvo dos hombres de una edad madura que yo estimé por sus consejeros y ayo, que vinieron y se assentaron a sus pies. Y de las viandas que yo le puse*

---

[1] Margin: Nunca este Baneque pareçió. Por ventura era la ysla de Jamayca.
[2] Margin: Vino el rey a la nao.

was more gold on Tortuga than on Española because it is nearer to Baneque. The Admiral said that he did not believe that there were gold mines on that island of Española nor on Tortuga, but that they brought it from Baneque, and that they bring very little of it, because these people have nothing to give for it. That land is so rich that they do not need to work very much to support themselves, nor to clothe themselves because they go naked. And the Admiral believed that he was very close to the source and that Our Lord would show him where the gold originates from. He had information that from there to Baneque was four days' journey, which could be 30 or 40 leagues, which could be sailed in one day of fine weather.[153]

Tuesday 18 December

On this day he lay anchored off that beach because there was no wind and also because the cacique had said that he would bring gold, not that the Admiral set much store by the gold that he could bring (he says), because there were no mines there, but in order to find out more about where they brought it from. At dawn, then, he ordered the flagship and the caravel to be decked with arms and banners for the feast that day, which was Santa María de la O, or the commemoration of the Annunciation. Many lombard shots were fired and the king of that island of Española (says the Admiral) had set out early from his house which was about 5 leagues away, as far as he could judge, and arrived at the hour of terce at that village where there were already some of the sailors whom the Admiral had sent to see if the gold was coming. They said that the king was on his way with more than 200 men, and that four men were carrying him on a litter and that he was a young man, as was said earlier. Today, while the Admiral was eating beneath the forecastle, he came to the ship with all his men. And the Admiral says to the Monarchs: *Your Highnesses would no doubt approve of the ceremony and respect with which they all treat him, although they all go naked. As soon as he came aboard the ship he found that I was eating at the table beneath the forecastle and he strode right up and sat down beside me and did not wish to give me the chance to go out to meet him nor rise from the table, but bade me continue my meal. I thought that he would be pleased to eat some of our food. I then ordered him to be brought something to eat. When he entered below the forecastle he gestured with his hand that his men should remain outside and so they did with the greatest readiness and respect in the world and they all sat on the deck except two men of mature age, whom I took to be his counsellors and tutor, who came and sat at his feet. And of the dishes which I put before him he took just enough*

delante, tomava de cada una tanto como se toma para hazer la salva, y después luego lo demás enbiávalo a los suyos y todos comían della y así hizo en el bever que solamente llegava a la boca y después así lo dava a los otros, y todo con un estado maravilloso y muy pocas palabras y aquellas que él dezía, según yo podía entender, eran muy assentadas y de seso y aquellos dos le miravan a la boca y hablavan por él y con él y con mucho acatamiento. Después de comido un escudero traya un çinto que es proprio como los de Castilla en la hechura, salvo que es de otra obra, que él tomó y me lo dio, y dos pedaços de oro labrados que eran muy delgados, que creo que aquí alcançan poco de él, puesto que tengo que están muy vezinos de donde naçe y ay mucho. Yo vide que le agradava un arambel que yo tenía sobre mi cama. Yo se lo di y unas cuentas muy buenas de ámbar que yo traya al pescueço, y unos çapatos colorados, y una almarraxa de agua de azahar de que quedó tan contento que fue maravilla. Y él y su ayo y consejeros llevan grande pena porque no me entendían ni yo a ellos. Con todo le cognoscí que me dixo, que si me compliese algo de aquí, que toda la ysla estava a mi mandar. Yo embié por unas cuentas mías adonde por un señal tengo un exçelente[1] de oro en que está esculpido Vuestras Altezas y se lo amostré, y le dixe otra vez como ayer que Vuestras Altezas mandavan y señoreavan todo lo mejor del mundo, y que no avía tan grandes prínçipes. Y le mostré las vanderas reales y las otras de la cruz, de que él tuvo en mucho y qué grandes señores serían Vuestras Altezas, dezía él contra sus consejeros, pues de tal lexos y del çielo me avían enbiado hasta aquí sin miedo y otras cosas muchas se passaron que yo no entendía, salvo que bien vía que todo tenía a grande maravilla. Después que ya fue tarde y él se quiso yr, el Almirante le enbió en la barca muy honrradamente y hizo tyrar muchas lombardas, y puesto en tierra subió en sus andas y se fue con sus más de dozientos hombres. Y su hijo le llevavan atrás en los hombros de un yndio, hombre muy honrrado. A todos los marineros y gente de los navíos dondequiera que lo topava les mandava dar de comer y hazer mucha honrra. Dixo un marinero que le avía topado en el camino y visto, que todas las cosas que le avía dado el Almirante y cada una dellas llevava delante del rey un hombre a lo que pareçía de los más honrrados. Yva su hijo atrás del rey buen rato con tanta compañía de gente como él. Y otro tanto un hermano del mismo rey, salvo que yva el hermano a pie, y llevávanlo de braço dos hombres honrrados. Este vino a la nao después del rey, al qual dio el Almirante algunas cosas de los dichos resgates, y allí supo el Almirante que al rey llamavan en su lengua caçique. En este día se resgató dizque poco oro, pero supo el Almirante de un hombre viejo, que avía muchas yslas

---

[1] Margin: Este exçelente era moneda que valía dos castellanos.

from each to sample them and then sent the rest to his men and they all ate it, and he did the same with the drink, which he merely raised to his lips and then gave to the others, and all with an amazing gravity and with few words, and those he did speak, as far as I could understand, were very wise and considered and those two men watched his mouth and spoke for him and with him and with great respect. After he had eaten, a page brought a belt just like those from Castile in manufacture although the workmanship is different, which he took and gave to me, and two pieces of worked gold which were very thin, because I believe that they get very little of it here, although I hold that they are very close to its source and there is a great deal of it. I saw that he liked a tapestry which I had over my bed. I gave it to him with some very good amber beads which I had around my neck, and some red slippers, and a flask of orange-flower water with which he was so pleased that it was amazing. He and his tutor and counsellors are very sad because they could not understand me nor I them. Nevertheless, I understood him to say that if I wanted anything from there, the whole island was at my disposal. I sent for some of my beads among which I have a gold coin on which Your Highnesses are portrayed and I showed it to him, and told him again, as I did yesterday, that Your Highnesses hold sway over and are masters of the greater part of the world, and that there were no princes as great. And I showed him the royal standards and the others with the cross, at which he was greatly impressed and said in reply to his counsellors that what great lords Your Highnesses must be since from so far away and from heaven you had sent me here without fear, and other things passed between them which I did not understand, except that I could see that he marvelled at everything. When it got late and he wished to leave, the Admiral sent him off in the boat with great honour and had many lombards fired, and when he was set down ashore he climbed onto his litter and went off with more than 200 of his people. And they carried his son on the shoulders of an Indian, a very honourable man. Wherever the sailors and the ships' crews came across him he ordered them to be given food and treated with great honour. A sailor said that he had met him on the road and seen a man, who appeared to be one of the highest ranking, carrying before the king every one of the things which the Admiral had given him. The king's son followed a good way behind him with as large an escort as he had. And the same distance behind again came a brother of the same king, except that the brother came on foot and two honourable men led him by the arms. This man came to the ship after the king and the Admiral gave him some of the things for barter, and the Admiral learned then that they called the king in their language 'cacique'. He says that on that day they traded very little gold, but the Admiral learned from an old man that there were many

comarcanas a çient leguas y más, según pudo entender, en las quales nasçe muy mucho oro, hasta dezirle que avía ysla que era toda oro, y en las otras que ay tanta cantidad que lo cogen y çiernen como con çedaço, y lo funden y hazen vergas y mill labores: figurava por señas la hechura. Este viejo señaló al Almirante la derrota y el paraje donde estava. Determinóse el Almirante de yr allá, y dixo que si no fuera el dicho viejo tan prinçipal persona de aquel rey, que lo detuviera y llevara consigo, o si supiera la lengua que se lo rogara y creya según estava bien con él y con los cristianos que se fuera con él de buena gana. Pero porque tenía ya aquellas gentes por de los Reyes de Castilla, y no era razón de hazelles agravio, acordó de dexallo. Puso una cruz muy poderosa en medio de la plaça de aquella poblaçión, a lo qual ayudaron los yndios mucho y hizieron dizque oraçión y la adoraron, y por la muestra que dan espera en Nuestro Señor el Almirante que todas aquellas yslas an de ser cristianos.

Miércoles 19 de diziembre

Esta noche se hizo a la vela por salir de aquel golpho que haze allí la Ysla de la Tortuga con la Española, y siendo de día tornó el viento levante, con el qual todo este día no pudo salir de entre aquellas dos yslas, y a la noche no pudo tomar un puerto que por allí pareçía. Vido por allí tres o quatro cabos de tierra y una grande baya y río, y de allí vido una angla muy grande y tenía una poblaçión, y a las espaldas un valle entre muchas montañas altíssimas llenas de árboles, que juzgó ser pinos, y sobre los Dos Hermanos ay una montaña muy alta y gorda que va de nordeste al sudueste, y del Cabo de Torres al lesueste está una ysla pequeña a la qual puso nombre Sancto Thomás porque es mañana su vigilia.[1] Todo el çerco de aquella ysla tiene cabos y puertos maravillosos, según juzgava él desde la mar. Antes de la ysla de la parte del güeste ay un cabo que entra mucho en la mar alto y baxo, y por eso le puso nombre Cabo Alto y Baxo. Del Cabo de Torres al leste quarta del sueste ay 60 millas hasta una montaña más alta que otra que entra en la mar y pareçe desde lexos ysla por sí por un degollado que tiene de la parte de tierra. Púsole nombre Monte Caribata, porque aquella provinçia se llamava Caribata. Es muy hermoso y lleno de árboles verdes y claros sin nieve y sin ñiebla, y era entonçes por allí el tiempo, quanto a los ayres y templança como por março en Castilla, y en quanto a los árboles y yervas como por mayo; las noches dizque eran de quatorze oras.

---

[1] Margin: Estos Dos Hermanos y el Cabo de Torres no lo a nombrado hasta agora.

neighbouring islands 100 leagues away and more, as far as he could understand, on which much gold is to be found, and he even said that there was one island which was all gold, and that on the others there was so much that they gather it and sift it with a kind of sieve and melt it and make it into bars and a thousand other objects: he illustrated the work by signs. This old man showed the Admiral the course and the position. The Admiral decided to go there and said that if that old man had not been such an eminent subject of that king he would have detained him and taken him with him, or if he had known the language he would have asked him to go, and believed that he and the Christians were on such good terms that he would have gone with them quite willingly. But because he already held those people to be subjects of the King and Queen of Castile, and that it was not right to do them an injury, he decided to leave him. He put a very imposing cross in the middle of the square of that village, in which the Indians helped greatly and he says that they offered up a prayer and worshipped it and from the signs that they give the Admiral hopes in Our Lord that all those islands will become Christian.

Wednesday 19 December

This night he set sail to leave that gulf which the Island of Tortuga forms with Española, and when day came the wind turned E, so that all this day he could not get out from between those two islands, and at night he could not make a harbour which appeared in that area. He sighted three or four capes there and a large bay and river, and from there he saw a very wide bay where there was a village, and behind it a valley between many very high mountains covered with trees, which he thought were pines, and above the Dos Hermanos[154] there is a very high, broad mountain which runs from NE to SW, and off the Cabo de Torres[155] to the ESE there is a small island to which he gave the name Santo Tomás[156] because tomorrow is his vigil. All around that island there are capes and marvellous harbours, as far as he could judge from the sea. Before reaching the island on the W side there is a cape which runs far out to sea and is both high and low, and for that reason he called it Cape Alto y Bajo.[157] From the Cabo de Torres to the SE by E it is 60 miles to a mountain, higher than any other, which juts into the sea and appears from a distance to be an island apart, owing to a ravine it has on the landward side. He gave it the name of Monte Caribata,[158] because that province was called Caribata. It is very beautiful and covered in light-green trees, and without snow or mists on it; and at that time the weather there was like March in Castile, as far as the breezes and the mild temperatures are concerned, and as for the trees and plants, like May. He says that the nights were fourteen hours long.[159]

Jueves 20 de diziembre

Oy al poner del sol entró en un puerto que estava entre la ysla de Sancto Thomás y el Cabo de Caribata y surgió. Este puerto es hermosíssimo y que cabrían en él quantas naos ay en cristianos. La entrada de él pareçe desde la mar impossible a los que no oviesen en él entrado, por unas restringas de peñas que passan desde el monte hasta quasi la ysla y no puestas por orden, sino unas acá y otra[s] acullá, unas a la mar y otras a la tierra; por lo qual es menester estar despiertos para entrar por unas entradas que tiene muy anchas y buenas para entrar sin temor y todo muy fondo de siete braças,[1] y passadas las restringas dentro ay doze braças. Puede la nao estar con una cuerda qualquiera amarrada contra qualesquiera vientos que aya. A la entrada deste puerto dizque avía un cañal,[2] que queda a la parte del güeste de una ysleta de arena, y en ella muchos árboles, y hasta el pie della ay siete braças. Pero ay muchas baxas en aquella comarca, y conviene abrir el ojo hasta entrar en el puerto; después no ayan miedo a toda la tormenta del mundo. De aquel puerto se pareçía un valle grandíssimo y todo labrado que desciende a él del sueste, todo çercado de montañas altíssimas que pareçe que llegan al çielo, y hermosíssimas, llenas de árboles verdes. Y sin duda que ay allí montañas más altas que la ysla de Tenerife en Canaria, que es tenida por de las más altas que puede hallarse. Desta parte de la ysla de Santo Thomás está otra ysleta a una legua, y dentro della otra, y en todas ay puertos maravillosos; mas cumple mirar por las baxas. Vido también poblaçiones y ahumadas que se hazían.

Viernes 21 de diziembre

Oy fue con las barcas de los navíos a ver aquel puerto el qual vido ser tal que afirmó que ninguno se le yguala de quantos aya jamás visto, y escúsase diziendo que a loado los passados tanto que no sabe cómo lo encareçer, y que teme que sea juzgado por manificador exçessivo más de lo que es la verdad. A esto satisfaze diziendo que él trae consigo marineros antiguos, y estos dizen y dirán lo mismo, y todos quantos andan en la mar, conviene a saber, todas las alabanças que a dicho de los puertos passados ser verdad, y ser éste muy mejor que todos ser asimismo verdad. Dize más desta manera: *Yo e andado veynte y tres años en la mar sin salir della tiempo que se aya de contar, y vi todo el levante y poniente*

---

[1] MS: braços.
[2] Margin: Creo quiere dezir cañaveral.

Thursday 20 December

Today at sunset he entered a harbour between the island of Santo Tomás and Cape Caribata, and anchored.[160] This harbour is very beautiful and one in which as many ships as there are in Christendom could lie. From the sea, the entrance looks impossible to anyone who has not entered it, owing to some reefs of rock which run from the mountain almost as far as the island, and which are not in any order but some here and others there, some more out to sea and others more inland. For this reason one has to be alert to enter by some good, wide passages which there are and which can be negotiated without fear as it is all seven fathoms deep, and once the reefs are passed it is twelve fathoms deep inside. The flagship can be moored with any piece of rope against whatever winds may blow. At the entrance to this harbour he says that there was a reed bed on the W side of a small sandy island on which there are many trees, with a depth of seven fathoms right up to the edge. But there are many shallows in that area, and it is important to keep one's eyes open until the harbour is entered. Then there is no need to fear the strongest storm. From that harbour could be seen a very wide valley, all cultivated, running down to the harbour from the SE, surrounded on all sides by very high mountains which seem to reach to the sky, and very beautiful, covered in green trees. Without doubt there are mountains there which are higher than the island of Tenerife in the Canaries, which is held to be one of the highest to be found.[161] From this side of the island of Santo Tomás there is another islet a league away, and within that another, and there are marvellous harbours on all of them; but it is essential to look out for the shallows. He also saw villages and smoke signals which they made.[162]

Friday 21 December

Today he went with the ships' boats to see that harbour. He saw that it was such that he affirmed that none he has seen so far can equal it, and he apologizes, saying that as he has praised the previous ones so much, he does not know how to commend this one, and that he fears that he may be thought to have built them up beyond what is the truth. He justifies this by saying that he has with him long-serving mariners and they say and will say the same, as will anyone who goes to sea, that is, that all the praise he has spoken of the previous harbours is true, and that it is equally true that this is very much better than all of them. He goes on in this manner: *I have spent twenty-three years at sea and have not left it for any length of time worth mentioning, and I have seen everything from east to west,* by which he means that he has been to the north, that is, to England,[163] *and*

que dize por yr al camino de septentrión que es Inglaterra, *y e andado la Guinea, mas en todas estas partidas no se hallará la perfeción de los puertos *** fallados siempre lo *** mejor que el otro que yo con buen tiento mirava mi escrevir, y torno a dezir que affirmo aver bien escripto y que agora éste es sobre todos. Y cabrían en él todas las naos del mundo, y çerrado, que con una cuerda la más vieja de la nao la tuviese amarrada. Desde la entrada hasta el fondo avrá çinco leguas.* Vido unas tierras muy labradas, aunque todas son así, y mandó salir dos hombres fuera de las barcas que fuesen a un alto para que viesen si avía población, porque de la mar no se vía ninguna, puesto que aquella noche çerca de las diez oras vinieron a la nao en una cano[a] çiertos yndios a ver al Almirante y a los cristianos por maravilla y les dio de los resgates, con que se holgaron mucho. Los dos cristianos bolvieron, y dixeron dónde avían visto una población grande un poco desviada de la mar. Mandó el Almirante remar hazia la parte donde la población estava hasta llegar çerca de tierra, y vio unos yndios que venían a la orilla de la mar y pareçía que venían con temor, por lo qual mandó detener las barcas y que les hablasen los yndios que traya en la nao, que no les haría mal alguno. Entonçes se allegaron más a la mar y el Almirante más a tierra, *y después que del todo perdieron el miedo, venían tantos que cobrían la tierra dando mill gracias*[1] *así hombres como mugeres y niños. Los unos corrían de acá y los otros de allá a nos traer pan que hazen de niames a que ellos llaman ajes, que es muy blanco y bueno, y nos trayan agua en calabaças y en cántaros de barro de la hechura de los de Castilla, y nos trayan quanto en el mundo tenían* y sabían que el Almirante quería, y todo con un coraçón tan largo y tan contento que era maravilla. *Y no se diga que porque lo que davan valía poco por eso lo davan liberalmente* (dize el Almirante) *porque lo mismo hazían y tan liberalmente los que davan pedaços de oro como los que davan la calabaça del agua; y fácil cosa es de cognoçer* (dize el Almirante) *quándo se da una cosa con muy deseoso coraçón de dar.* Estas son sus palabras. *Esta gente no tiene varas ni azagayas ni otras ningunas armas, ni los otros de toda esta ysla, y tengo que es grandíssima. Son así desnudos como su madre los parió, así mugeres como hombres, que en las otras tierras de la Juana y las otras de las otras yslas trayan las mugeres delante de sí unas cosas de algodón con que cobijan su natura tanto como una bragueta de calças de hombre. En espeçial después que passan de edad de doze años, mas aquí ni moça ni vieja. Y en los otros lugares todos los hombres hazían esconder sus mugeres de los cristianos por zelos, mas allí no. Y ay muy lindos cuerpos de*

---

[1] Margin: No[ta].

I have been to Guinea, but in all these parts you will not find harbours so perfect.
***
having always found ***
better than the other, so that I looked back over what I have written and I say again that I have written correctly and that this one is better than all of them. All the ships in the world would fit within it and it is so sheltered that the oldest rope on the ship would keep it moored. From the entrance to the end of the harbour must be five leagues.[164] He saw some lands which were well cultivated, although they are all like that, and he ordered two men to leave the boats and go to a high point to see if there were any village, since from the sea none could be seen, although that night around ten o'clock certain Indians came to the ship in a canoe to see and to marvel at the Admiral and the Christians and he gave them some of the things for barter, with which they were very pleased. The two Christians returned and told him where they had seen a large village a short distance away from the sea. The Admiral ordered them to row towards the shore in the direction of the village, and he saw some Indians coming down to the shore and, as they seemed to do so fearfully, he ordered the boats to stop and the Indians he had with him on the ship to tell them that he would not do them any harm. They then came closer to the sea and the Admiral drew closer to the land, *and when they had completely lost their fear, so many of them came, men and women and children alike, that they covered the land, giving many thanks, Some ran here and others ran there to bring us bread which they make out of 'niames' which they call 'ajes',*[165] *and which is very white and good, and they brought us water in gourds and in earthenware pots made like those in Castile, and they brought us everything they had in the world* and which they knew the Admiral wanted, and all with such a generous and happy heart that it was marvellous. *And let it not be said that because what they gave was worth little for that reason they gave generously* (says the Admiral) *because those who gave pieces of gold did so in the same way and just as generously as those who gave a gourd full of water; and it is easy to see* (says the Admiral) *when someone gives something with a glad heart.* These are his words. *These people have no staves or spears nor any other arms, nor have the others throughout the island, which I estimate to be very large. They are as naked as their mothers bore them, women and men alike, for in the other lands of Juana and the other islands the women wore a cotton garment in front of them with which they cover their genitals, somewhat like a pair of men's drawers, especially when they reach the age of twelve, but here neither young nor old women wear them. And in the other places all the men made their wives hide from the Christians out of jealousy, but not here. And there are some very shapely women and they were the first*

mugeres y ellas las primeras que venían a dar gracias al çielo, y traer quanto tenían, en espeçial cosas de comer, pan de ajes y gonça avellanada, y de çinco o seys maneras de fructas, de las quales mandó curar el Almirante para traer a los Reyes. No menos dizque hazían las mugeres en las otras partes antes que se ascondiesen. Y el Almirante mandava en todas partes estar todos los suyos sobre aviso que no enojasen a alguno en cosa ninguna, y que nada les tomassen contra su voluntad, y así les pagavan todo lo que dellos resçebían. Finalmente, dize el Almirante que no puede creer que hombre aya visto gente de tan buenos coraçones[1] y francos para dar, y tan temerosos que ellos se deshazían todos por dar a los cristianos quanto tenían, y en llegando los cristianos luego corrían a traerlo todo. Después enbió el Almirante seys cristianos a la población para que la viesen qué era, a los quales hizieron quanta honrra podían y sabían y les davan quanto tenían, porque ninguna duda les queda sino que creyan el Almirante y toda su gente aver venido del çielo. Lo mismo creyan los yndios que consigo el Almirante traya de las otras islas, puesto que ya se les avía dicho lo que devían de tener. Después de aver ydo[2] los seys cristianos, vinieron çiertas canoas con gente a rogar al Almirante de partes de un señor que fuese a su pueblo quan[do] de allí se partiese. Canoa es una barca en que navegan y son dellas grandes y dellas pequeñas. Y visto que el pueblo de aquel señor estava en el camino sobre una punta de tierra esperando con mucha gente al Almirante, fue allá. Y antes que se partiese vino a la playa tanta gente que era espanto, hombres y mugeres y niños, dando bozes que no se fuesse sino que se quedase con ellos. Los mensajeros del otro señor que avía venido a conbidar estavan aguardando con sus canoas porque no se fuese sin yr a ver al señor. Y así lo hizo, y en llegando que llegó el Almirante adonde aquel señor le estava esperando, y tenían muchas cosas de comer, mandó assentar toda su gente. Manda que lleven lo que tenían de comer a las barcas donde estava el Almirante junto a la orilla de la mar. Y como vido que el Almirante avía resçebido lo que le avían llevado, todos o los más de los yndios dieron a correr al pueblo que devía estar çerca para traerle más comida y papagayos y otras cosas de lo que tenían con tan franco coraçón que era maravilla. El Almirante les dio cuentas de vidro y sortijas de latón y cascaveles, no porque ellos demandassen algo, sino porque le pareçía que era razón, y sobre todo (dize el Almirante) porque los tiene ya por cristianos y por de los Reyes de Castilla más que las gentes de Castilla, y dize que otra cosa no falta salvo saber la lengua y mandarles, porque todo lo que se les mandare harán

---

[1] Margin: No[ta].
[2] MS: ydos.

to come and give thanks to heaven, and bring everything they need, especially things to eat, bread made from ajes, and chufa,[166] and five or six kinds of fruit, which the Admiral ordered to be dried and brought back to the Monarchs. He says that the women from the other parts did the same before hiding themselves away. Everywhere the Admiral ordered all his men to be careful not to offend anyone in any way, and to take nothing from them against their will, and so they paid for everything they received from them. Finally, the Admiral says that he cannot believe that anyone has seen such good-hearted people and so ready to give, and so timid that they did their utmost to give the Christians everything they had, and when they Christians arrived they ran to bring them everything. Later the Admiral sent six Christians to the village to see what it was like, and they paid them all the respect they could or knew how to and gave them everything they had, because there was no doubt that they believed that the Admiral and all his men had come from heaven. The Indians whom the Admiral had brought with him from the other islands believed the same, although he had already told them what they must believe. After the six Christians had left, some people in canoes came to ask the Admiral on behalf of a chieftain to go to his village when he left there. Canoe is a boat in which they travel, and some of them are large and others small. Seeing that that chief's village was on the way on a point of land and that he was waiting for the Admiral with many men, he went there. And before he set out so many people came down to the beach that it was frightening, men and women and children shouting to him not to go but to stay with them. The messengers of the other chief who had come to invite him were waiting with their canoes to make sure that he did not leave without going to see the chieftain. And he did so, and when the Admiral arrived at the place at which the chief was awaiting him with many things to eat, he ordered all his people to sit down. He ordered them to take the food to the boats where the Admiral was, close to the shore. And when he saw that the Admiral had accepted what they had taken to him, all or most of the Indians ran off to the village which must have been nearby to bring him more food and parrots and other things they had, with such a generous heart that it was marvellous. The Admiral gave them glass beads and brass rings and hawks' bells, not because they asked for anything, but because it seemed to him to be the right thing to do, and above all (says the Admiral) because he holds them to be Christians already, and subjects of the Monarchs of Castile, more so even than the people of Castile, and he says that the only thing needed is to know the language and give them orders, because

sin contradición alguna. Partióse de allí el Almirante para los navíos y los yndios davan bozes así hombres como mugeres y niños que no se fuessen y se quedasen con ellos los cristianos. Después que se partían venían tras ellos a la nao canoas llenas dellos, a los quales hizo hazer mucha honrra y dalles de comer y otras cosas que llevaron. Avía también venido antes otro señor de la parte del güeste y aun a nado venían muy mucha gente, y estava la nao más de grande media legua de tierra. *El señor que dixe se avía tornado; enbiéle çiertas personas para que le viesen y le preguntasen destas yslas; él los resçibió muy bien y los llevó consigo a su pueblo para dalles çiertos pedaços grandes de oro, y llegaron a un gran río, el qual los yndios passaron a nado; los cristianos no pudieron y así se tornaron. En toda esta comarca ay montañas altíssimas que pareçen llegar al çielo que la de la ysla de Tenerife pareçe nada en comparaçión dellas en altura y en hermosura, y todas son verdes, llenas de arboledas, que es una cosa de maravilla. Entremedias dellas ay vegas muy graçiosas, y al pie deste puerto al sur ay una vega tan grande que los ojos no pueden llegar con la vista al cabo, sin que tenga impedimento de montaña, que pareçe que deve tener quinze o veynte leguas. Por la qual viene un río y es toda poblada y labrada y está tan verde agora como si fuera en Castilla por mayo o por junio, puesto que las noches tienen catorze oras y sea la tierra tanto septentrional. Así, este puerto es muy bueno para todos los vientos que puedan ventar, çerrado y hondo, y todo poblado de gente muy buena y mansa y sin armas buenas ni malas, y puede qualquier navío estar sin miedo en él que otros navíos que vengan de noche a los saltear. Porque, puesto que la boca sea bien ancha, de más de dos leguas, es muy çerrada de dos restringas de piedra que escasamente la veen sobre agua, salvo una entrada muy angosta en esta restringa que no pareçe sino que fue hecho a mano y que dexaron una puerta abierta quanto los navíos puedan entrar. En la boca ay siete braças de hondo hasta el pie de una ysleta llana que tiene una playa y árboles al pie della; de la parte del güeste tiene la entrada y se puede llegar una nao sin miedo hasta poner el bordo junto a la peña. Ay de la parte del norueste ay tres yslas, y un gran río a una legua del cabo deste puerto. Es el mejor del mundo;* púsole nombre el Puerto de la Mar de Sancto Thomás porque era oy su día. Díxole mar por su grandeza.

Sábado 22 de diziembre

En amaneçiendo dio las velas para yr su camino a buscar las yslas que los yndios le dezían que tenían mucho oro y de algunas que tenían más oro que tierra; no le hizo tiempo y ovo de tornar a surgir, y enbió la barca a pescar con la red. El

whatever they are ordered they will do without demur. The Admiral left there to go back to the ships and the Indians, men and women and children alike, shouted that the Christians should not go and that they should stay with them. After they had left, canoes full of Indians came after them to the ship, and he had them treated with great respect and given something to eat and other things they took with them. Another chieftain had also previously come from the west, and many people even swam out, and the ship was a good half league from the shore. *The chieftain I mentioned had gone back; I sent some men to see him and ask him about these islands; he received them very well and took them with him to his village to give them some large pieces of gold, and they reached a large river which the Indians swam across; the Christians could not, and so turned back. In the whole of this area there are very high mountains which seem to reach to the sky, so that the mountain of Tenerife seems nothing in comparison with them in height or in beauty, and they are all green, covered in groves of trees, so that it is a thing of wonder.*[167] *Between them there are delightful plains*[168] *and at the end of this harbour to the S there is a plain so extensive that the eyes cannot see to the end; there are no mountains to interrupt it and it seems that it must be fifteen or twenty leagues long. Across it flows a river and it is all inhabited and cultivated and is as green at the moment as if it were in Castile in May or June, although the nights are fourteen hours long and the land is so far north. Thus, this harbour is very good for whatever wind might blow, sheltered and deep, and inhabited throughout by very good and gentle people without weapons, good or bad, and any ship can lie in it without fear that other ships might come in the night and attack them. Because, although the mouth may be good and wide, more than two leagues, it is closed off by reefs which can scarcely be seen above water, except for a very narrow entrance in this reef which looks exactly as if it had been made by hand leaving a door open just wide enough for ships to enter. At the mouth it is seven fathoms deep up to the edge of a flat islet which has a beach and trees at its foot; the entrance is on the western side and a ship can come close enough to put its flank against the rock without fear. To the NW there are three islands and a great river a league from the end of this harbour. It is the best in the world;* he gave it the name Puerto de la Mar de Santo Tomás because today was his day. He called it 'sea' on account of its great size.

Saturday 22 December

At day break he set sail to proceed on his way in search of the islands which the Indians told him had much gold, and some which, they said, had more gold than earth. The weather was not favourable and he had to turn back and anchor, and

señor de aquella tierra,[1] que tenía un lugar çerca de allí, le enbió una grande canoa llena de gente, y en ella un principal criado suyo a rogar al Almirante que fuese con los navíos a su tierra y que le daría quanto tuviese. Enbióle con aquél un çinto que en lugar de bolsa traya una carátula que tenía dos orejas grandes de oro de martillo, y la lengua y la nariz. *Y como sea esta gente de muy franco coraçón que quanto le piden dan con la mejor voluntad del mundo, que les pareçe que pidiéndoles algo les hazen grande merçed,* esto dize el Almirante, toparon la barca y dieron el çinto a un grumete, y vinieron con su canoa a bordo de la nao con su embaxada. Primero que los entendiese passó alguna parte del día; ni los yndios que él traya los entendían bien porque tienen alguna diversidad de vocablos en nombres de las cosas. En fin, acabó de entender por señas su conbite. El qual determinó de partir el domingo para allá, aunque no solía partir de puerto en domingo, sólo por su devoçión y no por superstición alguna; pero con esperança (dize él) que aquellos pueblos an de ser cristianos por la voluntad que muestran y de los Reyes de Castilla, y porque los tiene ya por suyos y porque le sirvan con amor, les quiere y trabaja hazer todo plazer. Antes que partiese oy, enbió seys hombres a una población muy grande tres leguas de allí de la parte del güeste, porque el señor della vino el día passado al Almirante y dixo que tenía çiertos pedaços de oro. En llegando allá los cristianos, tomó el señor de la mano al escrivano del Almirante, que era uno dellos el qual enbiava el Almirante para que no consintiese hazer a los demás cosa yndevida a los yndios, porque como fuessen tan francos los yndios y los españoles tan cudiçiosos y desmedidos, que no les basta que por un cabo de agujeta y aun por un pedaço de vidro y de escudilla y por otras cosas de no nada les davan los yndios quanto querían; pero, aunque sin dalles algo se lo querrían todo aver y tomar, lo que el almirante siempre prohibía, y aunque también eran muchas cosas de poco valor sino era el oro las que davan a los cristianos; pero el Almirante, mirando al franco coraçón de los yndios, que por seys contezuelas de vidro daría[n] y davan un pedaço de oro, por eso mandava que ninguna cosa se reçibiese dellos que no se les diese algo en pago. Así que tomó por la mano el señor al escrivano y lo llevó a su casa con todo el pueblo, que era muy grande, que le acompañava, y les hizo dar de comer, y todos los yndios les trayan muchas cosas de algodón labradas y en ovillos hilado. Después que fue tarde, dioles tres ánsares muy gordas el señor y unos pedaçitos de oro, y vinieron con ellos mucho número de gente y les trayan todas las cosas que allá avían resgatado, y a ellos mismos

---

[1] Margin: Éste era Guacanagarí, el señor del Marien donde el Almirante hizo la fortaleza, y dexó los treynta y nueve cristianos.

he sent the boat to do some fishing with the net. The chieftain of that land,[169] who had a village nearby, sent a large canoe full of people, including one of his principal servants, to ask the Admiral to go with the ships to his land and he would give him everything he had. He sent with him a belt which instead of a pouch had a mask with two large ears, and the tongue and mouth, of beaten gold. *And since these people have such generous hearts that they give whatever is asked of them with the best will in the world, and since they think that anyone asking for something is doing them a great honour*, so says the Admiral, they came up to the boat and gave the belt to a ship's boy, and came alongside the ship in their canoe to deliver their message. A good part of the day passed before he could understand them; neither could the Indians he had with him understand them, since they have somewhat different words for the names of things.[170] At length he managed to understand their invitation by signs. He decided to leave for that place on Sunday, although he used not to leave harbour on a Sunday, purely out of devotion and not from any superstition; but in the hope (he says) that those people will be Christians, on account of the good will they show, and subjects of the Monarchs of Castile as he holds them already to be, and so that they will serve them devotedly he wishes to do everything he can to please them. Before he set out today he sent six men to a very large village three leagues away to the W, because on the previous day the chief came to the Admiral and said that he had some pieces of gold. When the Christians arrived there, he took the Admiral's secretary[171] by the hand - he was one of those the Admiral sent to prevent the others from doing anything unworthy to the Indians, because the Indians were so generous and the Spaniards so grasping and undisciplined that they were not satisfied that the Indians gave them whatever they wanted for the end of a lace or even a piece of glass or earthenware and other things that were worth nothing, but wanted to take everything without giving them anything, which the Admiral always forbade, although many of the things they gave the Christians were of little value except the gold; but the Admiral, considering the generous hearts of the Indians, who would give a piece of gold for six glass beads, for this reason ordered that nothing should be accepted from them without something being given in payment. So the chief took the secretary by the hand and led him to his house accompanied by all the people from the village, which was very large, and he ordered them to be given food and all the Indians brought them articles of cotton, spun or woven and wound into balls. Later in the afternoon, the chief gave them three very fat geese and some small pieces of gold, and a large number of people came with them and carried all the things which they had bartered for, and they argued among themselves for the chance

porfiavan de traellos a cuestas,[1] y de hecho lo hizieron por algunos ríos y por algunos lugares lodosos. El Almirante mandó dar al señor algunas cosas, y quedó él y toda su gente con gran contentamiento, creyendo verdaderamente que avían venido del çielo y en ver los cristianos se tenían por bienaventurados. Vinieron este día más de çiento y veynte canoas a los navíos todas cargadas de gente y todos traen algo, espeçialmente de su pan y pescado y agua en cantarillos de barro, y simientes que son buenas especias. Echavan un grano en una escudilla de agua y bévenla y dezían los yndios que consigo traya el Almirante que era cosa saníssima.

Domingo 23 de diziembre

No pudo partir con los navíos a la tierra de aquel señor que lo avía enbiado a rogar y conbidar por falta del viento; pero enbió con los tres mensajeros que allí esperavan las barcas con gente y al escrivano. Entre tanto que aquellos yvan, enbió dos de los yndios que consigo traya a las poblaçiones que estavan por allí çerca del paraje de los navíos, y bolvieron con un señor a la nao con nuevas que en aquella ysla Española avía gran cantidad de oro,[2] y que a ella lo venían a comprar de otras partes, y dixéronle que allí hallaría quanto quisiese. Vinieron otros que confirmavan aver en ella mucho oro, y mostrávanle la manera que se tenía en cogello. Todo aquello entendía el Almirante con pena; pero todavía tenía por çierto que en aquellas partes avía grandíssima cantidad dello, y que, hallando el lugar donde se saca, avrá gran barato dello y según ymaginava que por no nada. Y torna a dezir que cree que deve aver mucho, porque en tres días que avía que estava en aquel puerto, avía avido buenos pedaços de oro, y no puede creer que allí lo traygan de otra tierra. *Nuestro Señor que tiene en las manos todas las cosas vea de me remediar y dar como fuere su servicio.* Estas son palabras del Almirante. Dize que aquella ora cree aver venido a la nao más de mill personas, y que todos trayan algo de lo que posseen y antes que lleguen a la nao con medio tyro de ballesta, se levantan en sus canoas en pie y toman en las manos lo que traen, diziendo: Tomad, tomad. También cree que más de quinientos vinieron a la nao nadando por no tener canoas y estava surta çerca de una legua de tierra. Juzgava que avían venido çinco señores y hijos de señores con toda su casa, mugeres y niños, a ver los cristianos. A todos mandava dar el Almirante porque todo dizque era bien enpleado, y dize: *Nuestro Señor me*

---

[1] Margin: No[ta].
[2] Margin: Tenían razón de dezirlo.

to carry the Christians on their backs, which they did across some rivers and muddy places. The Admiral ordered the chief to be given some things, and he and all his people were very contented, truly believing that they had come from heaven, and they held themselves fortunate to have seen the Christians. On this day more than one hundred and twenty canoes came to the ships, all full of people and each person bringing something, especially some of their bread and fish, and water in little earthenware pots, and seeds which are good spices. They put some grain in a dish of water and drink it, and the Indians the Admiral had with him said that it was very good for the health.[172]

Sunday 23 December

For lack of wind he could not leave with the ships for the land of that chief who had sent his men to invite him; but with the three messengers who were waiting there he sent the boats with some men and the secretary. While they were away, he sent two of the Indians he had with him to the neighbouring villages, near the ships' anchorage, and they returned to the flagship with a chief and with news that there was a great quantity of gold on that island of Española, and that people came there from other places to buy it, and they told him that he would find there as much as he wished. Others came who confirmed that there was much gold there, and showed him the method they used to collect it. The Admiral understood all this with difficulty; but he was already certain that in that region there was a very great quantity of it and that, once he had found the place where it was collected, there would be a great fortune to be had at no cost, as he imagined. And he says again that there must be a lot of it, because in three days since he had been in that harbour, he had acquired substantial pieces of gold, and he cannot believe that they should bring it from another land. *May Our Lord who has everything in his hands come to my aid and give me whatever may be in his service.* These are the words of the Admiral. He says that he believes that at that hour more than one thousand people had come to the ship, and that they all brought some of their possessions and before they come within half a crossbow-shot of the ship, they get up in their canoes and take what they bring in their hands, saying: 'Take, take'. He also believes that more than five hundred swam to the ship because they did not have canoes although it was anchored nearly a league from shore. He judged that five chiefs and sons of chiefs had come with all their household, women and children, to see the Christians. He ordered something to be given to all of them because he says that it was a good investment, and says: *May Our Lord take pity and guide me that I may find this gold,*

adereçe por su piedad que halle este oro, digo su mina, que hartos tengo aquí que dizen que la saben. Estas son sus palabras. En la noche llegaron las barcas y dixeron que avía gran camino hasta donde venían, y que al monte de Caribatán hallaron muchas canoas con muy mucha gente que venían a ver al Almirante y a los cristianos del lugar donde ellos yvan. Y tenía por çierto que si aquella fiesta de Navidad pudiera estar en aquel puerto, viniera toda la gente de aquella ysla, que estima ya por mayor que Inglaterra, por verlos. Los quales se bolvieron todos con los cristianos a la poblaçión, la qual dizque affirmavan ser la mayor y la más conçertada de calles que otra de las passadas y halladas hasta allí. La qual dizque es de parte de la Punta Sancta[1] al sueste quasi tres leguas. Y como las canoas andan mucho de remos, fuéronse delante a hazer saber al caçique que ellos llamavan allí. Hasta entonçes no avía podido entender el Almirante si lo dizen por rey o por governador. También dizen otro nombre por grande que llaman 'nitayno':[2] no sabía si lo dezían por hidalgo o governador o juez. Finalmente el caçique vino a ellos, y se ayuntaron en la plaça que estava muy barrida, todo el pueblo, que avía más de dos mill hombres. Este rey hizo mucha honrra a la gente de los navíos, y los populares cada uno les traya algo de comer y de bever. Después el rey dio a cada uno unos paños de algodón que visten las mugeres y papagallos para el Almirante y çiertos pedaços de oro; davan también los populares de los mismos paños y otras cosas de sus casas a los marineros por pequeña cosa que les davan. La qual según la reçibían pareçía que la estimavan por reliquias. Ya a la tarde, queriendo despedir, el rey les rogava que aguardasen hasta otro día, lo mismo todo el pueblo. Visto que determinavan su venida, venieron con ellos mucho del camino trayéndoles a cuestas lo que el cacique y los otros les avían dado hasta las barcas que quedavan a la entrada del río.

Lunes 24 de diziembre

Antes de salido el sol levantó las anclas con el viento terral. Entre los muchos yndios que ayer avían venido a la nao que les avían dado señales de aver en aquella ysla oro, y nombrado los lugares donde lo cogían, vido uno parece que más dispuesto y afiçionado o que con más alegría le hablava, y halagólo rogándole que se fuese con él a mostralle las minas del oro. Este truxo otro compañero o pariente consigo, los quales entre los otros lugares que nombravan

---

[1] Margin: Esta Punta Sancta no a nombrado.
[2] Margin: 'Nitayno' era prinçipal y señor después del rey, como grande del reyno.

that is, the mine, for I have many here who say that they know of it. These are his words. At night the boats returned and said that it was a long way to where they had returned from and that at Monte Caribatán[173] they found many canoes with very many people coming to see the Admiral and the Christians from the place where they were heading. And he was certain that if he could be in that harbour for the feast of the Nativity, all the people of that island, which he now thinks is larger than England, would come to see them. They all returned with the Christians to the village which he says they declared to be the largest and with the most organised street-plan of any they had previously found. He says it is almost three leagues to the SE of Punta Santa.[174] And as the canoes can be rowed at a good speed, they went ahead to tell the 'cacique', as they called him there. Until then the Admiral had not been able to understand whether by this they mean 'king' or 'governor'. They also use another word for lord, 'nitayno'. He did not know if they used it for 'nobleman' or 'governor' or 'judge'.[175] Eventually the cacique came to them and all the townspeople gathered together in the square which was swept very clean, and there were more than two thousand of them. This king did great honour to the men from the ships, and the common people all brought them something to eat and drink. Afterwards, the king gave each of them some cotton cloths which the women wear, and parrots for the Admiral, and some pieces of gold; the inhabitants also gave some of the same cloth and other things from their houses to the sailors, in return for any little thing they gave them. From the way they received these trifles it seemed that they valued them as holy relics. In the late afternoon, when they wished to leave, the king asked them to wait until the next day, as did all the people. When it was clear that they were determined to return, they came with them a good deal of the way, to the boats which were waiting at the entrance to the river, carrying on their shoulders what the cacique and the others had given them.

Monday 24 December

Before sunrise he weighed anchor with the wind off the land. Among the many Indians who had come to the ship yesterday and who had given them indications that there was much gold on that island, and had named places where they collected it, he saw one who was more amicably disposed or more inclined to speak to him, and he paid him the compliment of asking him to go with him and show him the gold mines. This man brought another companion or relative with him, and, among the other places they named where gold was collected, they spoke of

donde se cogía el oro, dixeron de Çipango al qual ellos llaman Çybao.[1] Y allí affirman que ay gran cantidad de oro, y que el caçique trae las vanderas de oro de martillo, salvo que está muy lexos al leste. El Almirante dize aquí estas palabras a los Reyes: *Crean Vuestras Altezas que en el mundo todo no puede aver mejor gente ni más mansa.*[2] *Deven tomar Vuestras Altezas grande alegría porque luego los harán cristianos y los avrán enseñado en buenas costumbres de sus reynos, que más mejor gente ni tierra puede ser, y la gente y la tierra en tanta cantidad que yo no sé ya cómo lo escriva. Porque yo e hablado en superlativo grado [de] la gente y la tierra de la Juana a que ellos llaman Cuba, mas ay tanta differençia dellos y della a ésta en todo como del día a la noche. Ni creo que otro ninguno que esto oviese visto oviese hecho ni dixesse menos de lo que yo tengo dicho y digo que es verdad que es maravilla las cosas de acá y los pueblos grandes desta ysla Española que así la llamé, y ellos le llaman Bohío; y todos de muy singularíssimo tracto amoroso y habla dulçe, no como los otros que pareçe quando hablan que amenazan; y de buena estatura hombres y mugeres y no negros. Verdad es que todos se tiñen, algunos de negro y otros de otra color y los más de colorado. He sabido que lo hazen por el sol, que no les haga tanto mal, y las casas*[3] *y lugares tan hermosos y con señorío en todos como juez o señor dellos; y todos le obedeçen que es maravilla. Y todos estos señores son de pocas palabras, y muy lindas costumbres, y su mando es lo más con hazer señas con la mano y luego es entendido que es maravilla.* Todas son palabras del Almirante. Quien oviere de entrar en la mar de Sancto Thomé se deve meter una buena legua sobre la boca de la entrada sobre una y[s]leta llana que en el medio ay, que le puso nombre la Amiga, llevando la proa en ella. Y después que llegare a ella con el [tiro] de una piedra passe de la parte del güeste y quédele ella al leste y se llegue a ella y no a la otra parte porque viene una restringa muy grande del güeste e aun en la mar fuera della ay unas tres baxas y esta restringa se llega a la Amiga un tyro de lombarda y entremedias passará y hallará a lo más baxo siete braças y casgajos abaxo. Y dentro hallará puerto para todas las naos del mundo y que estén sin amarras. Otra restringa y baxas vienen de la parte del leste a la dicha ysla Amiga y son muy grandes y salen en la mar mucho y llega hasta el cabo quasi dos leguas; pero entrellas pareçió que avía entrada a tiro de dos lombardas de la Amiga. Y al pie del Monte Caribatán de la parte del güeste ay un muy buen puerto y muy grande.

---

[1] Margin: Las minas de Cybao.
[2] Margin: El Almirante loa mucho los yndios.
[3] MS: ?cosas.

Cipango, which they call Cibao.[176] And they declare that there is a great quantity of gold there and that the cacique carries banners of beaten gold, but it is very far to the east. At this point the Admiral says the following words to the Monarchs: *Your Highnesses may believe that in the whole world there can be no better or more docile people. Your Highnesses should be very pleased because in due course you will make them Christians and will have taught them the good customs of your kingdoms, for there can be no better people or land, and in such quantity that I do not know how to describe it. For I have spoken in superlative terms of the people and the land of Juana which they call Cuba, but there is as much difference between those people and that land and all of this as there is between day and night. Nor do I believe that anyone else who had seen this would have done or said less than what I have said, and I say that the things here and the great peoples of this island of Española, as I have called it, and which they call Bohío, are truly marvellous. They all have a most exceptionally charming manner and are softly spoken, not like the others who seem to threaten when they speak; and they are of good stature, both men and women, and not black. It is true that they all paint themselves, some black and others another colour, mostly red. I have established that they do it because of the sun, so that it does not do them so much harm, and their houses and villages are so beautiful, and all well-regulated, with a kind of judge or overlord; and they all obey him marvellously. All these chieftains are of few words, and very good manners, and their way of governing is usually by gestures of the hand which are marvellously well understood.* These are all the Admiral's words. Anyone wishing to enter the sea of Santo Tomás should station himself a good league above the mouth of the entrance by a flat island which is in the centre, which he named Amiga,[177] and should steer for that. And after arriving within a stone's throw of it, he should pass to the W, leaving it to the E, and should not pass it on the other side because there is a very large reef to the W and even in the sea beyond it there are three shoals and this reef stretches to within a lombard-shot of Amiga. He will be able to pass between them and will find a minimum depth of seven fathoms and gravel beneath. Inside he will find a harbour for all the ships in the world and they may lie unmoored. There is another reef and shoals from the E towards the island of Amiga and they are very large and extend far out into the sea reaching almost within two leagues of the cape; but among them there seemed to be an entrance two lombard shots away from Amiga, and at the foot of Monte Caribatán to the W there is a very good, large harbour.

Martes 25 de diziembre, día de Navidad

Navegando con poco viento el día de ayer desde la mar de Sancto Thomé hasta la Punta Sancta sobre la qual a una legua estuvo así hasta passado el primer quarto que serían a las onze oras de la noche, acordó echarse a dormir porque avía dos días y una noche que no avía dormido. Como fuese calma, el marinero que governava la nao acordó yrse a dormir[1] y dexó el governario a un moço grumete, lo que mucho siempre avía el Almirante prohibido en todo el viaje, que oviese viento o que oviese calma, conviene a saber que no dexasen governar a los grumetes. El Almirante estava seguro de bancos y de peñas, porque el domingo quando enbió las barcas a aquel rey avían passado al leste de la dicha Punta Sancta bien tres leguas y media, y avía[n] visto los marineros toda la costa y los baxos que ay desde la dicha Punta Sancta al leste sueste bien tres leguas, y vieron por dónde se podía passar, lo que todo este viaje no hizo. Quiso Nuestro Señor que a las doze oras de la noche como avían visto acostar y reposar el Almirante y vían que era calma muerta y la mar como en una escudilla, todos se acostaron a dormir, y quedó el governallo en la mano de aquel muchacho, y las aguas que corrían llevaron la nao sobre uno de aquellos bancos. Los quales puesto que fuesse de noche sonavan que de una grande legua se oyeran y vieran, y fue sobre él tan mansamente que casi no se sentía. El moço que sintió el governalle y oyó el sonido de la mar dio bozes, a las quales salió el Almirante, y fue tan presto que aún ninguno avía sentido que estuviesen encallados. Luego el maestre de la nao cuya era la guardia salió, y díxoles el Almirante a él y a los otros que halasen el batel que trayan por popa, y tomasen un ancla, y la echasen por popa y él con otros muchos saltaron en el batel, y pensava el Almirante que hazían lo que les avía mandado; ellos no curaron sino de huyr a la caravela que estava a barlovento media legua. La caravela no los quiso resçebir haziéndolo virtuosamente y por esto bolvieron a la nao; pero primero fue a ella la barca de la caravela. Quando el Almirante vido que se huyan y que era su gente y las aguas menguavan y estava ya la nao la mar de través,[2] no viendo otro remedio, mandó cortar el mástel y alijar de la nao todo quanto pudieron para ver si podían sacarla; y como todavía las aguas menguassen, no se pudo remediar, y tomó lado hazia la mar traviesa puesto que la mar era poca o nada. Y entonçes se abrieron los conventos y no la nao. El Almirante fue a la caravela para poner en cobro la gente de la nao en la caravela, y como ventase ya ventezillo de la tierra,

---

[1] Margin: Por descuydo del marinero perdió el Almirante su nao.

[2] Margin: que quiere dezir que menguava el agua, o que corría hazia abaxo.

Tuesday 25 December, Christmas Day

Sailing yesterday with little wind from the sea of Santo Tomás to the Punta Santa a league off which he stood until after the first quarter, which would be at eleven o'clock at night, he decided to get some sleep because he had not slept for two days and a night. Although it was calm, the sailor who was steering the ship decided to go to sleep and left the tiller to a young ship's boy, which the Admiral had always strictly forbidden throughout the voyage, whether it was windy or calm; that is, they should not allow the boys to steer. The Admiral felt safe from sandbanks and rocks because on Sunday, when he sent the boats to that chief, they had passed a good three and a half leagues to the E of Punta Santa, and the sailors had seen the whole coast and the shoals from Punta Santa to the ESE for a good three leagues, and they saw where it was possible to pass; this was something he had not done on the whole of this voyage. Our Lord willed that at twelve o'clock at night, seeing that the Admiral had gone to bed to rest and that it was dead calm and the sea like water in a bowl, they all lay down to sleep, and the rudder was left in the hands of that boy, and the currents which were running took the ship on to one of those banks. Although it was night-time the sea sounded against them so that they could be heard and seen a good league away, and it went aground so gently that it was hardly noticed.[178] The boy who felt the rudder and heard the sound of the sea gave a shout, at which the Admiral came out, and happened so quickly that no one had yet realised that they were aground. Then the Master,[179] whose watch it was, came out and the Admiral told him and the others to launch the boat they carried astern, and take an anchor and drop it astern of the ship, and he and many others jumped into the boat, and the Admiral thought that they were doing what he had ordered; they were concerned only to escape to the caravel which stood half a league off to windward. The caravel with every good reason would not take them aboard and so they returned to the flagship; but the caravel's boat got to her first. When the Admiral saw that they were fleeing and that they were his crew and that the tide was ebbing and that the ship was broadside on to the sea, seeing no other remedy, he ordered the mast to be cut down and everything they could to be jettisoned from the ship to see if they could get her refloated; and as the waters were still receding, she could not be saved, and she settled on her side, broadside on to the sea, although there was little or no sea. And her seams opened though she stayed in one piece. The Admiral went to the caravel to put the ship's crew safely on the caravel, and as there was now a light breeze off the land, and much

y también aún quedava mucho de la noche, ni suppiesen quánto duravan los bancos, temporejó a la corda hasta que fue de día, y luego fue a la nao por de dentro de la restringa del banco. Primero avía enbiado el batel a tierra con Diego de Arana, de Córdova, alguazil del armada, y Pero Gutiérrez, respostero de la Casa Real, a hazer saber al rey que los avía enbiado a conbidar y rogar el sábado que se fuese con los navíos a su puerto, el qual tenía su villa adelante obra de una legua y media del dicho banco, el qual como lo supo dizen que lloró, y enbió toda su gente de la villa con canoas muy grandes y muchas a descargar todo lo de la nao.[1] Y así se hizo y se descargó todo lo de las cubiertas en muy breve espaçio: tanto fue el grande aviamiento y diligencia que aquel rey dio. Y él con su persona con hermanos y parientes estavan poniendo diligençia así en la nao como en la guarda de lo que se sacava a tierra para que todo estuvie[se] a muy buen recaudo. De quando en quando enbiava uno de sus parientes al Almirante llorando a lo consolar, diziendo que no rescibiese pena ni enojo que él le daría quanto tuviese. Çertifica el Almirante a los Reyes que en ninguna parte de Castilla tan buen recaudo en todas las cosas se pudiera poner sin faltar un agujeta. Mandólo poner todo junto con las casas entre tanto que se vaziavan algunas casas que quería dar donde se pusiese y guardase todo. Mandó poner hombres armados en rededor de todo que velasen toda la noche. *Él con todo el pueblo lloravan; tanto* (dize el Almirante) *son gente de amor y sin cudiçia y convenibles para toda cosa, que certifico a Vuestras Altezas que en el mundo creo que no ay mejor gente ni mejor tierra.*[2] *Ellos aman a sus próximos como a sí mismos, y tienen una habla la más dulçe del mundo, y mansa y siempre con risa. Ellos andan desnudos, hombres y mugeres, como sus madres los parieron. Mas crean Vuestras Altezas que entre sí tienen costumbres muy buenas, y el rey muy maravilloso estado de una çierta manera tan continente que es plazer de verlo todo, y la memoria que tienen y todo quieren ver y preguntan qué es y para qué.* Todo esto dize así el Almirante.

Miércoles 26 de diziembre

Oy a salir del sol vino el rey de aquella tierra que estava en aquel lugar a la caravela Niña donde estava el Almirante y quasi llorando le dixo que no tuviese pena que él le daría quanto tenía, y que avía dado a los cristianos que estavan en tierra dos muy grandes casas y que más les daría si fuesen menester, y quantas

---

[1] Margin: Nótese aquí la humanidad de los yndios contra los tyranos que los an estirpado.
[2] Margin: No[ta].

of the night still left, and they did not know how far the banks extended, he stood off until daybreak, and then went inside the reef to the flagship. First of all he had sent the boat ashore with Diego de Arana, from Córdoba, the bailiff of the fleet, and Pero Gutiérrez, chamberlain of the royal household, to inform the chieftain who had sent the invitation on Saturday to take the ships to his harbour and whose town was a matter of a league and a half beyond the sandbank. As soon as he heard the news they say that he wept, and he sent all his people from the town with many very large canoes to offload the ship. This was done and everything was offloaded from the holds in a very short space of time, so great was the expeditiousness and diligence that the king displayed. He in person, with his brothers and family, worked diligently both on the ship and on safeguarding what was taken off so that everything would be fully secure. From time to time he sent one of his relatives weeping to the Admiral to console him, saying that they should not be upset or distressed because he would give him everything he had. The Admiral assures the Monarchs that nowhere in Castile would such good care have been taken about everything that not a lace was missing. He ordered everything to be put next to the houses while some which he wanted to make available were emptied and everything could be put there and safeguarded. He ordered armed men to be placed around everything and to guard it all night. He and all his people were crying; *they are* (says the Admiral) *so loving a people and so lacking in cupidity and so willing to do anything, that I assure Your Highnesses that I believe that there are no better people in the world, and no better land. They love their neighbours as themselves, and have the softest speech in the world, and are docile and always laughing. They go naked, men and women, as their mothers bore them. But Your Highnesses may believe that their dealings with each other are very good, and the king has a most marvellous bearing and such a sober manner that it is a pleasure to see it all, and their memory, and they want to see everything and ask what it is and what it is for.* The Admiral says all this in these words.

Wednesday 26 December

Today at sunrise the king of that land who was in that village came to the caravel Niña where the Admiral was and almost in tears told him not to be upset because he would give him everything he had, and that he had given the Christians on shore two very large houses and that he would give them more if

canoas pudiesen cargar y descargar la nao y poner en tierra quanta gente quisiese; y que así lo avía hecho ayer, sin que se tomase una migaja de pan ni otra cosa alguna, *tanto* (dize el Almirante) *son fieles y sin cudiçia de lo ageno;* y así era sobre todos aquel rey virtuoso.[1] En tanto que el Almirante estava hablando con él, vino otra canoa de otro lugar que traya çiertos pedaços de oro los quales quería dar por un cascavel, porque otra cosa tanto no deseavan como cascaveles, y que aún no llega la canoa a bordo quando llamavan y mostravan los pedaços de oro diziendo chuq chuque, por cascaveles; que están en puntos de se tornar locos por ellos. Después de aver visto esto y partiéndose estas canoas que eran de los otros lugares, llamaron al Almirante y le rogaron que les mandase guardar un cascavel hasta otro día, porque él traería quatro pedaços de oro tan grandes como la mano. Holgó el Almirante de oyr esto, y después un marinero que venía de tierra dixo al Almirante que era cosa de maravilla las pieças de oro que los cristianos que estavan en tierra resgatavan por no nada; por una agujeta davan pedaços que serían más de dos castellanos, y que entonçes no era nada, al respeto de lo que sería dende a un mes. El rey se holgó mucho con ver al Almirante alegre, y entendió que deseava mucho oro, y díxole por señas que él sabía çerca de allí adónde avía dello muy mucho en grande suma, y que estuviese de buen coraçón que él daría quanto oro quisiese. Y dello dizque le dava razón y en espeçial que lo avía en Çipango a que ellos llamavan Çybao en tanto grado que ellos no lo tienen en nada y que él lo trahería allí aunque también en aquella ysla Española a quien llaman Bohío y en aquella provinçia Caribata lo avía mucho más.[2] El rey comió en la caravela con el Almirante, y después salió con él en tierra, donde hizo al Almirante mucha honrra, y le dio colación de dos o tres maneras de ajes y con camarones y caça y otras viandas que ellos tenían y de su pan que llamavan caçabi. Dende lo llevó a ver unas verduras de árboles junto a las casas, y andavan con él bien mill personas, todos desnudos. El señor ya traya camisa y guantes que el Almirante le avía dado, y por los guantes hizo mayor fiesta que por cosa de las que le dio. En su comer con su honestidad y hermosa manera de limpieza, se mostrava bien ser de linaje. Después de aver comido que tardó buen rato estar a la mesa, truxeron çiertas yervas con que se fregó mucho las manos; creyó el Almirante que lo hazía para ablandarlas, y diéro[n]lo aguamanos. Después que acabaron de comer, llevó a la playa al Almirante y el Almirante enbió por un arco turquesco y un manojo de flechas, y el Almirante hizo tyrar a un hombre de su compañía que sabía dello; y el señor

---

[1] MS: era corrected from es; aquel corrected from este.
[2] Margin: Cybao era provinçia de la misma ysla españ[ol]a donde avía las minas muy ricas.

necessary, and as many canoes as could load and unload the ship and put ashore as many men as he wished; and that he had done so yesterday, without a crumb of bread or anything else being taken, *so trustworthy are they* (says the Admiral) *and respectful of other people's property*; and that king was even more honest than the others. While the Admiral was talking with him, another canoe came from another village carrying certain pieces of gold which they wanted to give for a hawk's bell, for they desired nothing so much as hawks' bells, and the canoe scarcely reached the shore before they called out and showed the pieces of gold, saying 'chuq chuque', meaning 'hawks' bells'; they nearly go mad over them. Having seen this, and with the canoes from the other villages about to leave, they called the Admiral and asked him to have a hawk's bell kept for them until the following day, because they would bring four pieces of gold as large as a hand. The Admiral was pleased to hear this, and afterwards a sailor returning from the shore told the Admiral that it was amazing to see the pieces of gold which the Christians on shore were bartering for nothing; for a shoe-lace they were giving pieces which would be worth more than two castellanos, and that this was nothing to what it would be like within a month. The king was very pleased to see the Admiral happy, and understood that he wanted much gold, and told him by signs that he knew where there was a very great deal nearby, and that he should be of good heart because he would give him as much gold as he wished. The Admiral says that he gave him information about it and told him in particular that there was so much gold in Cipango, which they called Cibao, that they attach no value to it at all and that he would bring some of it even though there was much more on that island of Española, which they call Bohío, and in that province of Caribata. The king ate on board the caravel with the Admiral, and afterwards went ashore with him, where he did the Admiral great honour, and gave him a meal of two or three types of ajes and shrimps and game and other kinds of food which they had, and some of their bread which they call cassava. Then he took him to see some groves of trees near the houses, and a good thousand people, all naked, went with him. The chief now wore the tunic and gloves which the Admiral had given him, and was more excited by the gloves than by anything else he had given him. From his eating habits, his decency and delightful cleanliness, he showed clearly that he was of good birth. After eating, for they remained at table for a fair while, they brought certain herbs which he thoroughly rubbed into his hands; the Admiral thought that he did so to soften them, and they gave him water for his hands. After they had finished eating, he took the Admiral to the beach and the Admiral sent for a Turkish bow and a handful of arrows, and the Admiral had one of his men who was skilled in archery shoot some arrows; and the chief thought it was wonderful since he does

como no sepa qué sean armas porque no las tienen ni las usan, le pareçió gran cosa; aunque dizque el comienço fue sobre el habla de los de Caniba que ellos llaman caribes que los vienen a tomar y traen arcos y flechas sin hierro que en todas aquellas tierra[s] no avía memoria de él y de azero ni de otro metal salvo de oro y de cobre aunque cobre no avía visto sino poco el Almirante. El Almirante le dixo por señas que los Reyes de Castilla mandarían destruyr a los Caribes y que a todos se los mandarían traer las manos atadas. Mandó el Almirante tyrar una lombarda y una espingarda, y viendo el effecto que su fuerça hazían y lo que penetravan, quedó maravillado. Y quando su gente oyó los tiros cayeron todos en tierra. Truxeron al Almirante una gran carátula que tenía grandes pedaços de oro en las orejas y en los ojos y en otras partes, la qual le dio con otras joyas de oro que el mismo rey avía puesto al Almirante en la cabeça y al pescueço; y a otros cristianos que con él estavan dio también muchas. El Almirante resçibió mucho plazer y consolaçión destas cosas que vía, y se le templó el angustia y pena que avía resçebido y tenía de la pérdida de la nao, y cognosçió que Nuestro Señor avía hecho encallar allí la nao porque hiziese allí asiento. *Y a esto* (dize él) *vinieron tantas cosas a la mano, que verdaderamente no fue aquél desastre salvo gran ventura. Porque es çierto* (dize él) *que si yo no encallara que yo fuera de largo sin surgir en este lugar; porque él está metido acá dentro en una grande baya y en ella dos o más restringas de baxas. Ni este viaje dexara aquí gente, ni aunque yo quisiera dexarla no les pudiera dar tan buen aviamiento ni tantos pertrechos ni tantos mantenimientos ni adereço para fortaleza. Y bien es verdad que mucha gente desta que va aquí me avían rogado y hecho rogar que les quisiese dar licencia para quedarse.*[1] *Agora tengo ordenado de hazer una torre y fortaleza todo muy bien y una grande cava, no porque crea que aya esto menester por esta gente,*[2] *porque tengo por dicho que con esta gente que yo traygo sojugaría toda esta ysla, la qual creo que es mayor que Portugal y más gente al doblo; mas son desnudos y sin armas y muy cobardes fuera de remedio. Mas es razón que se haga esta torre y se esté como se a de estar, estando tan lexos de Vuestras Altezas; y porque cognozcan el ingenio de la gente de Vuestras Altezas y lo que pueden hazer porque con amor y temor le obedezcan.*[3] *Y así ternán tablas para hazer toda la fortaleza dellas y mantenimientos de pan y vino para más de un año y simientes para sembrar y la barca de la nao y un calafate y un carpintero y un lombardero, y un tonelero y muchos entre ellos hombres que desean mucho,*

---

[1] Margin: No[ta].
[2] Margin: No[ta].
[3] Margin: No[ta].

not know what weapons are as they neither possess or use them; he says, however, that what started it all was some talk about the people from Caniba whom they call Caribs and who come to capture them and carry bows and arrows without iron tips, for in all those lands there was no knowledge of iron or steel nor of any other metal except gold and copper, although the Admiral had only seen a little copper. The Admiral told him by signs that the Monarchs of Castile would order the Caribs to be destroyed and brought to them with their hands tied. The Admiral ordered a lombard and a musket to be fired, and seeing the effect of their force and the extent to which they penetrated, he was amazed. And when his people heard the shots they all fell to the ground. They brought the Admiral a great mask which had large pieces of gold in the ears and eyes and in other places, which they gave him, together with other gold ornaments, which the king himself placed on the Admiral's head and neck; and he also gave many ornaments to other Christians who were with him. The Admiral took great pleasure and consolation from these things which he saw, and the anguish and pain he had suffered from the loss of the flagship was mitigated, and he knew that Our Lord had caused the ship to run aground there so that he might establish a settlement.[180] *So many things came to hand* (he says) *that in truth it was not a disaster but a great piece of good fortune. Because it is certain* (he says) *that if I had not run aground I would have kept out to sea without anchoring in this place; because it is tucked away inside a great bay with two or more sandbanks. Nor would I have left men here on this voyage, and if I had wanted to leave them I could not have provided them with such thorough preparation, nor so much equipment or supplies or materials for a fort. And it is true that many of the men with me had pleaded for permission to stay. Now I have ordered a tower and fortress to be built, all in good order, with a large moat, not that I think this is needed for these Indians, (because I am sure that with these men I have with me I could subdue the whole of this island, which I believe is larger than Portugal and with twice the population) and they are naked, unarmed and timid beyond redress. But it is right that this tower should be built as it should, so that the Indians, although so far away from Your Highnesses, will realise the skill of Your Highnesses' subjects and see what they can do, so that with love and fear they will obey you. And so they will have planks from which to build the whole fortress and supplies of bread and wine for more than a year, and seed to sow, and the ship's boat, and a caulker and a carpenter and a gunner and a cooper, and many among those men who desire to*

por servicio de Vuestras Altezas y me hazer plazer, de saber la mina adonde se coge el oro. Así que, que todo es venido mucho a pelo, para que se faga este comienço. Y sobre todo que quando encalló la nao fue tan passo que quasi no se sintió ni avía ola ni viento. Todo esto dize el Almirante. Y añade más para mostrar que fue gran ventura y determinada voluntad de Dios que la nao allí encallase porque dexase allí gente, que si no fuera por la trayción del maestre y de la gente que eran todos o los más de su tierra, de no querer echar el ancla por popa para sacar la nao como el Almirante les mandava, la nao se salvara y así no pudiera saberse la tierra (dize él) como se supo aquellos días que allí estuvo, y adelante por los que allí entendía dexar porque él yva siempre con intención de descubrir y no parar en parte más de un día, si no era por falta de los vientos, porque la nao dizque era muy pesada y no para el officio de descubrir. Y llevar tal nao (dizque) causaron los de Palos que no cumplieron con el Rey e la Reyna lo que le avían prometido: dar navíos convenientes para aquella jornada, y no lo hizieron. Concluye el Almirante diziendo que de todo lo que en la nao avía no se perdió una agujeta, ni tabla ni clavo, porque ella quedó sana como quando partió, salvo que se cortó y rajó algo para sacar la vasija y todas las mercaderías y pusiéronlas todas en tierra y bien guardadas como está dicho. Y dize que espera en Dios que, a la buelta que él entendía[1] hazer de Castilla, avía[2] de hallar un tonel de oro que avrían resgatado los que avía de dexar, y que avrían hallado la mina del oro y la espeçería; y aquello en tanta cantidad, que los Reyes antes de tres años enprendiesen y adereçasen para yr a conquistar la casa sancta, *que así* (dize él) *protesté a Vuestras Altezas que toda la ganançia desta mi empresa se gastase en la conquista de Hierusalém y Vuestras Altezas se rieron y dixeron que les plazía y que sin esto tenían aquella gana.* Estas son palabras del Almirante.

Jueves 27 de diziembre

En saliendo el sol vino a la caravela el rey de aquella tierra y dixo al Almirante que avía embiado por oro, y que lo quería cobrir todo de oro antes que se fuesse, antes le rogava que no se fuese y comieron con el Almirante el rey e un hermano suyo y otro su pariente muy privado, los quales dos le dixeron que querían yr a Castilla con él. Estando en esto vinieron [nuevas] cómo la caravela Pinta estava en un río al cabo de aquella ysla; luego enbió el caçique allá una canoa, y en ella el Almirante un marinero, porque amava tanto al Almirante que era maravilla.

---

[1] MS: entiende corrected to entendía.
[2] MS: a corrected to avía.

serve Your Highnesses and to please me by finding the mine where the gold is collected. Thus everything has happened so that this beginning can be made, particularly since, when the ship went aground, it was so gently that no one felt it and there was neither wave nor wind. All this the Admiral says. And he adds more to show that it was good fortune and the determined will of God that the ship should run aground there so that he should leave men behind, and that if it were not for the treachery of the Master and the crew who were all or mostly from his region not wanting to drop the anchor astern to pull off the ship as the Admiral had ordered, the ship would have been saved and he would not have learned about the land (he says) all that he learned in those days that he was there, and will learn from those whom he intended to leave there, because it was always his intention to explore and not stop anywhere more than a day, unless through lack of wind, because he says that the ship was very heavy and not suited to exploration. And (he says) the people of Palos caused him to take such a ship by not fulfilling the promise they had made to the King and Queen: to give him suitable ships for that expedition, and they did not do so. The Admiral concludes by saying that not a shoe-lace was lost from what was on the ship, nor a plank nor a nail, because she remained as sound as when she left, except that she was somewhat cut and split in order to get at the barrels and the cargo which they put ashore and kept under guard as has been said. And he says that he hopes to God that, on the return journey which he intended to make from Castile, he would find a barrel of gold which those he was to leave behind would have bartered for, and that they would have found the gold mine and the spices; and in such quantity that the Monarchs would be able in three years to undertake preparations for the conquest of the Holy Land, *just as* (he says) *it was my declared intention to Your Highnesses that the whole of the profit from this my enterprise should be spent on the conquest of Jerusalem and Your Highnesses laughed and said you were pleased and that, even without this expedition, that was your intention.* These are the Admiral's words.

Thursday 27 December

At sunrise the king of that land came to the caravel and told the Admiral that he had sent for gold, and that he wanted to cover him with gold before he left, but that he would rather he did not leave, and the king ate with the Admiral together with a brother of his and another close relative, both of whom told him that they wanted to go with him to Castile. Meanwhile, there came news that the caravel Pinta was in a river at the end of that island;[181] the cacique immediately dispatched a canoe, because he was so wonderfully fond of the Admiral, and the

Ya entendía el Almirante con quanta priesa podía por despacharse para la buelta de Castilla.

Viernes 28 de diziembre

Para dar orden y priesa en el acabar de hazer la fortaleza, y en la gente que en ella avía de quedar, salió el Almirante en tierra y pareçióle que el rey le avía visto quando yva en la barca, el qual se entró presto en su casa dissimulando, y enbió a un su hermano que resçibiese al Almirante. Y llevólo a una de las casas que tenía dadas a la gente del Almirante, la qual era la mayor y mejor de aquella villa. En ella le tenían aparejado un estrado de camisas de palma donde le hizieron asentar. Después el hermano enbió un escudero suyo a dezir al rey que el Almirante estava allí como que el rey no sabía que era venido, puesto que el Almirante creya que lo dissimulava por hazelle mucha más honrra. Como el escudero se lo dixo, dio el caçique dizque a correr para el Almirante, y púsole al pescueço una gran plasta de oro que traya en la mano. Estuvo allí con él hasta la tarde, deliberando lo que avía de hazer.

Sábado 29 de diziembre

En saliendo el sol vino a la caravela un sobrino del rey muy moço y de buen entendimiento y buenos hygados (como dize el Almirante). Y como siempre trabajase por saber adónde se cogía el oro, preguntava a cada uno porque por señas ya entendía algo; y así aquel mançebo le dixo que a quatro jornadas avía una ysla al leste que se llamava Guarionex, y otras que se llamavan Macorix, y Mayonic y Fuma y Çybao y Coroay,[1] en las quales avía ynfinito oro, los quales nombres escrivió el Almirante. Y supo esto que le avía dicho un hermano del rey e riñó con él según el Almirante entendió. También otras vezes avía el Almirante entendido que el rey trabajava porque no entendiese dónde nasçía y se cogía el oro, porque no lo fuese a resgatar o comprar a otra parte. *Mas es tanto y en tantos lugares y en esta misma ysla Española* (dize el Almirante) *que es maravilla.* Siendo ya de noche le embió el rey una gran carátula de oro, y enbióle a pedir un baçín de aguamanos y un jarro; creyó el Almirante que lo pedía para mandar hazer otro y así se lo enbió.

---

[1] Margin: Estas no eran yslas sino provinçias de la ysla española.

Admiral sent a sailor off in it. The Admiral was now preparing with all possible speed for the return to Castile.

Friday 28 December

In order to direct and hasten the completion of the fortress, and to brief the men who were to remain, the Admiral went ashore and it appeared to him that the king had seen him in the boat, and pretending not to have done so, quickly went into his house and sent a brother of his to receive the Admiral. And he took him to one of the houses which he had given to the Admiral's men, which was the biggest and best in that town. In it they had prepared a platform made from the inner bark of the palm tree, where they bade him sit down. Then the brother sent a page to tell the king that the Admiral was there, as if the king did not know that he had come, although the Admiral believed that he was pretending so that he could do him much greater honour. When the page told him, as he says, the king came running towards the Admiral, and put a great plate of gold which he was carrying around his neck. He stayed there with him until the afternoon, deciding what needed to be done.

Saturday 29 December

At sunrise one of the king's nephews came to the caravel, very young and intelligent and full of beans (as the Admiral puts it). And as he always strove to find out where the gold came from, he asked everyone, for he could now understand something by signs; and so this young man told him that four days' journey away to the east there was an island called Guarionex, and others called Macorix, and Mayonic and Fuma and Cibao and Coroay,[182] where there was gold without end. The Admiral wrote the names down, and when one of the king's brothers found out what he had told him, he was very angry with him, as far as the Admiral could tell. The Admiral had understood at other times that the king was trying to prevent him from finding out where the gold came from so that he would not go and barter or buy elsewhere. *But there is so much of it and in so many places, and on this island of España itself* (says the Admiral), *that it is amazing.* After dark, the king sent him a great mask of gold with a request for a jug and bowl; the Admiral believed that he wanted them to have another set made, and so he sent them to him.

Domingo 30 de diziembre

Salió el Almirante a comer a tierra y llegó a tiempo que avían venido çinco reyes subjectos a aqueste que se llamava Guacanagarí, todos con sus coronas representando muy buen estado que dize el Almirante a los Reyes que Sus Altezas ovieran plazer de ver la manera dellos. En llegando en tierra, el rey vino a rescebir al Almirante y lo llevó de braços a [la] misma casa de ayer a do tenía un estrado y sillas en que asentó al Almirante y luego se quitó la corona de la cabeça y se la puso al Almirante y el Almirante se quitó del pesqüeço un collar de buenos alaqueques y cuentas muy hermosas de muy lindos colores que pareçía muy bien en toda parte, y se lo puso a él, y se desnudó un capuz de fina grana, que aquel día se avía vestido, y se lo vistió, y enbió por unos borzeguíes de color que le hizo calçar, y le puso en el dedo un grande anillo de plata, porque avían dicho que vieron una sortija de plata a un marinero y que avía hecho mucho por ella. Quedó muy alegre y muy contento, y dos de aquellos reyes que estavan con él vinieron adonde el Almirante estava con él,[1] y truxeron al Almirante dos grandes plastas de oro cada uno la suya. Y estando así vino un yndio diziendo que avía dos días que dexara la caravela Pinta al leste en un puerto. Tornóse el Almirante a la caravela, y Viçeynte Anes, capitán della, affirmó que avía visto ruybarbo,[2] y que lo avía en la ysla Amiga que está a la entrada de la mar de Sancto Thomé que estava seys leguas de allí, e que avía cognosçido los ramos y rayz. Dizen que el ruybarbo echa unos ramitos fuera de tierra y unos frutos que pareçen moras verdes quasi secas y el palillo que está çerca de la rayz es tan amarillo y tan fino, como la mejor color que puede ser para pintar, y debaxo de la tierra haze la rayz como una grande pera.

Lunes 31 de diziembre

Aqueste día se ocupó en mandar tomar agua y leña para la partida a España, por dar noticia presto a los Reyes para que enbiasen navíos que descubriesen lo que quedava por descubrir, porque ya *el negoçio pareçía tan grande y de tanto tomo que es maravilla* (dixo el Almirante). Y dize que no quisiera partirse hasta que oviera visto toda aquella tierra que yva hazia el leste y andarla toda por la costa por saber también (dizque) el tránsito de Castilla a ella, para traer ganados y otras cosas. Mas como oviese[3] quedado con un solo navío, no le pareçía

---

[1] Margin: No[ta].
[2] Margin: Ruybarbo.
[3] MS: aya corrected to oviese.

Sunday 30 December

The Admiral went ashore to eat and arrived just as five kings, who were subjects of this one called Guacanagarí, had arrived, all with their crowns indicating their high rank, and the Admiral tells the Monarchs that Their Highnesses would have been very pleased at their bearing. On reaching the shore, the king came to meet the Admiral and took him by the arm to the same house as yesterday where he had a dais and chairs on which the Admiral sat and then he took off his crown and put it on the Admiral's head, and the Admiral took off a collar of good bloodstones[183] and very beautiful beads of very fine colours which looked good in every way, and put it around the king's neck, and took off a cloak of fine scarlet cloth which he had worn that day, and put it on him, and sent for some coloured boots which he made him put on, and placed on his finger a large silver ring, because they said that they saw a sailor with a silver jewel and he had tried hard to obtain it. He was very happy and content and two of the kings who were with him came to where the Admiral was beside him and brought the Admiral two large plates of gold, one each. At which there came an Indian saying that two days before he had left the caravel Pinta in a harbour to the east. The Admiral returned to the caravel and Vicente Yáñez, her captain, declared that he had seen rhubarb,[184] and that there was some on the island of Amiga which is in the entrance to the Mar de Santo Tomás six leagues away, and that he had recognized the stems and root. They say that rhubarb sends out small branches above ground and bears fruits like green mulberries, almost dry, and the stem near the root is yellow and as fine as the best possible colour for painting, and under the soil the root grows like a large pear.

Monday 31 December

This day was spent in getting water and wood fetched for the departure for Spain, to inform the Monarchs quickly so that they could send ships to discover what remained to be discovered, because *now the enterprise seemed so great and of such importance that it is a marvel* (said the Admiral). And he says that he did not wish to leave until he had seen all that land to the east and sailed the whole coast (he says) to be sure also of the route from Castile to there, so as to bring cattle and other things. But since he had been left with only one ship, it did not

razonable cosa ponerse a los peligros que le pudieran ocurrir descubriendo. Y quexávase que todo aquel mal e inconveniente [provenía de] averse apartado de él la caravela Pinta.

Martes 1º de enero

A media noche despachó la barca que fuese a la ysleta Amiga para traer el ruybarbo. Bolvió a bísperas con un serón dello; no truxeron más porque no llevaron açada para cavar; aquello llevó por muestra a los Reyes. El rey de aquella tierra dizque avía embiado muchas canoas por oro. Vino la canoa que fue a saber de la Pinta y el marinero y no la hallaron. Dixo aquel marinero que a veynte leguas de allí avían visto un rey que traya en la cabeça dos grandes plastas de oro, y luego que los yndios de la canoa le hablaron se las quitó, y vido también mucho oro a otras personas. Creyó el Almirante que el rey Guacanagarí devía de aver prohibido a todos que no vendiesen oro a los cristianos, porque passasse todo por su mano. Mas él avía sabido los lugares como dixo antier donde lo avía en tanta cantidad que no lo tenían en preçio. También la espeçería que comen (dize el Almirante) es mucha y más vale que pimienta y manegueta. Dexava encomendados a los que allí quería dexar que oviesen quanta pudiesen.

Miércoles 2 de enero

Salió de mañana en tierra para se despedir del rey Guacanagarí e partirse en el nombre del Señor e diole una camisa suya. Y mostróle la fuerça que tenían y effecto que hazían las lombardas. Por lo qual mandó armar una y tyrar al costado de la nao que estava en tierra, porque vino a propósito de platica[r] sobre los caribes con quien tienen guerra. Y vido hasta dónde llegó la lombarda y cómo passó el costado de la nao y fue muy lexos la piedra por la mar. Hizo hazer también un escaramuça con la gente de los navíos armada diziendo al caçique que no oviese miedo a los caribes aunque viniesen. Todo esto dizque hizo el Almirante porque tuviese por amigos a los cristianos que dexava, y por ponerle miedo que los temiese. Llevólo el Almirante a comer consigo a la casa donde estava aposentado y a los otros que yvan con él. Encomendóle mucho el Almirante a Diego de Arana, y a Pero Gutiérrez, y a Rodrigo Escobedo que dexava juntamente por sus tenientes de aquella gente que allí dexava, porque todo fuese bien regido y governado a servicio de Dios y de Sus Altezas. Mostró mucho amor el caçique al Almirante y gran sentimiento en su partida

seem to him sensible to risk the dangers which exploration might entail. And he complained that all those problems and difficulties resulted from the caravel Pinta having left him.

Tuesday 1 January

At midnight he sent the boat to the island of Amiga to fetch the rhubarb. It returned at vespers with a large basketful; they did not bring more because they had no spade for digging; he took it as a sample for the Monarchs. The king of that land had sent, he says, many canoes for gold. The canoe that went to enquire after the Pinta returned, together with the sailor, but they did not find her. The sailor said that twenty leagues from there they had seen a chief with two large plates of gold on his head, and as soon as the Indians in the canoe spoke to him he took them off; he also saw other people with much gold. The Admiral believed that the king Guacanagarí must have forbidden anyone to sell gold to the Christians, so that it would all pass through his hands. But he had learned of the places where, as he said the day before yesterday,[185] there was so much gold that they attach no value to it at all. Also, the spices they eat (says the Admiral) are many and worth more than pepper and allspice. He had ordered those whom he wanted to leave behind that they should obtain as much as they could.

Wednesday 2 January

In the morning he went ashore to take his leave of Guacanagarí, and to set out in the name of the Lord, and he gave him one of his shirts. He also showed him the power of the lombards and the effect they produced. For this purpose, and arising out of a conversation about the Caribs with whom they are at war, he ordered a lombard to be loaded and fired at the side of the flagship which was aground. And he saw the range of the lombard and how the shot passed through the side of the ship and went into the sea some way beyond. He also had some men from the ships arm themselves and stage a mock battle, telling the cacique that he should not be afraid of the Caribs even if they did come. The Admiral says that he did all this so that the king would treat the Christians he was leaving behind as friends, and to inspire fear of them. The Admiral took him and the others who were with him to eat with the Admiral at the house in which he was lodging. The Admiral entrusted him to Diego de Arana, Pedro Gutiérrez and Rodrigo de Escobedo whom he left in joint charge of the men he was leaving behind, so that everything would be well administered in the service of God and Their Highnesses. The cacique showed the Admiral great affection and great

mayormente quando lo vido yr a embarcarse. Dixo al Almirante un privado de aquel rey que avía mandado hazer un estatua de oro puro tan grande como el mismo Almirante y que de desde a diez días la avían de traer. Enbarcóse el Almirante con propósito de se partir luego mas el viento no le dio lugar. Dexó en aquella ysla Española que los yndios dizque llamavan Bohío treynta y nueve hombres con la fortaleza y dizque mucho amigos de aquel rey Guacanagarí e sobre aquellos por sus tenientes a Diego de Arana natural de Córdova, y a Pero Gutiérrez respostero de estrado del Rey criado del despensero mayor, e a Rodrigo de Escobedo natural de Segovia sobrino de fray Rodrigo Pérez con todos sus poderes que de los Reyes tenía. Dexóles todas las mercaderías que los Reyes mandaron comprar para los resgates que eran muchas, para que las trocasen y resgatasen por oro con todo lo que traya la nao. Dexóles también pan vizcocho para un año y vino y mucha artillería, y la barca de la nao para que ellos como marineros que eran los más fuesen quando viessen que convenía a descubrir la mina del oro, porque a la buelta que bolviese el Almirante hallase mucho oro; y lugar donde se assentasse una villa porque aquél no era puerto a su voluntad. Mayormente que el oro que allí trayan venía dizque del leste, y quanto más fuesen al leste tanto estavan çercanos de España. Dexóles tanbién simientes para sembrar, y sus officiales escrivano y alguazil, y entre aquellos un carpintero de naos y calafate y un buen lombardero que sabe bien de ingenios y un tonelero y un phísico, y un sastre y todos dizque hombres de la mar.

Jueves 3 de enero

No partió oy porque anoche dizque vinieron tres de los yndios que traya de las yslas que se avían quedado, y dixéronle que los otros y sus mugeres vernían al salir del sol. La mar también fue algo alterada y no pudo la barca estar en tierra; determinó partir mañana mediante la gracia de Dios. Dixo que si él tuviera consigo la caravela Pinta, tuviera por çierto de llevar un tonel de oro porque osara seguir las costas destas yslas, lo que no osava hazer por ser solo, porque no le acaeçiese algún ynconveniente. Y se impidiese su buelta a Castilla y la notiçia que devía dar a los Reyes de todas las cosas que avía hallado. Y si fuera çierto que la caravela Pinta llegara a salvamento en España con aquel Martín Alonso Pinçón, dixo que no dexara de hazer lo que deseava. Pero porque no sabía de él, y porque ya que vaya podrá ynformar a los Reyes de mentiras porque no le manden dar la pena que él mereçía como a quien tanto mal avía hecho y hazía

regret at his leaving, particularly when he saw him going to embark. One of the king's counsellors told that Admiral that he had ordered a statue of pure gold to be made, as large as the Admiral himself, and that it would be brought in ten days' time. The Admiral embarked with the intention of leaving directly but the wind did not allow him to do so. He left on that island, which he says the Indians called Bohío, thirty-nine men with the fort and, he says, they were very friendly with king Guacanagarí. In charge of them were Diego de Arana from Córdoba, Pedro Gutiérrez the King's chamberlain and servant of the chief steward, and Rodrigo de Escobedo from Segovia, nephew of Fr. Rodrigo Pérez, entrusted with all the powers given to him by the Monarchs. He left them all the cargo which the Monarchs ordered to be bought for trading, of which there was a substantial amount, so that they could exchange and barter it for gold, along with everything from the flagship. He also left them a year's supply of biscuit and wine and much artillery, and the ship's boat so that, being sailors as most of them were, they could go when the time seemed right, in search of the gold mine, so that on his return the Admiral would find much gold, and a place where they could establish a settlement, because that harbour was not to his liking, especially as the gold they brought there came, as he says, from the east, and the further east they went the nearer they were to Spain. He also left them seed to sow and his officials, the secretary and the bailiff, and among the company a ship's carpenter and caulker and a good gunner with a knowledge of machinery and a cooper and a doctor, and a tailor, all of whom he says were seamen.

Thursday 3 January

He did not set out today because he says that last night three of the remaining Indians he had brought from the islands came to him and said that the others and their wives would come at sunrise.[186] Moreover, the sea was rather rough and the boat could not go to the shore; he decided to leave tomorrow, the grace of God permitting. He said that if he had had the caravel Pinta with him, he would certainly have taken back a barrel of gold, because he could have risked following the coasts of these islands, which he dare not do alone in case any accident should befall him and prevent his return to Castile with the news which he had to give to the Monarchs of all the things that he had found. And if he were certain that the caravel Pinta would arrive safely in Spain under Martín Alonso Pinzón's command, he said that he would not abandon what he wished to do. But because he had no news of him, and because, if he were to go, he could lie to the Monarchs so that they would not punish him as he deserved for all the harm he had done and was still doing in having gone off without permission and

en averse ydo sin liçençia y estorvar los bienes que pudieran hazerse y saberse de aquella vez, dize el Almirante, confiava que Nuestro Señor le daría buen tiempo y se podía remediar todo.

Viernes 4 de enero

Saliendo el sol levantó las anclas con poco viento con la barca por proa el camino del norueste para salir fuera de la restringa por otra canal más ancha de la que entró. La qual y otras son muy buenas para yr por delante de la Villa de la Navidad.[1] Y por todo aquello el más baxo fondo que halló fueron tres braças hasta nueve y estas dos van de norueste al sueste según aquellas restringas eran grandes que duran desde el Cabo Sancto hasta el Cabo de Sierpe que son más de seys leguas y fuera en la mar bien tres y sobre el Cabo Sancto[2] a una legua no ay más de ocho braças de fondo, y dentro del dicho cabo de la parte del leste ay muchos baxos y canales para entrar por ellos y toda aquella costa se corre norueste sueste y es toda playa y la tierra muy llana hasta bien quatro leguas la tierra adentro. Después ay montañas muy altas y es toda muy poblada de poblaçiones grandes y buena gente según se mostravan con los cristianos. Navegó así al leste camino de un monte muy alto que quiere pareçer ysla pero no lo es porque tiene participaçión con tierra muy baxa, el qual tiene forma de un alfaneque muy hermoso al qual puso nombre Monte Cristo. El qual está justamente al leste de el Cabo Santo y avrá diez y ocho leguas. Aquel día por ser el viento muy poco no pudo llegar al Monte Cristi con seys leguas. Halló quatro ysletas de arena muy baxas con una restringa que salía mucho al norueste y andava mucho al sueste. Dentro ay un grande golpho que va desde el dicho monte al sueste bien veynte leguas, el qual deve ser todo de poco fondo y muchos bancos. Y dentro de él en toda la costa muchos ríos no navegables aunque aquel marinero que el Almirante enbió con la canoa a saber nuevas de la Pinta dixo que vido un río en el qual podían entrar naos. Surgió por allí el Almirante seys leguas de Monte en diez y nueve braças dando la buelta a la mar por apartarse de muchos baxos y restringa[s] que por allí avía, donde estuvo aquella noche. Da el Almirante aviso que el que oviere de yr a la Villa de la Navidad que cognosciere a Monte Cristo deve meterse en la mar dos leguas, etc., pero porque ya se sabe la tierra y más por allí no se pone aquí. Concluye

---

[1] Margin: Llamó la Villa de la Navidad la fortaleza y el asiento que allí hizo porque llegó allí día de la Navidad como pareçe por lo de arriba.
[2] MS: Cabo Sancto bien tres y sobre el Cabo Santo a una legua

preventing all the benefits and knowledge which could be had at once, says the Admiral, he was confident that Our Lord would give him fair weather and everything could be put right.

Friday 4 January

At sunrise he weighed anchor with a light wind on a NW course; the boat went ahead of him to find a way out through the shoals by a wider channel than that through which he entered. This channel and others are fine for approaching the town of Navidad. The least depth he found throughout was from three to nine fathoms. These two channels run from NW to SE through the large shoals that stretch from Cabo Santo[187] to the Cabo de Sierpe,[188] more than six leagues, and a good three leagues[189] into the sea. A league above Cabo Santo the bottom is no more than eight fathoms, and inside that cape to the E there are many shoals and channels by which to pass through them. All that coast runs NW to SE and is all beach and the terrain is very flat for a good four leagues inland. Then there are very high mountains, and it is all thickly populated with large villages and good people, to judge from their treatment of the Christians. Thus he sailed E on course for a very high hill which looks almost like an island but is not because it is joined to the land by a very low-lying isthmus. This hill has the shape of a very beautiful pavilion and he named it Monte Cristo. It is about 18 leagues due E of Cabo Santo. Because the wind was very light he could not get within six leagues of Monte Cristo that day. He found four very low sandy islets[190] with a sandbank which stretched far to the NW and to the SE. Inside there is a large gulf[191] which runs a good twenty leagues SE from the hill, which must be all very shallow and with many banks. Inside the gulf the whole coast has many rivers, none of them navigable, although that sailor whom the Admiral sent in the canoe to seek news of the Pinta said that he had seen a river which ships could enter.[192] The Admiral anchored six leagues from Monte Cristo in nineteen fathoms, having put out to sea in order to steer clear of the many shoals and sandbanks in that area, and there he stayed that night. The Admiral advises that anyone wanting to go to the town of Navidad should first sight Monte Cristo and stay two leagues offshore, etc., but since the land is well known up there, I will not write it

que Çipango estava en aquella ysla, y que ay mucho oro y espeçería y almáçiga y ruybarbo.

Sábado 5 de enero

Quando el sol quería salir dio la vela con el terral; después ventó leste, y vido que de la parte del susueste del Monte Cristo entre él y una ysleta pareçía ser buen puerto para surgir esta noche, y tomó el camino al lessueste y después al sursueste bien seys leguas açerca del Monte y halló andadas la[s] seys leguas diez y siete braças de hondo y muy limpio y anduvo así tres leguas con el mismo fondo. Después abaxó a doze braças hasta el morro del Monte, y sobre el morro del Monte a una legua halló nueve, y limpio todo, arena menuda. Siguió así el camino hasta que entró entre el Monte y la ysleta adonde halló tres braças y media de fondo con baxamar muy singular puerto adonde surgió. Fue con la barca a la ysleta donde halló huego y rastro que avían estado allí pescadores. Vido allí muchas piedras pintada[s] de colores o cantera de piedras tales de labores naturales muy hermosas dizque para edifiçios de iglesia o de otras obras reales como las que halló en la ysleta de Sant Salvador. Halló también en esta ysleta muchos pies de almáçiga. Este Monte Cristo dizque es muy hermoso y alto y andable de muy linda hechura, y toda la tierra çerca de él es baxa, muy linda campiña, y él queda así alto que viéndolo de lexos pareçe ysla que no comunique con alguna tierra.[1] Después del dicho monte al leste vido un cabo a xxiiii millas al qual llamó Cabo del Bezerro desde el qual hasta el dicho monte passa[n] en la mar bien dos leguas unas restringas de baxos aunque le pareçió que avía entre ellas canales para poder entrar; pero conviene que sea de día, y vaya soldando con la barca primero. Desde el dicho monte al leste hazia el Cabo del Bezerro las quatro leguas es todo playa y tierra muy baxa y hermosa y lo otro es todo tierra muy alta y grandes montañas labradas y hermosas y dentro de la tierra va una sierra de nord[u]este al sueste, la más hermosa que avía visto que pareçe propria como la sierra de Córdova. Pareçen también muy lexos otras montañas muy altas hazia el sur y el sueste, y muy grandes valles y muy verdes y muy hermosos y muy muchos ríos de agua; todo esto en tanta cantidad apazible que no creya encareçerlo la milléssima parte. Después vido al leste del dicho monte una tierra que pareçía otro monte así como aquel de Cristo en grandeza y hermosura. Y dende a la quarta del leste al nordeste es tierra no tan alta y avría bien çien millas, o çerca.

---

[1] Margin: Dize verdad que por mar y por tierra pareçe ysla como un montón de trigo.

here. He concludes that Cipango was on that island, and that there is much gold and spices and mastic and rhubarb.

Saturday 5 January

As the sun was about to rise he set sail with a land breeze; then the wind turned E and he saw that to the SSE of Monte Cristo, between it and an islet, there seemed to be a good anchorage for that night. He steered ESE and then SSE for a good six leagues towards Monte Cristo and after covering six leagues he found a depth of seventeen fathoms, very clear, and he sailed three leagues with the same depth. Then it decreased to twelve fathoms as far as the head of Monte Cristo, and one league beyond the head he found nine fathoms, all very clear and with fine sand. He followed this course until he was between Monte Cristo and the islet[193] where he found three and a half fathoms at low tide and an exceptional harbour where he anchored. He took the boat to the islet where he found a fire and signs that fishermen had been there. He saw there a sort of quarry of many coloured stones of various hues, all shaped by nature in a very beautiful way, suitable, he says, for church buildings or other official buildings, and like those he found on the islet of San Salvador. He also found on this islet many mastic trees.[194] He says that this Monte Cristo is very beautiful and high and accessible with a very pretty shape, and all the land around it is low, making a very lovely plain, and the hill itself is high so that from a distance it appears to be an island cut off from any other land. 24 miles beyond this hill to the E he saw a cape which he named Cabo del Bezerro[195] and, between them both, shoals and sandbanks extend into the sea for two leagues, although it seemed to him that there were channels among them which would allow entry; but it had best be by day with the boat going ahead to take soundings. Four leagues E of Monte Cristo towards Cabo del Bezerro it is all beach and very beautiful lowland and the rest is very high land with large mountains with beautifully cultivated slopes and inland a mountain range runs from NW to SE, the most beautiful he had seen and just like the sierra in Córdoba. In the far distance there also appear to be very high mountains to the S and SE and very large valleys, very green and very beautiful and with many streams of water; all of this is so pleasant that he did not believe he was exaggerating by a thousandth part. Then he saw to the E of the mountain some land which looked like another hill just like Monte Cristo in size and beauty. From there E by N to the NE the land is not so high and must extend for a good hundred miles, or thereabouts.

Domingo 6 de enero

Aquel puerto es abrigado de todos los vientos salvo de norte y norueste y dize que poco reynan por aquella tierra;[1] y aun destos se pueden guareçer detrás de la ysleta. Tiene tres hasta quatro braças. Salido el sol dio la vela por yr la costa delante, la qual toda corría al leste. Salvo que es menester dar reguardo a muchas restringas de piedra y arena que ay en la dicha costa; verdad es que dentro della ay buenos puertos y buenas entradas por sus canales. Después de mediodía ventó leste rezio, y mandó sobir a un marinero al topo del mástel para mirar los baxos, y vido venir la caravela Pinta[2] con leste a popa, y llegó al Almirante, y porque no avía donde surgir por ser baxo bolvióse el Almirante al Monte Cristi a desandar diez leguas atrás que avía andado, y la Pinta con él. Vino Martín Alonso Pinçón a la caravela Niña donde yva el Almirante a se escusar diziendo que se avía partido de él contra su voluntad, dando razones para ello; pero el Almirante dize que eran falsas todas y que con mucha sobervia y cudiçia se avía apartado aquella noche que se apartó de él. Y que no sabía (dize el Almirante) de dónde le oviese venido las sobervias y deshonestidad que avía usado con él aquel viaje. Las quales quiso el Almirante dissimular, por no dar lugar a las malas obras de Sathanás que deseava impedir aquel viaje como hasta entonçes avía hecho. Sino que por dicho de un yndio de los que el Almirante le avía encomendado con otros que lleva en su caravela, el qual le avía dicho que en una ysla que se llamava Baneque avía mucho oro, y como tenía el navío sotil y ligero se quiso apartar y yr por sí dexando al Almirante. Pero el Almirante quísose detener y costear la ysla Joana y la Española pues todo era un camino del leste. Después que Martín Alonso fue a la ysla Baneque dizque y no halló nada de oro, se vino a la costa de la Española, por información de otros yndios que le dixeron aver en aquella ysla Española que los yndios llamavan Bohío mucha cantidad de oro y muchas minas, y por esta causa llegó çerca de la Villa de la Navidad, obra de quinze leguas y avía entonçes más de veynte días; por lo qual pareçe que fueron verdad las nuevas que los yndios davan por las quales enbió el rey Guacanagarí la canoa y el Almirante el marinero; y devía de ser yda quando la canoa llegó. Y dize aquí el Almirante que resgató la caravela mucho oro que por un cabo de agujeta le davan buenos pedaços de oro del tamaño de dos dedos y a vezes como la mano, y llevava el Martín Alonso la mitad y la otra mitad se repartía por la gente. Añide el Almirante diziendo a los

---

[1] Margin: No avía experimentado la yra destos dos vientos.
[2] Margin: Vieron la caravela Pinta.

Sunday 6 January

That harbour is sheltered from all winds except from the N and NW which, he says, do not prevail in that land;[196] and it is possible to shelter even from these behind the islet. The depth is from three to four fathoms. At sunrise he set sail to go along the coast which stretched to the east. It is necessary, however, to look out for many reefs and sandbanks along that coast; although it is true that inside them there are good harbours and good approaches through their channels. After midday a strong easterly wind blew, and he ordered a sailor to climb to the top of the mast to look out for shallows, and he saw the caravel Pinta approaching from the E, and she reached the Admiral, and as there was nowhere to anchor because the water was shallow, the Admiral returned to Monte Cristi, undoing the ten leagues he had sailed, and the Pinta went with him. Martín Alonso Pinzón came to the caravel Niña where the Admiral was and made his excuses, saying that he had become separated from him against his will, giving reasons; but the Admiral says that they were all untrue and that he had acted out of great pride and greed on the night that he had gone off and left him. And the Admiral says that he had no idea where he had got the arrogance and disloyalty with which he had treated him on that voyage. The Admiral decided to turn a blind eye, so as not to give Satan a chance to do his evil deeds by hindering the voyage as he had done up till then. One of the Indians whom the Admiral had entrusted to him together with others he carried on his caravel had told him that there was much gold on an island called Baneque, and as his ship was light and fast he decided to go off on his own, leaving the Admiral behind. But the Admiral decided to stay and coast the island of Juana and Española because it was all in the same easterly direction. After Martín Alonso went to the island of Baneque he says that he found no sign of gold there and came to the coast of Española, acting on information from other Indians who told him that there was a great quantity of gold on that island of Española which the Indians called Bohío, and for this reason he came within fifteen leagues of the town of Navidad more than twenty days before; from which it seems that the reports which the Indians gave, and which king Guacanagarí had sent the canoe and the Admiral the sailor to investigate, were correct; and she must have gone by the time the canoe got there. And the Admiral says here that the caravel bartered for much gold, for they would give good pieces of gold, the size of two fingers and sometimes a hand, in exchange for a piece of leather thong, and Martín Alonso took half and the other half was divided among the men. The Admiral adds, address-

Reyes: *Así que, Señores Príncipes, que yo cognozco que milagrosamente mandó quedar allí aquella nao Nuestro Señor, porque es el mejor lugar de toda la ysla para hazer el assiento y más açerca de las minas de oro.* También dizque supo que detrás de la ysla Joana de la parte del sur ay otra ysla grande[1] en que ay muy mayor cantidad de oro que en ésta en tanto grado que cogían los pedaços mayores que havas y en la ysla Española se cogían los pedaços de oro de las minas como granos de trigo.[2] Llamávase dizque aquella ysla Yamaye. También dizque supo el Almirante que allí hazia el leste avía una ysla adonde no avía sino solas mugeres, y esto dizque de muchas personas lo sabía. Y que aquella ysla Española o la otra ysla Y[a]maye estava çerca de tierra firme diez jornadas de canoa que podía ser sesenta o setenta leguas y que era la gente vestida allí.

Lunes 7 de enero

Este día hizo tomar una agua que hazía la caravela; calafetalla; y fueron los marineros en tierra a traer leña y dizque hallaron muchos almáçigos y lignáloe.

Martes 8 de enero

Por el viento leste y sueste mucho que ventava no partió este día; por lo qual mandó que se guarneçiese la caravela de agua y leña y de todo lo nesçessario para todo el viaje. Porque aunque tenía voluntad de costear toda la costa de aquella Española que andando al camino pudiese; pero porque los que puso en las caravelas por capitanes que eran hermanos, conviene a saber Martín Alonso Pinçón y Viceynte Anes, y otros que les seguían con sobervia y cudiçia estimando que todo era ya suyo, no mirando la honrra que el Almirante les avía hecho y dado, no avían obedeçido ni obedeçían sus mandamientos, antes hazían y dezían muchas cosas no devidas contra él, y el Martín Alonso lo dexó desde 21 de noviembre hasta seys de enero sin causa ni razón sino por su desobediençia; todo lo qual el Almirante avía çufrido y callado por dar buen fin a su viaje; así que por salir de tan mala compañía con los quales dize que cunplía dissimular aunque gente desmandada y aunque tenía dizque consigo muchos hombres de bien, pero no era tiempo de entender en castigo, acordó bolverse y no parar

---

[1] Margin: Dize verdad pero es tierra firme no isla.
[2] Margin: Y aun como una gran hogaça de pan de Alcalá o como un quartal de Valladolid se halló grano de oro en la Española, e yo lo vi, y otros muchos de libra y de dos y de tres y de ocho libras se hallaron en la Española.

ing the Monarchs: *And so, Sovereign Princes, I realize that it was a miracle that Our Lord commanded that ship to remain here, because it is the best place on the whole island to make a settlement and the closest to the gold mines.* He also says that he learned that on the other side of the island of Juana to the south there is another large island on which there is a much greater quantity of gold than on this one, so much so that they would collect nuggets larger than beans and on the island of Española the nuggets they collected from the mines were the size of grains of wheat. He says that that island was called Yamaye.[197] The Admiral also says that he learned that to the E there was an island where there were only women,[198] and he says that he heard this from many people. And that the island of Española or the other island of Yamaye were ten days by canoe from the mainland, which would be sixty or seventy leagues away, and that the people there wore clothes.

Monday 7 January

Today he had the caravel, which was leaking, pumped out and caulked; and the sailors went ashore to fetch firewood and he says that they found many mastic trees and aloe.

Tuesday 8 January

Because the wind blew strongly from the E and SE he did not set out today; he therefore ordered the caravel to be provisioned with water and firewood and everything necessary for the complete voyage. For although he wished to coast as much of that island of Española as possible on his homeward course, because the men he had put in charge of the caravels, who were brothers, that is Martín Alonso Pinzón and Vicente Yáñez, and others who supported them in their arrogance and greed, believing that everything was already theirs, and not heeding the honour which the Admiral had done to them, had not obeyed nor were obeying his orders, but rather were doing and saying many unjust things against him, and Martín Alonso had left him from 21 November till 6 January without cause or reason except his own disobedience, all of which the Admiral had suffered in silence to bring his voyage to a successful conclusion, in order to be rid of such bad company on whom he says it is best to turn a blind eye even though they are a rabble and even though he says he had with him many good men, but now was not the time to worry about punishment, he decided to return home as

más, con la mayor priesa que le fuese possible. Entró en la barca y fue al río que es allí junto, hazia el sursueste del Monte Cristo una grande legua donde yvan los marineros a tomar agua para el navío, y halló que el arena de la boca del río el qual es muy grande y hondo era dizque toda llena de oro y en tanto grado que era maravilla puesto que era muy menudo.[1] Creya el Almirante que por venir por aquel río abaxo se desmenuzava por el camino. Puesto que dize que en poco espacio halló muchos granos tan grandes como lantejas; mas de lo menudito dize que avía mucha cantidad. Y porque la mar era llena y entrava la agua salada con la dulce, mandó subir con la barca el río arriba un tiro de piedra: hincheron los barriles desde la barca, y bolviéndose a la caravela hallavan metidos por los aros de los barriles pedaçitos de oro y lo mismo en los aros de la pipa. Puso por nombre el Almirante al río, el Río del Oro. El qual de dentro passada la entrada muy hondo aunque la entrada es baxa y la boca muy ancha y de él a la Villa de la Navidad diez y siete leguas. Entremedias ay otros muchos ríos grandes, en especial tres, los quales creya que devían tener mucho más oro que aquél porque son más grandes, puesto que éste es quasi tan grande como Guadalquivir por Córdova;[2] y dellos a las minas de oro no ay veynte leguas.[3] Dize más el Almirante que no quiso tomar de la dicha arena que tenía tanto oro, pues sus Altezas lo tenía[n] todo en casa y a la puerta de su Villa de la Navidad, sino venirse a más andar, por llevalles las nuevas; y por quitarse de la mala compañía que tenía y que siempre avía dicho que era gente desmandada.

Miércoles 9 de enero

A medianoche levantó las velas con el viento sueste y navegó al lesnordeste. Llegó a una punta que llamó Punta Roxa que está justamente al leste del Monte Cristo sesenta millas y al abrigo della surgió a la tarde, que serían tres oras antes que anocheçiese. No osó salir de allí de noche porque avía muchas restringas, hasta que se sepan, porque después serán provechosas si tienen como deven tener canales y tienen mucho fondo y buen surgidero seguro de todos vientos. Estas tierras desde Monte Cristo hasta allí donde surgió son tierras altas y llanas

---

[1] Margin: Este río es Yaqui muy poderoso y de mucho oro y podía ser que lo hallase entonçes el Almirante como dize porque entonçes estava virgen como dizen. Pero todavía creo que mucho dello devía ser magasita porque allí ay mucha y pensava quiçá el Almirante que era oro todo lo que reluzía.

[2] Margin: Mayor es éste que todos aquellos, yo lo sé.

[3] Margin: Ni quatro leguas ay dellos a las minas.

fast as possible and delay no longer. He got into the boat and went to the river nearby,[199] a league SSE of Monte Cristo where the sailors went to fetch water for the ship, and he found that the sand in the river mouth, which was wide and deep, was he says all full of gold, so much so that it was a wonder even though the grains were very fine. The Admiral believed that as it came down that river the gold was broken into pieces. He says, however, that in a small area he found many grains as large as lentils, but he says that there was a great deal of the smaller size.[200] And because the tide was full and the salt water mingled with the fresh, he ordered them to go a stone's throw up-river in the boat; they filled the barrels from the boat and on returning to the caravel they found small pieces of gold caught in the hoops and the same in the hoops of the cask. The Admiral gave the river the name Río del Oro. It is very deep inside the entrance although the entrance itself is shallow and the mouth very wide; it is seventeen[201] leagues from the town of Navidad. In between there are many other large rivers, three in particular which he believed must contain more gold than that one because they are larger, although this one is almost as large as the Guadalquivir at Córdoba and from them to the gold mines is less than twenty leagues.[202] The Admiral also says that he did not want to take back any of the sand which contained so much gold, since their Highnesses had it all at their disposal and at the gate of their town of Navidad, but preferred to return at all speed to give them the news; and to be rid of the bad company he was keeping and whom he had always said were a rabble.

Wednesday 9 January

At midnight he set sail with a SE wind on a course ENE. He arrived at a point which he called Punta Roja[203] which is sixty miles due E of Monte Cristo, and he anchored in the shelter of it that afternoon, about three hours before sunset. He dare not proceed at night because there were many reefs, at least not until they are charted, but afterwards they will be of use if there are channels through them as there must be and a good depth and a good anchorage, secure from all winds. The land from Monte Cristo to the place where he anchored is high and flat with

y muy lindas campiñas y a las espaldas muy hermosos montes que van de leste a güeste y son todos labrados y verdes que es cosa de maravilla ver su hermosura y tienen muchas riberas de agua. En toda esta tierra ay muchas tortugas de las quales tomaron los marineros en el Monte Cristi que venían a desovar en tierra, y eran muy grandes como una grande tablachina. El día passado quando el Almirante yva al Río del Oro, dixo que vido tres serenas[1] que salieron bien alto de la mar pero no eran tan hermosas como las pintan, que en alguna manera tenían forma de hombre en la cara. Dixo que otras vezes vido algunas en Guinea en la costa de la Manegueta. Dize que esta noche con el nombre de Nuestro Señor partiría a su viaje sin más detenerse en cosa alguna, pues avía hallado lo que buscava[2] porque no quiere más enojo con aquel Martín Alonso hasta que Sus Altezas supiesen las nuevas de su viaje y de lo que a hecho. *Y después no çufriré* (dize él) *hechos de malas personas y de poca virtud. Las quales contra quien les dio aquella honrra presumen hazer su voluntad con poco acatamiento.*

Jueves 10 de enero

Partióse de donde avía surgido y al sol puesto llegó a un río al qual puso nombre Río de Graçia; está de la parte del sueste tres leguas. Surgió a la boca que es buen surgidero a la parte del leste;[3] para entrar dentro tiene un banco que no tiene sino dos braças de agua y muy angosto; dentro es buen puerto çerrado sino que tiene mucha bruma. Y della yva la caravela Pinta donde yva Martín Alonso muy maltratada, porque dizque estuvo allí resgatando diez y seys días donde resgataron mucho oro que era lo que deseava Martín Alonso. El qual después que supo de los yndios que el Almirante estava en la costa de la misma ysla Española y que no lo podía errar, se vino para él. Y dizque quisiera que toda la gente del navío jurara que no avían estado allí sino seys días. Mas dizque era cosa tan pública su maldad que no [se] podía encobrir. El qual (dize el Almirante) tenía hechas leyes que fuese para él la mitad del oro que se resgatase o se oviese. Y quando ovo de partirse de allí, tomó quatro hombres yndios y dos moços por fuerça. A los quales el Almirante mandó dar de vestir y tornar en tierra que se fuesen a sus casas. *Lo qual* (dize) *es servicio de Vuestras Altezas, porque hombres y mugeres son todos de Vuestras Altezas así desta ysla en especial como de las*

---

[1] Margin: Vido tres serenas.
[2] Margin: No[ta].
[3] Margin: Este río es el que dizen de Martín Alonso Pinçón que está çinco leguas de Puerto de Plata.

very lovely fields flanked by very beautiful hills which run from E to W and are all cultivated and green and it is marvellous to see their beauty and they have many streams of water. In all this land there are many turtles some of which the sailors captured at Monte Cristi as they came ashore to lay their eggs, and they were the size of a large wooden shield. The previous day when the Admiral went to the Río del Oro, he said that he saw three sirens which rose high above the water but they were not as beautiful as they are depicted, and in some ways they had the face of a man. He said that he had seen some on other occasions in Guinea and the coast of Manegueta.[204] He says that tonight he would set out in the name of Our Lord without further delay for any reason, as he had found what he was looking for and he wanted no more trouble with Martín Alonso until Their Highnesses knew the news of his voyage and what he has done. *Then (he says) I will not suffer the deeds of evil men of little worth who presume to have their own way with scant regard for him who gave them that honour.*

Thursday 10 January

He left his anchorage and at sunset reached a river to which he gave the name Río de Gracia[205] three leagues to the SE. He anchored to the east of the river mouth which is a good anchorage; entering the river, there is a very narrow sand bank with only two fathoms of water; once inside it is a good safe harbour except that there are many shipworms. The caravel Pinta in which Martín Alonso sailed had been badly attacked by them because, he says, she was trading there for sixteen days and traded for much gold, which is what Martín Alonso wanted. Once he found out from the Indians that the Admiral was on the coast of that same island of Española and that he could not miss him, he came to meet him. And he says that he wanted the men to swear that they had only been there six days. But he says that his misdeeds were so manifest that he could not conceal them. He had (says the Admiral) made a rule that half the gold they traded for or obtained should be his. And when he went to leave he took four Indian men and two boys by force. The Admiral ordered them to be given clothes and put ashore so that they could return home. *I did this* (he says) *in the interests of Your Highnesses, because these men and women are all Your Highnesses's subjects, particularly*

otras.[1] *Mas aquí donde tiene[n] ya asiento Vuestras Altezas se deve hazer ho[n]rra y favor a los pueblos pues que en esta ysla ay tanto oro y buenas tierra y espeçería.*

Viernes 11 de enero

A medianoche salió del Río de Graçia con el terral; navegó al leste hasta un cabo que llamó Belprado quatro leguas y de allí al sueste está el monte a quien puso Monte de Plata[2] y dize que ay ocho leguas. De allí del Cabo de Belprado al leste quarta del sueste está el cabo que dixo del Angel y ay diez y ocho leguas; y deste cabo al Monte de Plata ay un golfo y tierras las mejores y más lindas del mundo todas campiñas altas y hermosas que van mucho la tierra dentro; y después ay una sierra que va de leste a güeste muy grande y muy hermosa y al pie del monte ay un puerto muy bueno y en la entrada tiene quatorze braças. Y este monte es muy alto y hermoso, y todo esto es poblado mucho. Y creya el Almirante devía aver buenos ríos y mucho oro. Del Cabo del Angel al leste quarta del sueste ay quatro leguas a una punta que puso del Hierro y al mismo camino, quatro leguas, está una punta que llamó la Punta Seca. Y de allí al mismo camino a seys leguas está el cabo que dixo Redondo; y de allí al leste está el Cabo Françés, y en este cabo de la parte del leste ay una angla grande mas no le pareçió aver surgidero. De allí una legua está el Cabo del Buen Tiempo; de éste al sur quarta del sueste ay un cabo que llamó Tajado, una grande legua. De éste hazia el sur vido otro cabo y pareçióle que avría quinze leguas. Oy hizo gran camino por[que] el viento y las corrientes yvan con él. No osó surgir por miedo de los baxos y así estuvo a la corda toda la noche.

Sábado 12 de enero

Al quarto del alva navegó al leste con viento fresco y anduvo así hasta el día y en este tiempo veynte millas y en dos oras después andaría veynte y quatro millas. De allí vido al sur tierra y fue hazia ella y estaría della 48 millas y dize que dado reguardo al navío andaría esta noche 28 millas al nornordeste. Quando vido la tierra llamó a un cabo que vido el Cabo de Padre y Hijo porque a la punta de la

---

[1] Margin: No[ta].
[2] Margin: Este monte llamó de plata porque es muy alto y está siempre sobre la cumbre una ñiebla que lo haze blanco, o plateado, y al pie de él está el puerto que se dize por aquel Monte de Plata.

those of this island as of the others. But here, where Your Highnesses already have a settlement the people should be treated with honour and respect since on this island there is so much gold and good lands and spices.

Friday 11 January

At midnight he left the Río de Gracia with the land breeze; he steered E to a cape which he called Belprado,[206] a distance of four leagues, and from there to the SE is the mountain which he called Monte de Plata[207] and he says it is eight leagues. From Cape Belprado E by S is the cape he called del Angel,[208] eighteen leagues away; and between this cape and Monte de Plata there is a gulf and the best and most beautiful lands in the world, all beautiful high fields stretching well inland; then there is a very large and very beautiful sierra running E to W and at the foot of the hill there is a very good harbour forty fathoms deep at the entrance.[209] And this hill is very high and beautiful and is all well populated. And the Admiral believed that there must be good rivers and much gold. From Cabo del Angel four leagues E by S there is a point which he called del Hierro,[210] and four leagues in the same direction is a point which he called the Punta Seca.[211] And six leagues beyond in the same direction is the cape he called Redondo;[212] and from there to the east is the Cabo Francés,[213] and to the east of this cape there is a large bay but it did not appear to him to have an anchorage. A league from there is the Cabo del Buen Tiempo;[214] from this a good league S by E is a cape he called Tajado.[215] To the south of this cape he saw another at a distance of what seemed to him to be fifteen leagues. Today he made good progress because the wind and the currents were with him. He did not dare to anchor for fear of shoals and so stood off all night.

Saturday 12 January

At the dawn watch he steered E with a fresh wind and proceeded on that course until daybreak during which time he made twenty miles and in the next two hours he made about another twenty-four miles. From there he saw land to the south at a distance of about 48 miles and made for it; he says that by keeping out to sea he made about 28 miles NNE that night. When he sighted land he named a cape he saw Cabo del Padre e Hijo[216] because on the eastern end it has two

parte del leste tiene dos farallones, mayor el uno que el otro. Después al leste dos leguas vido una grande abra y muy hermosa entre dos grandes montañas, y vido que era grandíssimo puerto bueno y de muy buena entrada; pero por ser muy de mañana y no perder camino, porque por la mayor parte del tiempo haze por allí lestes y entonces le lleva nornorueste, no quiso detenerse más. Siguió su camino al leste hasta un cabo muy alto y muy hermoso y todo de piedra tajado a quien puso por nombre Cabo del Enamorado el qual estava al leste de aquel puerto a quien llamó Puerto Sacro 32 millas. Y en llegando a él descubrió otro muy más hermoso y más alto y redondo de peña todo así como el Cabo de Sant Viçeynte en Portugal y estava del Enamorado al leste 12 millas. Después que allegó a emparejarse con el del Enamorado vido, entremedias de él y de otro, vido que se hazía una grandíssima baya que tiene de anchor tres leguas y en medio della está una ysleta pequeñuela, el fondo es mucho a la entrada hasta tierra. Surgió allí en doze braças; enbió la barca en tierra por agua y por ver si avían lengua; pero la gente toda huyó. Surgió también por ver si toda era aquella una tierra con la Española; y lo que dixo ser golpho sospechava no fuese otra ysla por sí. Quedava espantado de ser tan grande la ysla Española.

Domingo 13 de enero

No salió deste puerto por no hazer terral con que saliese; quisiera salir por yr a otro mejor puerto porque aquél era algo descubierto, y porque quería ver en qué parava la conjunción de la luna con el sol que esperava a 17 deste mes y la opposición della con Júpiter y conjunción con Mercurio y el Sol en oppósito con Júpiter, que es causa de grandes vientos.[1] Enbió la barca a tierra en una hermosa playa para que tomasen de los ajes para comer, y hallaron çiertos hombres con arcos y flechas con los quales se pararon a hablar y les compraron dos arcos y muchas flechas y rogaron a uno dellos que fuese a hablar al Almirante a la caravela y vino, el qual dizque era muy disforme en el acatadura más que otros que oviese visto. Tenía el rostro todo tyznado de carbón, puesto que en todas partes acostumbran de se teñir de diversas colores. Traya todos los cabellos muy largos[2] y encogidos y atados atrás, y después puestos en una redezilla de plumas de papagayos y él así desnudo como los otros. Juzgó el

---

[1] Margin: Por aquí parece que el Almirante sabía algo de astrología, aunque estos planetas parecen que no están bien puestos por falta del mal escrivano que lo trasladó.
[2] Margin: Estos devían ser los que llamavan ciguayos que todos traían lo[s] cabellos así muy largos.

rocky outcrops, one larger than the other. Then two leagues further E he saw a large and very beautiful inlet between two large mountains, and saw that it was a very fine large harbour with a very good entrance; but because it was very early in the morning and so as not to lose time, because for the most part the winds in that region are easterlies and at that time he had a NNW wind, he did not want to delay any longer. He followed his course E to a very high and very beautiful cape, all of jagged rock, to which he gave the name Cabo del Enamorado[217] and which was 32 miles E of that harbour he called Puerto Sacro.[218] Arriving at this cape, he discovered another, higher and more beautiful,[219] all of rock and with a rounded top like Cape St Vincent in Portugal, and this was 12 miles E of the Enamorado. After he had come abreast of the Enamorado he saw between it and the other cape a huge bay three leagues wide,[220] and in the middle there is a tiny little island, and there is a good depth of water right up to the shore. He anchored there in 12 fathoms; he sent the boat ashore for water and to see if they could make contact; but all the people fled. He anchored also to find out if all that land was part of Española, for he suspected that what he called a gulf might have made a separate island. He was amazed to find that the island of Española was so large.

Sunday 13 January

He did not leave this harbour because there was no land breeze with which to do so; he would have liked to go to a better harbour because that one was somewhat exposed, and because he wanted to see the conjunction of the Moon with the Sun and Mercury[221] which was expected on the 17th of this month and the opposition of the Moon and the Sun with Jupiter, which is a cause of strong winds. He sent the boat ashore to a beautiful beach so that they could gather ajes to eat, and they found some men with bows and arrows, with whom they stopped to talk and with whom they traded two bows and many arrows, and asked one of them to go to talk to the Admiral on the caravel, and he came. He says he was very ugly to look at, more so than others he had seen. His face was all blackened with charcoal,[222] although everywhere they are accustomed to paint their faces in various colours. His hair was very long and drawn back and tied behind and gathered in a small net of parrot feathers,[223] and he was as naked as the others.

Almirante que devía de ser de los caribes[1] que comen los hombres, y que aquel golfo que ayer avía visto, que hazía apartamiento de tierra y que sería ysla por sí. Preguntóle por los caribes y señalóle al leste çerca de allí la qual dizque ayer vio el Almirante antes que entrase en aquella baya, y díxole el yndio que en ella avía muy mucho oro, señalándole la popa de la caravela que era bien grande y que pedaços avía tan grandes. Llamava al oro tuob y no entendía por caona[2] como le llaman en la primera parte de la ysla, ni por noçay como lo nombravan en San Salvador y en las otras yslas. Al alambre, o a un oro baxo llaman en la Española tuob. De la ysla de Matinino dixo aquel yndio que era toda poblada de mugeres sin hombres[3] y que en ella ay muy mucho tuob que es oro o alambre y que es más al leste de Carib. También dixo de la ysla de Goanin[4] adonde ay mucho tuob. Destas yslas dize el Almirante que avía por muchas personas días avía notiçia. Dize más el Almirante que en las yslas passadas estavan con gran temor de Carib y en algunas le llamavan Caniba, pero en la Española Carib, *y que deve de ser gente arriscada, pues andan por todas estas yslas y comen la gente que pueden aver.* Dize que entendía algunas palabras, y por ellas dizque saca otras cosas, y que los yndios que consigo traya entendían más puesto que fallava differençia de lenguas por la gran distançia de las tierras. Mandó dar al yndio de comer, y diole pedaços de paño verde y colorado y cuentezuelas de vidro a que ellos son muy affiçionados, y tornóle a embiar a tierra, y díxole que truxese oro si lo avía lo qual creya por algunas cositas suyas que él traya. En llegando la barca a tierra, estavan detrás los árboles bien çinquenta y çinco hombres desnudos con los cabellos muy largos así como las mugeres los traen en Castilla.[5] Detrás de la cabeça trayan penachos de plumas de papagayos y de otras aves y cada uno traya su arco. Desçendió el yndio en tierra y hizo que los otros dexasen sus arcos y flechas y un pedaço de palo que es como un *** muy pesado que traen en lugar de espada.[6] Los quales después se llegaron a la barca y la gente de la barca

---

[1] Margin: No eran caribes ni los ovo en la Española jamás.

[2] Margin: Caona llamavan al oro en la mayor parte de la ysla Española, pero avía otras dos o tres lenguas.

[3] Margin: No entendía el Almirante aqueste yndio.

[4] Margin: Este guanin no era ysla según yo creo, sino el oro baxo que según los yndios de la Española tenía un olor porque lo preçiavan mucho y a éste llamavan guanin.

[5] Margin: Estos çierto eran los que se llamavan çyguayos en las sierras y costa del norte de la Española desde quasi Puerto de Plata hasta Higuay inclusive.

[6] Margin: Este es del árbol de palma que es duríssimo hecho a manera de una paleta de hierro que fazen para freyr uuevos o pescado, grande de quatro palmos boto por todas partes. Llámanle macana.

The Admiral thought he must be one of the man-eating Caribs, and that the gulf he had seen yesterday divided the land and made a separate island. He asked him about the Caribs and the Indians pointed out some land nearby to the E which the Admiral says he saw yesterday before entering that bay, and the Indian told him that there was much gold there, and pointing to the poop of the caravel which was very large he said that there were pieces just as large. He called gold 'tuob' and did not understand 'caona' as they call it in the first part of the island, nor 'nozay' as they call it in San Salvador and the other islands. On Española they call copper or low-grade gold 'tuob'. Of the island of Matinino[224] the Indian said that it was entirely populated by women without men and that there is a great deal of 'tuob', that is, gold or copper, and that it is further E than Carib. He also spoke of the island of Goanin[225] where there is much 'tuob'. The Admiral says that he had information about these islands from many people days before. The Admiral further says that the people on the previous islands were very afraid of the Carib and some called them 'Caniba', but 'Carib' on Española, and that *they must be a daring people for they roam these islands eating anyone they can capture*. He says that he understood a few words and from them he says he gathers other things, and that the Indians he had with him understood more although they found the languages different due to the great distance between the lands. He ordered food to be given to the Indian and gave him pieces of green and red cloth and small glass beads, of which they are very fond, and sent him ashore again, and told him to bring gold if there was any, which the Admiral believed to be so from some trinkets the Indian had with him. When the boat reached the shore there were a good fifty-five men behind the trees, naked and with very long hair just like women wear in Castile. At the back of the head they wore bunches of parrots' and other birds' feathers, and each one carried a bow. The Indian landed and made the others lay down their bows and arrows and a piece of wood which is like a very heavy [club] which they carry instead of a sword. They then came to the boat and the sailors went ashore and began to

salió a tierra y començáronles a comprar los arcos y flechas y las otras armas porque el Almirante así lo tenía ordenado. Vendidos dos arcos no quisieron dar más; antes se aparejaron de arremeter a los cristianos y prendellos. Fueron corriendo a tomar sus arcos y flechas donde los tenían apartados, y tornaron con cuerdas en las manos para dizque atar los cristianos. Viéndolos venir corriendo a ellos, estando los cristianos apercebidos, porque siempre los avisava desto el Almirante, arremetieron los cristianos a ellos,[1] y dieron a un yndio una gran cuchillada en las nalgas y a otro por los pechos hirieron con una saetada lo qual visto que podían ganar poco aunque no eran los cristianos sino siete y ellos çinquenta y tantos, dieron a huyr que no quedó ninguno, dexando uno aquí los flechas y otro allí los arcos. Mataran dizque los cristianos muchos dellos, si el piloto que yva por capitán dellos no lo estorvara. Bolviéronse luego a la caravela los cristianos con su barca, y sabido por el Almirante dixo que por una parte le avía plazido y por otra no,[2] *porque ayan miedo a los cristianos, porque sin duda* (dize él) *la gente de allí es* dizque *de mal hazer* y que creya que eran los de Carib, y que comiesen los hombres, y porque viniendo por allí la barca que dexó a los xxxix hombres en la fortaleza y Villa de la Navidad, *tengan miedo de hazerles algún mal.* Y que *si no son de los caribes al menos deven ser fronteros y de las mismas costumbres, y gente sin miedo, no como los otros de las otras yslas que son cobardes y sin armas fuera de razón.* Todo esto dize el Almirante y que querría tomar algunos dellos. Dizque hazían muchas ahumadas como acostumbrava[n] en aquella ysla Española.

Lunes 14 de enero

Quisiera enbiar esta noche a buscar las casas de aquellos yndios por tomar algunos dellos, creyendo que eran caribes, y por el mucho leste y nordeste y mucha ola que hizo en la mar; pero ya de día vieron mucha gente de yndios en tierra, por lo qual mandó el Almirante yr allá la barca con gente bien adereçada. Los quales luego vinieron todos a la popa de la barca, y especialmente el yndio que el día antes avía venido a la caravela[3] y el Almirante le avía dado las cosillas de resgate. Con este dizque venía un rey el qual avía dado al yndio dicho unas cuentas que diese a los de la barca en señal de seguro y de paz. Este rey con tres de los suyos entraron en la barca y viniero[n] a la caravela. Mandóles el

---

[1] Margin: La primera pelea que se ovo entre yndios y cristianos en la ysla Española.
[2] Margin: No[ta].
[3] Margin: Tornaron los yndios de paz a contratar.

trade for the bows and arrows and other weapons from them as the Admiral had ordered. Having sold two bows they did not wish to sell any more; instead they made ready to attack the Christians and capture them. They ran to pick up their bows and arrows where they had left them and came back with ropes in their hands, he says, to tie up the Christians. Seeing them running towards them, and being ready because the Admiral had always warned about this, the Christians attacked them and gave one Indian a great gash on the buttocks and wounded another in the chest with an arrow. When they saw from this that they could gain little even though there were only seven of the Christians and more than fifty of them, they ran off leaving their bows and arrows scattered about, and none remained behind. He says that the Christians would have killed many of them if the pilot who was in charge of them had not prevented it. The Christians then returned to the caravel in the boat and when the Admiral learned what had happened he said that in one sense he was sorry but in another not; *because they will fear the Christians, because without doubt* (he says) *those people are*, he says, *evildoers* and he thought they were from the Carib and that they were man-eaters,[226] and because if the boat he left for the thirty-nine men in the fort and town of Navidad should come that way, *they will be afraid to do them any harm. And if they are not caribs they must at least be neighbours and have the same customs, and they are without fear, not like the others on the other islands who are cowardly beyond reason and unarmed.* All this the Admiral says, and that he would like to take some of them back. He says they made many smoke signals as was the custom on that island of Española.

Monday 14 January

He wanted to send men out tonight to look for the houses of those Indians to capture some of them, believing them to be caribs, and because the wind was from the E and NE and the sea very rough; but at daybreak they saw many Indian people on shore, and the Admiral therefore sent out the boat with some well-armed men. The Indians then all came to the stern of the boat, especially the one who had come to the caravel the day before and to whom the Admiral had given the trinkets for barter. With him, he says, came a king who had given that Indian some beads to give to the sailors as a token of security and peace. This king got into the boat with three of his men and came to the caravel. The

Almirante dar de comer vizcocho y miel, y diole un bonete colorado y cuentas y un pedaço de paño colorado, y a los otros también pedaços de paño el qual dixo que traería mañana una carátula de oro afirmando que allí avía mucho, y en Carib y en Matinino. Después los enbió a tierra bien contentos. Dize más el Almirante que le hazían agua mucha las caravelas por la quilla y quéxase mucho de los calafates que en Palos las calafatearon muy mal y que quando vieron que el Almirante avía entendido el defecto de su obra y los quisiera constreñir a que la emendaran, huyeron. Pero no obstante la mucha agua que las caravelas hazían, confía en Nuestro Señor que lo truxo, lo tornara por su piedad y misericordia que bien sabía su Alta Magestad quánta controversia tuvo primero antes que se pudiese expedir de Castilla,[1] que ningún otro fue en su favor sino Él porque Él sabía su coraçón y después de Dios Sus Altezas, y todo lo demás le avía sido contrario sin razón alguna.[2] Y dize más así: *Y an seydo causa que la corona real de Vuestras Altezas no tenga çient cuentos de renta más de la que tiene después que yo vine a les servir que son siete años agora a veynte días de henero este mismo mes*[3] *y más lo que acreçentado sería de aquí en adelante. Mas aquel poderoso Dios remediará todo.* Estas son sus palabras.

Martes 15 de enero

Dize que se quiere partir porque ya no aprovecha nada detenerse por aver passado aquellos desconciertos; deve dezir del escándalo de los yndios. Dize también que oy a sabido que toda la fuerça del oro estava en la comarca de la Villa de la Navidad de Sus Altezas y que en la ysla de Carib avía mucho alambre y en Matinino puesto que será dificultoso en Carib porque aquella gente dizque come carne humana; y que de allí se parecía la ysla dellos y que tenía determinado de yr a ella, pues está en el camino y a la de Matinino que dizque era poblada toda de mugeres, sin hombres, y ver la una y la otra y tomar dizque algunos dellos. Embió el Almirante la barca a tierra, y el rey de aquella tierra no avía venido porque dizque la población estava lexos; mas enbió su corona de oro como avía prometido. Y vinieron otros muchos hombres con algodón y con pan y ajes todos con sus arcos y flechas. Después que todo lo ovieron resgatado, vinieron dizque

---

[1] Margin: No[ta].
[2] Margin: Acuérdase el Almirante de las dificultades que tuvo en la corte quando propuso su descubrimiento.
[3] Margin: A xx de enero año de 1485 entró en la corte el Almirante a proponer su descubrimiento.

Admiral ordered them to be given biscuit and honey, and gave the king a red cap and beads and a piece of red cloth; to the others he also gave pieces of cloth, and the king said that the next day he would bring a mask of gold, saying that there was a great deal of gold there, and on Carib and Matinino also. Then he sent them ashore well pleased. The Admiral also says that the caravels were leaking badly at the keel and complains bitterly about the caulkers in Palos who caulked them very badly and who fled when they saw that the Admiral had noticed their defective work and intended to make them put it right. But in spite of the amount of water the caravels were taking in, he trusts that Our Lord who brought him will, out of pity and mercy, get him back again, for His High Majesty well knew how much trouble he had had before he could set out from Castile, for no one showed him any favour except Him, because He knew his heart, and after God, Their Highnesses; and everything else had been against him without reason. And he goes on : *And they have been the reason why Your Highnesses' royal crown does not have a hundred million more in revenue than it has since I came to serve you, now seven years ago on 20 January this present month,*[227] *plus the extra which would accrue from now on. But that all-powerful God will set everything to rights.* These are his words.

Tuesday 15 January

He says that he wishes to depart today because there was no longer anything to be gained from remaining after those disagreements; he must mean the commotion with the Indians. He also says that today he has learned that the bulk of the gold was in the area of Their Highnesses' town of Navidad and that there was a great deal of copper on the island of Carib and on Matinino, although there would be difficulties on Carib because of those people who, he says, eat human flesh; from there he could see their island[228] and had decided to go there, since it is on his route, and to go to Matinino which, he says, was inhabited entirely by women,[229] without men, and see them both and take back some of the inhabitants. The Admiral sent the boat ashore and the king of that land had not come because, he says, the village was a long way off; but he sent his crown of gold as he had promised, and many other men came with cotton and with bread and ajes, and all with their bows and arrows. After everything had been traded,

quatro mancebos a la caravela, y pareçiéronle al Almirante dar tan buena cuenta de todas aquellas yslas que estavan hazia el leste en el mismo camino que el Almirante avía de llevar, que determinó de traer a Castilla consigo.[1] Allí dizque no tenían hierro ni otro metal que se oviese visto, aunque en pocos días no se puede saber de una tierra mucho, así por la dificultad de la lengua que no entendía el Almirante sino por discreçión, como porque ellos no saben lo que él pretendía en pocos días. Los arcos de aquella gente dizque eran tan grandes como los de Françia e Inglaterra; las flechas son proprias como las azagayas de las otras gentes que hasta allí avía visto, que son de los pinpollos de las cañas, quando son simiente, que quedan muy derechas y de longura de una vara y media y de dos, y después ponen al cabo un pedaço de palo agudo de un palmo y medio y ençima deste palillo algunos le inxieren un diente de pescado, y algunos y los más le pone[n] allí yerva, y no tyran como en otras partes, salvo por una çierta manera que no pueden mucho offender. Allí avía muy mucho algodón y muy fino y luengo y ay muchas almáçigas, y pareçíale que los arcos eran de texo y que ay oro y cobre. También hay mucho axí que es su pimienta, della que vale más que pimienta y toda la gente[2] no come sin ella que la halla muy sana; puédense cargar çinquenta caravelas cada año en aquella Española. Dize que halló mucha yerva en aquella baya de la que hallavan en el golpho quando venía al descubrimiento, por lo qual creya que avía yslas al leste hasta en derecho de donde las començó a hallar; porque tiene por çierto que aquella yerva nasce en poco fondo junto a tierra. Y dize que si así es, muy çerca estavan estas yndias de las yslas de Canaria, y por esta razón creya que distavan menos de quatrocientas leguas.[3]

Miércoles 16 de enero

Partió antes del día tres oras del golfo que llamó el Golfo de las Flechas[4] con viento de la tierra; después con viento güeste llevando la proa al leste quarta del nordeste para yr dizque a la ysla de Carib donde estava la gente a quien todas aquellas yslas y tierras tanto miedo tenían, porque dizque con sus canoas sin número andavan todas aquellas mares y dizque comían los hombres que pueden

---

[1] Margin: Fue muy mal hecho traerlos contra su voluntad.
[2] Margin: Esta gente deve dezir por los cristianos.
[3] Margin: Bien juzgava.
[4] Margin: Sospecho que éste era el golfo de Samaná donde salen los ríos Yuna y Tamo ríos poderosos de la ysla Española.

he says that four young men came to the caravel, and they appeared to the Admiral to give such a good account of all those islands that were to the E on the same route that the Admiral was to take, that he decided to take them to Castile with him. He says that they had no iron, and no other metal had been seen there, although in a few days it is not possible to learn much about a country, because of the difficulty of the language which the Admiral did not understand except by conjecture and because in a few days they could not understand what he meant. He says that the bows of those people were as large as those of France and England; the arrows are the same as the spears of the other peoples he had seen previously, and are made from the stalks of canes which have gone to seed, which are very straight and about a yard and a half or two yards long, and they fix in the end a piece of sharp stick about one and a half palms long and into this stick some insert a fish tooth, and some, the majority, put poison on it. They do not shoot them as in other areas, but in a peculiar way which cannot do much harm. There was a great deal of very fine, long cotton there and much mastic, and it seemed to him that the bows were made of yew and that there is gold and copper. There is also much 'ají' which is their pepper, some of which is worth more than pepper, and all the men eat it with everything and find it very healthy; fifty caravels a year could be loaded with it on Española. He says that he found a lot of seaweed in that bay, of the same kind as they found in the gulf when he was on his journey of discovery, and for this reason he believed that there were islands stretching due E from where he began to find them; because he is certain that that weed grows in shallow water near to land. And he says that if that is the case, these Indies were very close to the Canary Islands, and for this reason he thought they were less than four hundred leagues away.

Wednesday 16 January

Three hours before daybreak he left that gulf which he called the Golfo de las Flechas[230] with the land breeze, and later with a W wind he steered E by N to go, he says, to the island of Carib where there lived the people of whom all those islands and lands were so afraid, because, he says, that with their innumerable canoes they roamed those seas and he says that they would eat anyone they

aver.[1] La derrota dizque le avía[n] mostrado unos yndios de aquellos quatro que tomó ayer en el Puerto de las Flechas. Después de aver andado a su pareçer 64 millas señaláronle los yndios quedaría la dicha ysla al sueste; quiso llevar aquel camino y mandó templar las velas, y después de aver andado dos leguas, refrescó el viento muy bueno para yr a España. Notó en la gente que començó a entristeçerse por desviarse del camino derecho por la mucha agua que hazían ambas caravelas y no tenían algún remedio salvo el de Dios. Ovo de dexar el camino que creya que lleva de la ysla, y bolvió al derecho de España, nordeste quarta del leste, y anduvo así hasta el sol puesto 48 millas, que son doze leguas.[2] Dixéronle los yndios que por aquella vía hallaría la ysla de Matinino que dizque era poblada de mugeres sin hombres, lo qual el Almirante mucho quisiera [ver] por llevar dizque a los Reyes çinco o seys dellas; pero dudava que los yndios supiesen bien la derrota, y él no se podía detener por el peligro del agua que cogían las caravelas; mas dizque era çierto que las avía,[3] y que a çierto tiempo del año venían los hombres a ellas de la dicha ysla de Carib que dizque estava dellas diez o doze leguas, y si parían niño enbiávanlo a la ysla de los hombres, y si niña, dexávanla consigo. Dize el Almirante que aquellas dos yslas no devían distar de donde avía partido xv o xx leguas y creya que eran al sueste y que los yndios no le supieron señalar la derrota. Después de perder de vista el cabo que nombró de Sant Theramo[4] de la ysla Española, que le quedava al güeste diez y seys leguas, anduvo doze leguas al leste quarta del nordeste. Llevava muy buen tiempo.

Jueves 17 de enero

Ayer al poner del sol calmóle algo el viento; andaría 14 ampolletas que tenía cada una media ora o poco menos hasta el rendir del primer quarto, y andaría quatro millas por ora que son 28 millas. Después refrescó el viento y anduvo así todo aquel quarto que fueron diez ampolletas y después otras seys hasta salido el sol ocho millas por ora y así andaría por todas ochenta y quatro millas que son 21 leguas al nordeste quarta del leste, y hasta el sol puesto andaría unas quarenta y quatro millas que son onze leguas al leste. Aquí vino un alcatraz a la ca~~la, y después otro, y vido mucha yerva de la que está en la mar.

---

[1] Margin: Oy dexó del todo la ysla Española.
[2] Margin: Buelta a España.
[3] Margin: Nunca esto después se averiguó que oviese tales mugeres.
[4] Margin: Este cabo de Sant Theramo creo cierto que es el que llaman agora el Cabo del Engaño.

could capture. He says that some of the four Indians he had taken yesterday in the Puerto de las Flechas showed him the course. After having sailed what he estimated to be 64 miles, the Indians indicated to him that the island in question would lie to the SE; he decided to take that route and ordered the sails trimmed, and after sailing two leagues, the wind got up fresh, ideal for returning to Spain. He noticed that the crew was beginning to get unhappy about leaving the direct route because both caravels were leaking badly and they could expect help from no one but God. He was forced to leave the route which he believed leads to the island, and returned to the direct route to Spain NE by E and sailed on for 48 miles, that is 12 leagues, till sunset. The Indians told him that on that route he would find the island of Matinino, which he says was inhabited by women without men. The Admiral would very much like to see it in order, he says, to take five or six of them back to the Monarchs; but he doubted that the Indians knew the course well, and he could not delay because of the danger from the water which the caravels were shipping; but he says that he was certain that they existed, and that at a certain time of the year the men came to them from the island of Carib, which he says was ten or twelve leagues away, and if they gave birth to a boy they sent him to the men's island, and if a girl, they kept her with them.[231] The Admiral says that those two islands cannot be more than 15 or 20 leagues from his point of departure and he believed that they were to the SE and that the Indians did not know how to show him the bearing. After losing sight of the cape on Española which he named San Theramo,[232] which lay sixteen leagues W, he sailed twelve leagues E by N. He had very good weather.

Thursday 17 January

Yesterday at sunset the wind fell somewhat; he sailed for about 14 half-hour sand-glasses or a little less until the end of the first quarter watch, and made about 4 miles an hour, which is 28 miles. Then the wind freshened and he sailed the whole of that watch, or ten glasses, and then another six until sunrise at eight miles an hour, and so in total he sailed about 84 miles, that is 21 leagues, NE by E, and by sunset he made about another forty-four miles, which is eleven leagues, E. A tern came to the caravel, and then another, and he saw a lot of weed of the kind that is found in the sea.

Viernes 18 de enero

Navegó con poco viento esta noche al leste quarta del sueste quarenta millas que son 10 leguas y después al sueste quarta del leste 30 millas que son 7 leguas y media hasta salido el sol. Después de salido [el] sol navegó todo el día con poco viento lesnordeste y nordeste y con leste más y menos puesta la proa a vezes al norte y a vezes a la quarta del nordeste y al nornordeste; y así contando lo uno y lo otro creyó que andaría sesenta millas que son 15 leguas. Pareçió poca yerva en la mar pero dize que ayer y oy pareçió la mar quajada de atunes y creyó el Almirante que de allí devían de yr a las almadravas del Duque de Conil y de Cáliz. Por un pescado que se llama rabiforcado que anduvo alrededor de la caravela y después se fue la vía del sursueste creyó el Almirante que avía por allí algunas yslas. Y al lessueste de la ysla Española dixo que quedava la ysla de Carib y la de Matinino y otras muchas.

Sábado 19 de enero

Anduvo esta noche çinquenta y seys millas al norte quarta del nordeste y 64 al nordeste quarta del norte. Después del sol salido navegó al nordeste con el viento lessueste con viento fresco y después a la quarta del norte, y andaría 84 millas que son veynte y una leguas. Vido la mar quajada de atunes pequeños; ovo alcatrazes, rabos de juncos, y rabiforcados.

Domingo 20 de enero

Calmó el viento esta noche y a rratos ventava unos balços de viento y andaría por todo veynte millas al nordeste. Después del sol salido andaría onze millas al sueste, después al nornordeste 36 millas que son nueve leguas. Vido infinitos atunes pequeños; los ayres dizque muy suaves y dulçes, *como en Sevilla por abril o mayo, y la mar*, dize, *a Dios sean dadas muchas gracias, siempre muy llana*. Rabiforcados y pardelas y otras aves muchas pareçieron.

Lunes 21 de enero

Ayer después del sol puesto navegó al norte quarta del nordeste con el viento leste y nordeste; andaría 8 millas por ora hasta media noche que serían çinquenta y seys millas. Después anduvo al nornordeste 8 millas por ora, y así serían en toda la noche çiento y quatro millas que son xxvi leguas a la quarta del norte de la parte del nordeste. Después del sol salido navegó al nornordeste con

Friday 18 January

This night he steered E by S with little wind for forty miles, that is 10 leagues, and then SE by E for 30 miles, which is 7 and a half leagues, until sunrise. After sunrise he sailed all day with little wind ENE and NE and E, more or less, steering sometimes N and at others N by E or NNE; and so in all he believed that he must have sailed sixty miles, which is 15 leagues. Very little weed appeared on the sea but he says that yesterday and today the sea appeared to be thick with tunny fish and the Admiral believed that they must go from there to the tunny fisheries of the Duke of Conil and Cádiz.[233] Judging by a [bird] called a frigate bird which flew around the caravel and then headed SSE the Admiral believed that there were some islands in the area. And to the ESE of Española he said there lay the islands of Carib and Matinino and many others.

Saturday 19 January

This night he sailed fifty-six miles N by E and 64 NE by N. After sunrise he steered NE with a fresh ESE wind and later NE by N, and he made about 84 miles, which is twenty-one leagues. He saw the sea thick with small tunny fish; there were gannets, reed-tails and frigate birds.

Sunday 20 January

This night the wind dropped and there were occasional gusts and in all he must have sailed twenty miles NE. After sunrise he made about eleven miles SE, then 36 miles, or nine leagues, NNE. He saw a huge number of small tunny; the breezes he says are very gentle and sweet, *as in Seville in April and May, and the sea*, he says, *is always calm, thanks be to God.* Frigate birds and petrels and many other birds appeared.

Monday 21 January

Yesterday after sunset he steered N by E with the wind E and NE; he made about 8 miles an hour until midnight, that is fifty-six miles. Then he sailed NNE at 8 miles an hour, and so in the whole night one hundred and four miles, which is 26 leagues, NE by N. After sunrise he steered NNE with the same E wind and

el mismo viento leste y a vezes a la quarta del nordeste, y andaría 88 millas en onze oras que tenía el día que son 21 leguas sacada una que perdió porque arribó sobre la caravela Pinta por haballe. Hallava los ayres más fríos y pensava dizque hallarlos más cada día quanto más se llegase al norte, y también por las noches ser más grandes por el angostura de la esp[h]era. Pareçieron muchos rabos de juncos y pardelas y otras aves pero no tantos peçes dizque por ser el agua más fría; vido mucha yerva.

Martes 22 de enero

Ayer después del sol puesto navegó al nornordeste con viento leste y tomava del sueste; andava 8 millas por ora hasta passadas çinco ampolletas y tres de antes que se començase la guardia que eran ocho ampolletas. Y así avría andado setenta y dos millas que son diez [y] ocho leguas. Después anduvo a la quarta del nordeste al norte seys ampolletas que serían otras 18 millas. Después quatro ampolletas de la segunda guarda al nordeste seys millas por ora que son tres leguas al nordeste. Después hasta el salir del sol anduvo al lesnordeste onze ampolletas seys leguas por ora que son siete leguas. Después al lesnordeste hasta las onze oras del día 32 millas. Y así calmó el viento y no anduvo más en aquel día. Nadaron los yndios, vieron rabos de juncos y mucha yerva.

Miércoles 23 de enero

Esta noche tuvo muchos mudamientos en los vientos; tanteado todo y dados los reguardos que los marineros buenos suelen y deven dar, dize que andaría esta noche al nordeste quarta del norte 84 millas que son 21 leguas. Esperava muchas vezes a la caravela Pinta porque andava mal de la bolina porque se ayudava poco de la mezana por el mástel no ser bueno. Y dize que si el capitán della, que [era] Martín Alonso Pinçón, tuviera tanto cuydado de proveerse de un buen mástel en las yndias donde tantos y tales avía, como fue cudiçioso de se apartar de él pensando de hinchir el navío de oro, él lo pusiera bueno. Pareçieron muchos rabos de juncos, y mucha yerva; el çielo todo turbado estos días pero no avía llovido y la mar siempre muy llana como en un río *a Dios sean dadas muchas gracias*. Después del sol salido andaría al nordeste franco çierta parte del día 30 millas que son siete leguas y media, y después lo demás anduvo al lesnordeste otras treynta millas que son siete leguas y media.

occasionally NE by N, and he made about 88 miles in eleven hours of daylight, which is 21 leagues after discounting one hour which he lost because he fell off towards the Pinta for a conference with her. He found the winds colder and thought, he says, that he would find them colder by the day the further north he went, and also because the nights were longer because of the shape of the Earth. Many reed-tails and petrels and other birds appeared but not so many fish, he says, because the water was colder; he saw a lot of weed.

Tuesday 22 January

Yesterday after sunset he steered NNE with the wind E and veering SE; he made 8 miles an hour until five sand-glasses had passed and three before the beginning of the next watch, which was eight half-hour glasses. So he must have made seventy-two miles, which is eighteen leagues.[234] Then he sailed N by E for six glasses, which would have been another 18 miles. Then four glasses of the second watch NE at six miles an hour, which is three leagues NE. Then until sunrise he sailed ENE for eleven glasses at six [miles] an hour, which is seven leagues.[235] Then 32 miles ENE until eleven o'clock in the morning. Then the wind fell and he made no more progress that day. The Indians went swimming, they saw reed-tails and a lot of weed.

Wednesday 23 January

This night the wind was very changeable; being on the alert for everything and taking the precautions which good sailors must and do take, he says that tonight he sailed about 84 miles NE by N, which is 21 leagues. He had to wait many times for the caravel Pinta because she was having trouble sailing close to the wind and was getting little help from the mizzen because the mast was not sound. And he says that if her captain, Martín Alonso Pinzón, had shown as much care in providing himself with a good mast in the Indies, where there were so many good ones, as he showed greed in sailing away thinking to fill the ship with gold, he would have put it to rights. Many reed-tails appeared, and much weed; the sky was all overcast these days but it had not rained and the sea was still as flat as a river *many thanks be to God*. After sunrise he made about 30 miles due NE for part of the day, that is seven and a half leagues, and then for the rest he sailed ENE for another thirty miles, which is seven and a half leagues.

Jueves 24 de enero

Andaría esta noche toda, consideradas muchas mudanças que hizo el viento, al nordeste 44 millas que fueron onze leguas. Después de salido el sol hasta puesto andaría al lesnordeste quatorze leguas.

Viernes 25

Navegó esta noche al lesnordeste un pedaço de la noche que fueron treze ampolletas nueve leguas y media; después anduvo al nornordeste otras seys millas. Salido el sol todo el día, porque calmó el viento, andaría al lesnordeste 28 millas que son 7 leguas. Mataron los marineros una tonina y un grandíssimo tiburón, y dizque lo avían bien menester porque no trayan ya de comer sino pan y vino y ajes de las yndias.

Sábado 26 de enero

Esta noche anduvo al leste quarta del sueste 56 millas que son quatorze leguas. Después del sol salido navegó a las vezes al lessueste y a las vezes al sueste; andaría hasta las onze oras del día quarenta millas. Después hizo otro bordo y después anduvo a la relinga y hasta la noche anduvo hazia el norte 24 millas que son seys leguas.

Domingo 27 de enero

Ayer después del sol puesto anduvo al nordeste y al norte y al norte quarta del nordeste y andaría çinco millas por ora y en treze oras serían 65 millas que son 16 leguas y media. Después del sol salido anduvo hazia el nordeste 24 millas que son seys leguas hasta mediodía y de allí hasta el sol puesto andaría tres leguas al lesnordeste.

Lunes 28 de enero

Esta noche toda navegó al lesnordeste; andaría 36 millas que son 9 leguas. Después del sol salido anduvo hasta el sol puesto al lesnordeste 20 millas que son çinco leguas. Los ayres halló templados y dulçes; vido rabos de juncos y pardelas, y mucha yerva.

Thursday 24 January

All this night, taking into account the very changeable wind, he made about 44 miles NE, which was eleven leagues. After sunrise until sunset he made about fourteen leagues ENE.

Friday 25

This night he steered ENE for part of the night, that is 13 sand-glasses, and made nine and a half leagues; then he sailed another six miles NNE. After sunrise he made about 28 miles, that is 7 leagues, ENE during the whole day because the wind died down. The sailors killed a porpoise and a very large shark, and he says they really needed them because they no longer had anything to eat but bread and wine and ajes from the Indies.

Saturday 26 January

This night he sailed 56 miles E by S, which is fourteen leagues. After sunrise he steered sometimes ESE and sometimes SE; he made about forty miles by eleven o'clock in the morning. Then he set another tack and sailed close to the wind and by nightfall he sailed 24 miles N, which is six leagues.

Sunday 27 January

Yesterday after sunset he sailed NE and N and N by E and made about five miles an hour or 65 miles in thirteen hours, which is 16 and a half leagues. After sunrise he sailed 24 miles NE, which is six leagues, by midday and from then until sunset he made about three leagues ENE.

Monday 28 January

All this night he steered ENE; he made about 36 miles, which is 9 leagues. After sunrise he sailed ENE until sunset for 20 miles, which is five leagues. He found the breezes temperate and sweet; he saw reed-tails and petrels, and much weed.

Martes 29 de enero

Navegó al lesnordeste y andaría en la noche con sur y sudueste 39 millas que son 9 leguas y media. En todo el día andaría 8 leguas. Los ayres muy templados como en abril en Castilla, la mar muy llana. Peçes que llaman dorados vinieron a bordo.

Miércoles 30 de enero

En toda esta noche andaría 7 leguas al lesnordeste. De día corrió al sur quarta al sueste treze leguas y media. Vido rabos de juncos y mucha yerva y muchas toninas.

Jueves 31 de enero

Navegó esta noche al norte quarta del nordeste treynta millas, y después al nordeste treynta y çinco millas que son diez y seys leguas. Salido el sol hasta la noche anduvo al lesnordeste 13 leguas y media. Vieron rabo de junco y pardelas.

Viernes 1º de hebrero

Anduvo esta noche al lesnordeste 16 leguas y media. El día corrió al mismo camino 29 leguas y un quarto. La mar muy llana a Dios gracias.

Sábado 2 de hebrero

Anduvo esta noche al lesnordeste quarenta millas que son 10 leguas. De día con el mismo viento a popa corrió 7 millas por ora por manera que en onze oras anduvo 77 millas que son 19 leguas y quarta. La mar muy llana gracias a Dios y los ayres muy dulçes. Vieron tan quajada la mar de yerva que si no la ovieran visto temieran ser baxos. Pardelas vieron.

Domingo 3 de hebrero

Esta noche yendo a popa con la mar muy llana a Dios gracias andaría 29 leguas. Parecióle la estrella del norte muy alta como en el Cabo de Sant Viçeynte. No pudo tomar el altura con el astrolabio ni quadrante porque la ola no le dio lugar. El día navegó al lesnordeste su camino y andaría diez millas por ora y así en onze oras 27 leguas.

Tuesday 29 January

He steered ENE and made about 39 miles, which is 9 leagues and a half, with the wind S and SW. He made about 8 leagues all day. The breezes very mild, like April in Castile, the sea very flat. Fish they call dorados came aboard.

Wednesday 30 January

All this night he made about 7 leagues ENE. By day he ran thirteen leagues and a half S by E. He saw reed-tails and much weed and many porpoises.

Thursday 31 January

This night he steered thirty miles N by E and then thirty-five miles NE, which is sixteen leagues. Between sunrise and nightfall he sailed 13 leagues and a half ENE. They saw a reed-tail and some petrels.

Friday 1 February

On this night he sailed 16 and a half leagues ENE. By day he ran the same course for 29 leagues and a quarter. The sea was very calm, thanks be to God.

Saturday 2 February

On this night he sailed ENE for forty miles, which is 10 leagues. By day with the same wind astern he made 7 miles an hour so that in eleven hours he sailed 77 miles, which is 19 leagues and a quarter. The sea was very calm, thanks be to God, and the breezes very gentle. They saw the sea so choked with weed that, had they not already met it, they would have feared that there were shoals. They saw some petrels.

Sunday 3 February

On this night with the wind astern and the sea very calm, thanks be to God, he made about 29 leagues. The north star seemed to him to be as high as at Cape St. Vincent.[236] He could not measure its elevation with the astrolabe nor the quadrant because the waves would not let him. By day he sailed on his course ENE and made about ten miles an hour and so in eleven hours, 27 leagues.

Lunes 4 de hebrero

Esta noche navegó al leste quarta del nordeste; parte anduvo 12 millas por ora y parte diez y así andaría 130 millas que son 32 leguas y media. Tuvo el çielo muy turbado y llovioso y hizo algún frío por lo qual dizque cognoscía que no avía llegado a las yslas de los Açores. Después del sol levantado mudó el camino y fue al leste. Anduvo en todo el día 77 millas que son 19 leguas y quarta.

Martes 5 de hebrero

Esta noche navegó al leste; andaría toda ella 54 millas que son quatorze leguas menos media. El día corrió 10 millas por ora y así en onze oras fueron 110 millas que son 27 leguas y media. Vieron pardelas y unos palillos que era señal que estavan çerca de tierra.

Miércoles 6 de hebrero

Navegó esta noche al leste; andaría onze millas por ora. En treze oras de la noche andaría 143 millas que son 35 leguas y quarta. Vieron mucha[s] aves y pardelas. El día corrió 14 millas por ora y así anduvo aquel día 154 millas que son 38 leguas y media. De manera que fueron entre día y noche 74 leguas poco más o menos. Viceynte Anes [halló] que oy por la mañana le quedava la ysla de Flores al norte, y la de la Madera al leste. Roldán[1] dixo que la ysla del Fayal o la de Sant Gregorio le quedava al nornordeste y el Puerto Sancto al leste. Pareció mucha yerva.

Jueves 7 de hebrero

Navegó esta noche al leste; andaría 10 millas por ora y así en treze oras 130 millas que son 32 leguas y media. El día, ocho millas por ora, en onze oras, 88 millas que son 22 leguas. En esta mañana estava el Almirante al sur de la ysla de Flores 75 leguas y el piloto Pero Alonso yendo al norte passava entre la Terçera y la de Sancta María, y a[l] leste passava de barlovento de la ysla de la Madera doze leguas de la parte del norte. Vieron los ma[rineros] yerva de otra manera de la passada de la que ay mucha en las yslas de los Açores. Después se vido de la passada.

---

[1] Margin: Este devía ser piloto.

Monday 4 February

Tonight he steered E by N; for a time he made 12 miles an hour and then ten, and so sailed about 130 miles, which is 32 leagues and a half. The sky was overcast and rainy, and it was rather cold, from which he says he realised that he had not reached the islands of the Azores. After sunrise he changed course to the E. Throughout the day he sailed 77 miles, which is 19 leagues and a quarter.

Tuesday 5 February

Tonight he steered E; he made about 54 miles during the night, which is fourteen leagues less a half. By day he made 10 miles an hour and so in eleven hours, 110 miles, which is 27 leagues and a half. They saw some petrels and some small sticks, which was a sign that they were near land.

Wednesday 6 February

Tonight he steered E; he made about eleven miles an hour. In thirteen hours of night he sailed about 143 miles, which is 35 leagues and a quarter. They saw many birds and petrels. During the day he made 14 miles an hour and so that day sailed 154 miles, which is 38 leagues and a half. So by day and night they went 74 leagues more or less. Vicente Yáñez[237] estimated that this morning the island of Flores lay to the N, and that of Madeira to the E. Roldán[238] reckoned that the island of Faial, or San Gregorio, lay to his NNE and Porto Santo to the E. A lot of weed appeared.

Thursday 7 February

Tonight he steered E; he made about 10 miles an hour and so in thirteen hours, 130 miles, which is 32 leagues and a half. During the day, at eight miles an hour for eleven hours, he made 83 miles, which is 22 leagues. This morning the Admiral was 75 leagues S of the island of Flores and the pilot Pero Alonso[239] estimated that by steering N he would pass between Terceira and Santa María, and by steering E he would pass to windward of the island of Madeira, 12 leagues off the north coast. The sailors saw a different kind of weed from that they had seen before, of which there is a great deal in the islands of the Azores. Later they saw the same kind as before.

Viernes 8 de hebrero

Anduvo esta noche tres millas por ora al leste por un rato y después caminó a la quarta del sueste; anduvo toda la noche 12 leguas. Salido el sol hasta mediodía corrió 27 millas; después hasta el sol puesto otras tantas que son treze leguas al sursueste.

Sábado 9 de hebrero

Un rato desta noche andaría tres leguas al sursueste y después al sur quarta del sueste; después al nordeste hasta las diez oras del día otras çinco leguas; y después hasta la noche anduvo 9 leguas al leste.

Domingo 10 de hebrero

Después del sol puesto navegó al leste toda la noche 130 millas que son 32 leguas y media. Al sol salido hasta la noche anduvo 9 millas por ora y así anduvo en onze oras 99 millas que son 24 leguas y media y una quarta.

En la caravela del Almirante carteavan o echavan punto Viçeynte Yanes y los dos pilotos Sancho Royz y Pero Alonso Niño y Rondán, y todos ellos passavan mucho adelante de las yslas de los Açores al leste por sus cartas y navegando al norte ninguno tomara la ysla de Sancta María que es la postrera de todas las de los Açores; antes serían delante con çinco leguas e fueran en la comarca de la ysla de la Madera o en el Puerto Sancto. Pero el Almirante se hallava muy desviado de su camino hallándose mucho más atrás que ellos. Porque esta noche le quedavan la ysla de Flores al norte y al leste yva en demanda a Nafe en Africa y pasava a barlovento de la ysla de la Madera de la parte del norte *** leguas. Así que ellos estavan más çerca de Castilla que el Almirante con 150 leguas. Dize que mediante la gracia de Dios desque vean tierra se sabrá quién andava más çierto. Dize aquí también que primero anduvo 263 leguas de la ysla del Hierro a la venida que viese la primera yerva, etc.

Lunes 11 de hebrero

Anduvo esta noche doze millas por ora a su camino y así en toda ella contó 39 leguas, y en todo el día corrió 16 leguas y media. Vido muchas aves de donde creyó estar çerca de tierra.

Friday 8 February

This night he sailed E for a while at three miles an hour and then went E by S; throughout the night he made 12 leagues. Between sunrise and midday he sailed 27 miles; then until sunset as many again, which makes thirteen leagues SSE.

Saturday 9 February

For part of this night he made about three leagues SSE and then S by E; then NE until ten o'clock in the morning for another five leagues; and then until nightfall he sailed 9 leagues to the E.

Sunday 10 February

After sunset he steered E all night for 130 miles, which is 32 leagues and a half. From sunrise until night he made 9 miles an hour, and so in eleven hours made 99 miles, which is 24 leagues and a half and a quarter.

On the Admiral's caravel Vicente Yáñez and the two pilots Sancho Ruiz[240] and Pero Alonso Niño and Roldán were plotting the course, and according to their charts they were all well to the E of the islands of the Azores and if they had sailed N none would have made the island of Santa María which is the easternmost island of all the Azores; rather, they would have been five leagues beyond and in the vicinity of the island of Madeira or Porto Santo. But the Admiral found himself way off their course and well behind them, for on this night he reckoned that the island of Flores lay to the N, and to the E he was heading for Nafe[241] in Africa and would have passed to windward of the island of Madeira ***[242] leagues to the N. So they were 150 leagues nearer Castile than the Admiral. He says that God willing as soon as they sight land they will know whose position was the most accurate. He also says here that on the outward journey he sailed 263 leagues from the island of Ferro before he saw the first weed, etc.

Monday 11 February

Tonight he sailed twelve miles an hour on his course and so made 39 leagues in all, and during the day he ran another 16 leagues and a half. He saw many birds, from which he believed that he was near land.

Martes 12 de hebrero

Navegó al leste seys millas por ora esta noche y andaría hasta el día 73 millas que son 18 leguas y un quarto. Aquí começó a tener grande mar y tormenta[1] y si no fuera la caravela dizque muy buena y bien adereçada temiera perderse. El día correría onze o doze leguas con mucho trabajo y peligro.

Miércoles 13 de hebrero

Después del sol puesto hasta el día tuvo gran trabajo del viento y de la mar muy alta y tormenta; relampagueó hazia el nornordeste tres vezes;[2] dixo ser señal de gran tempestad que avía de venir de aquella parte, o de su contrario. Anduvo a árbol seco lo más de la noche; después dio una poca de vela y andaría 52 millas que son treze leguas. En este día blandeó un poco el viento; pero luego creçió, y la mar se hizo terrible y cruzavan las olas que atormentavan los navíos. Andaría 55 millas que son treze leguas y media.

Jueves 14 de hebrero

Esta noche creció el viento y las olas eran espantables, contraria una de otra, que cruzavan y enbaraçavan el navío que no podía passar adelante ni salir de entremedias dellas y quebravan en él.[3] Llevava el papahígo muy baxo para que solamente lo sacase algo de las ondas; andaría así tres oras y correría 20 millas. Creçía mucho la mar y el viento, y viendo el peligro grande, començó a correr a popa donde el viento le llevase, porque no avía otro remedio. Entonçes començó a correr también la caravela Pinta en que yva Martín Alonso y desapareçió[4] aunque toda la noche hizo faroles el Almirante y el otro le respondía hasta que parez que no pudo más por la fuerça de la tormenta y porque se hallava muy fuera del camino del Almirante. Anduvo el Almirante esta noche al nordeste quarta del leste 54 millas que son 13 leguas. Salido el sol fue mayor el viento y la mar cruzando más terrible; llevava el papahígo solo y baxo para que el navío saliese de entre las ondas que cruzavan porque no lo hundiesen. Andava el camino del lesnordeste y después a la quarta hasta el nordeste; andaría seys oras así y en ellas 7 leguas y media.

---

[1] Margin: Començó a tener tormenta.
[2] Margin: Señal de mucho viento.
[3] Margin: Padeçió gran tormenta.
[4] Margin: Desapareçió la Pinta.

Tuesday 12 February

Tonight he steered E at six miles an hour and by daybreak sailed about 73 miles, which is 18 leagues and a quarter. Here he began to experience heavy seas and stormy weather and he says that if the caravel had not been very good and well equipped he feared that he would have been lost. By day he made about eleven or twelve leagues with great effort and at great risk.

Wednesday 13 February

Between sunset and daybreak he had a lot of trouble with the wind and the high waves and the stormy sea; three times there was lightning to the NNE; he said that it was a sign of a fierce storm coming from that quarter, that is, against him. He proceeded with bare masts for most of the night; then he put on a little sail and made about 52 miles, which is thirteen leagues. Today the wind abated a little; but later it strengthened and the sea became terrible with the waves crashing into each other and pounding the ships. He made about 55 miles, which is thirteen leagues and a half.

Thursday 14 February

Tonight the wind grew stronger and the waves were terrifying, crashing into each other and impeding the ship which could neither make headway nor extricate itself from them as they broke over it. He kept the mainsail very low simply to avoid the waves as far as possible; he must have sailed like that for three hours, making about 20 miles. The sea and the wind became much heavier, and seeing the great danger, he began to run before the wind wherever it took him, for there was nothing else he could do. Then the caravel Pinta with Martín Alonso aboard began to run before the wind and then disappeared even though all night the Admiral had flares burning and the other ship responded until it seems she could do so no longer because of the strength of the storm and because she was way off the Admiral's course.[243] Tonight the Admiral sailed 54 miles NE by E, which is 13 leagues. At sunrise the wind blew stronger and the crashing waves grew more terrible; he carried only the mainsail and kept it low so that the ship would escape from the waves breaking over her and not be sunk by them. He was following a course ENE and then NE by E; he sailed about six hours like that and during that time made 7 leagues and a half.

Él ordenó que se echase un romero[1] que fuese a Sancta María de Guadalupe y llevase un cirio de çinco libras de çera y que hiziesen voto todos que al que cayesse la suerte cumpliese la romería. Para lo qual mandó traer tantos garvanços quantas personas en el navío venían y señalar uno con un cuchillo haziendo una cruz y metellos en un bonete bien rebueltos. El primero que metió la mano fue el Almirante y sacó el garbanço de la cruz[2] y así cayó sobre él la suerte y desde luego se tuvo por romero y deudor de yr a complir el voto. Echóse otra vez la suerte[3] para enbiar romero a Santa María de Loreto que está en la marca de Ancona tierra del Papa que es casa donde Nuestra Señora ha hecho y haze muchos y grandes milagros y cayó la suerte a un marinero del Puerto de Sancta María que se llamava Pedro de Villa, y el Almirante le prometió de le dar dineros para las costas. Otro romero[4] acordó que se enbiase a que velase una noche en Sancta Clara de Moguer y hiziese dezir una missa para lo qual se tornaron a echar los garvanços con el de la cruz, y cayó la suerte al mismo Almirante. Después desto el Almirante y toda la gente hizieron voto[5] de en llegando a la primera tierra yr todos en camissa en proçessión a hazer oración en una iglesia que fuese de la invocaçión de Nuestra Señora.

Allende los votos generales, o comunes, cada uno hazía en espeçial su voto[6] porque ninguno pensava escapar, teniéndose todos por perdidos según la terrible tormenta que padeçían. Ayudava a acreçentar el peligro que venía el navío con falta de lastre por averse alivianado la carga siendo ya comidos los bastimentos y el agua y vino bevido. Lo qual por cudiçia del próspero tiempo que entre las yslas tuvieron no proveyó el Almirante, teniendo propósito de lo mandar lastrar en la ysla de las mugeres adonde lleva propósito de yr. El remedio que para esta neçessidad tuvo fue quando hazerlo pudieron henchir las pipas que tenían vazías de agua y vino, de agua de la mar y con esto en ella se remediaron.

Escrive aquí el Almirante las causas[7] que le ponían temor de que allí Nuestro Señor no quisiese que pereciese y otras que le davan esperança de que Dios lo

---

[1] Margin: Echan romeros y hazen voto.
[2] Margin: Cayó la suerte sobre el Almirante.
[3] Margin: Otro romero.
[4] Margin: Otro romero y cayó la suerte al Almirante.
[5] Margin: Otro voto.
[6] Margin: Hazían votos particulares.
[7] Margin: Pone las causas que le augmentavan el miedo de se perder y las que le davan esperança de salir a salvamento.

He declared that a pilgrim should be sent to Santa María de Guadalupe[244] with a candle made from five pounds of wax, and that everyone should make a vow that whoever was chosen by lot should make the pilgrimage. To this end he ordered as many chickpeas as there were people on board to be brought, and one of them to be marked by a cross made with a knife, and the chickpeas to be placed in a cap and well shaken. The Admiral was the first to put his hand in and he pulled out the bean with the cross and so he drew the lot and naturally took himself to be the pilgrim bound to go and fulfil the vow. Lots were drawn again for a pilgrim to be sent to Santa María de Loreto[245] which is in the province of Ancona, one of the papal lands,[246] and which is a house where Our Lady has performed and still performs many great miracles, and the lot was drawn by a sailor from Puerto de Santa María called Pedro de Villa, and the Admiral promised to give him money for his expenses. It was decided that another pilgrim should be sent to keep vigil for one night in Santa Clara de Moguer[247] and to have a mass said, and for this they again drew from the chickpeas including the one with the cross, and the lot fell to the Admiral again. After this the Admiral and all the men vowed that on the first land they reached they would all go in their shirts in a procession to pray in a church dedicated to Our Lady.

Apart from the general or communal vows, each man made his own pledge, for none thought he would escape, and all thought they were lost, so terrible was the storm they were suffering. The danger was made worse by the fact that the ship was short of ballast since the cargo had become lighter as the food was eaten and the water and wine drunk. The Admiral did not take on enough ballast, because he wanted to make the best of the fine weather they had among the islands, and intended to take on ballast on the Island of Women, which he intended to visit. The solution he found for this need was to fill, when they could, the empty water and wine casks with sea water, and with this they solved the problem.

At this point the Admiral writes about the reasons why he was afraid that Our Lord wanted him to perish there, and about others which made him hopeful that God would carry him to safety so that the news he was bringing to the Monarchs

avía de llevar en salvamento para que tales nuevas como llevava a los Reyes no pereçiesen. Pareçíale que el deseo grande que tenía de llevar estas nuevas tan grandes y mostrar que avía salido verdadero en lo que avía dicho y proferídose a descubrir, le ponía grandíssimo miedo de no lo conseguir y que cada mosquito dizque le podía perturbar e impedir. Atribúyelo esto a su poca fe y desfalleçimiento de confiança de la providencia divina. Confortáva[n]le por otra parte las merçedes que Dios le avía hecho en dalle tanta victoria descubriendo lo que descubierto avía y complídole Dios todos sus deseos, aviendo passado en Castilla en sus despachos muchas adversidades y contrariedades. Y que como antes oviese puesto su fin y endereçado todo su negoçio a Dios, y le avía oydo y dado todo lo que le avía pedido, devía creer que le daría complimiento de lo començado y le llevaría en salvamento. Mayormente que pues le avía librado a la yda quando tenía mayor razón de temer de los trabajos que [tenía] con los marineros y gente que llevava, los quales todos a una boz estavan determinados de se bolver y alçarse contra él haziendo protestaçiones,[1] y el eterno Dios le dio esfuerço y valor contra todos y otras cosas de mucha maravilla que Dios avía mostrado en él y por él en aquel viaje, allende aquellas que Sus Altezas sabían de las personas de su casa; así que (dize) que no deviera temer la dicha tormenta. Mas su flaqueza y congoxa (dize él) *no me dexava asensar la ánima*. Dize más que también le dava gran pena dos hijos que tenía en Córdova al estudio que los dexava güérfanos de padre y madre en tierra estraña, y los Reyes no sabían los serviçios que les avía en aquel viaje hecho y nuevas tan prósperas que les llevava, para que se moviesen a los remediar. Por esto y porque supiesen Sus Altezas cómo Nuestro Señor le avía dado victoria de todo lo que deseava de las yndias y suppiesen que ninguna tormenta avía en aquellas partes lo qual dize que se puede cognosçer por la yerva y árboles que están nacidos y creçidos hasta dentro en la mar; y porque si se perdiese con aquella tormenta los Reyes oviesen notiçia de su viaje,[2] tomó un pargamino y escrivió en él todo lo que pudo de todo lo que avía hallado, rogando mucho a quien lo hallase que lo llevase a los Reyes. Este pargamino enbolvió en un paño ençerado atado muy bien, y mandó traer un gran barril de madera y púsolo en él sin que ninguna persona supiese qué era, sino que pensaron todos que era alguna devoçión y así lo mandó echar en la mar. Después con los aguaçeros y turvionadas se mudó el viento al güeste y andaría así a popa sólo con el triquete çinco oras con la mar muy desconçertada y andaría dos leguas y media al nordeste. Avía quitado el papahígo de la vela mayor, por miedo que alguna onda de la mar no se lo llevase del todo.

---

[1] Margin: Las angustias y turbaçiones que padeçió a la yda de la gente que consigo llevava.
[2] Margin: Una yndustria que tuvo para que supiesen los reyes su viaje si se perdiese.

would not be lost. It seemed to him that the great desire he had to deliver such good news, and to show that he had been right in what he had said and undertaken to discover, made him very afraid that he would not manage to do so, and that the merest mosquito could upset his plans and prevent him. He attributes this to little faith and failing trust in Divine Providence. On the other hand he found comfort in the favour which God had shown him by giving him such a great victory in discovering what he had discovered and by fulfilling all his wishes, and having overcome many adversities and obstacles during his negotiations in Castile. And since he had always in the past entrusted the outcome of all his affairs to God's will, and He had heard him and given him all he had asked for, he must believe that He would allow him to complete what he had undertaken and would bring him to safety. Especially since He had delivered him on the outward voyage when he had greater reason to fear the problems he had with his crew and the men with him, all of whom were determined to a man to turn round and rebel against him in protest, and God eternal gave him strength and courage to face them all, and many other marvellous things which God had manifested in him and by him on that voyage, besides those things which Their Highnesses knew from the members of their household; so that (he says) he should not fear that storm. But his weakness and anguish (he says) *would not let me set my mind at rest*. He says further that he was also distressed about the two sons he had in school in Córdoba and of whom he was about to make fatherless and motherless in a strange land,[248] and the Monarchs did not know the services he had rendered them on that voyage nor the marvellous news that he was bringing them, which might move them to see that the boys were cared for. For this reason and so that Their Highnesses would know how Our Lord had given him in triumph everything he desired from the Indies and so that they would know that there were no storms in those regions, which, he says, is shown by the grass and trees which spring up and grow even in the sea; and so that, if he were to perish in that storm, the Monarchs would have news of his voyage, he took a piece of parchment and wrote on it everything he could about everything he had found, beseeching whomsoever might find it to take it to the Monarchs. He wrapped the parchment tightly in a waxed cloth and called for a large wooden barrel and put it in the barrel without anyone knowing what it was, for they all thought it was some act of devotion, and then ordered it to be thrown into the sea.[249] Then with the rain and the squalls the wind changed to the W, and he sailed before it with only the foresail set for about five hours with the sea very unsettled and he made about two leagues and a half NE. He had taken down the mainsail, for fear that a wave from the sea might carry it away altogether.

Viernes 15 de hebrero

Ayer después del sol puesto començó a mostrarse claro el çielo de la vanda del güeste y mostrava que quería de hazia allí ventar. Dio la boneta a la vela mayor; todavía la mar era altíssima aunque yva algo baxándose. Anduvo al lesnordeste quatro millas por ora y en treze oras de noche fueron treze leguas. Después del sol salido vieron tierra;[1] parecíales por proa al lesnordeste. Algunos dezían que era la ysla de la Madera, otros que era la Roca de Sintra en Portugal, junto a Lisboa. Saltó luego el viento por proa lesnordeste, y la mar venía muy alta del güeste; avría de la caravela a la tierra 5 leguas. El Almirante por su navegaçión se hallava estar con las yslas de los Açores,[2] y creya que aquella era una dellas; los pilotos y marineros se hallavan ya con tierra de Castilla.

Sábado 16 de hebrero

Toda esta noche anduvo dando bordos por encavalgar la tierra que ya se cognoscía ser ysla; a vezes yva al nordeste, otras al nornordeste, hasta que salió el sol, que tomó la buelta del sur por llegar a la ysla que ya no vían por la gran cerrazón, y vido por popa otra ysla que distaría 8 leguas. Después del sol salido hasta la noche anduvo dando bueltas por llegarse a la tierra con el mucho viento y mar que llevava. Al dezir de la Salve que es a boca de noche algunos vieron lumbre de sotavento y pareçía que devía ser la ysla que vieron ayer primero; y toda la noche anduvo barloventeando y allegándose lo más que podía para ver si al salir del sol vía alguna de las yslas. Esta noche reposó el Almirante algo porque desde el miércoles no avía dormido ni podido dormir y quedava muy tollido de las piernas por estar siempre desabrigado al frío y al agua y por el poco comer. El sol salido, navegó al sursudueste y a la noche llegó a la ysla, y por la gran cerrazón no pudo cognosçer qué ysla era.

Lunes 18 de hebrero

Después ayer del sol puesto anduvo rodeando la ysla para ver dónde avía de surgir y tomar lengua; surgió con una ancla que luego perdió; tornó a dar la vela y barloventeó toda la noche. Después del sol salido, llegó otra vez de la parte del

---

[1] Margin: Esta tierra era la ysla de Santa María en los Açores.
[2] Margin: El Almirante andava muy çierto en lo que avía andado, y los pilotos y marineros erravan.

Friday 15 February

Yesterday after sunset the sky began to clear in the W, showing that the wind was about to come from that direction. He had the bonnet added to the mainsail; the sea was still very rough but was beginning to grow calmer. He sailed ENE at four miles an hour and in thirteen hours of night made thirteen leagues. After sunrise they sighted land which appeared ahead of them to the ENE. Some said that it was the island of Madeira, others that it was the Rock of Sintra in Portugal, near Lisbon. The wind then veered sharply to a head wind from the ENE and the sea to the W became very high; the caravel was about 5 leagues from land. By the Admiral's reckoning he was off the Azores and he believed the land ahead was one of them; the pilots and sailors thought they were already off the coast of Castile.

Saturday 16 February

He spent all this night beating against the wind in order to reach the land which they now saw was an island; at times he went NE, at others NNE, until sunrise, when he turned S to reach the island which they could no longer see because of the dark cloud, and he saw astern another island about 8 leagues off. Between sunrise and nightfall he tacked back and forth trying to reach the land in the face of the strong wind and heavy seas. At the singing of the *Salve,* which is at nightfall, some saw a light to leeward and it seemed that it must be the island they first saw yesterday; he spent all night beating about and getting as close as possible so that at sunrise he might see one of the islands. This night the Admiral rested a little because he had not been able to sleep since Wednesday and his legs were troubling him from being constantly exposed to the cold and the wet, and he had had very little to eat. After sunrise he steered SSW and at nightfall reached the island, but could not tell which island it was for the dense low cloud.

Monday 18 February

After sunset yesterday he sailed round the island to see where he could anchor and talk to someone; he cast one anchor which he promptly lost; he set sail again and beat about all night. After sunrise he approached the island again from the

norte de la ysla y donde le pareció surgió con un ancla y enbió la barca en tierra y ovieron habla con la gente de la ysla y supieron cómo era la ysla de Sancta María,[1] una de las de los Açores, y enseñáronles el puerto dónde avían de poner la caravela; y dixo la gente de la ysla que jamás avían visto tanta tormenta como la que avía hecho los quinze días passados, y que se maravillavan cómo avían escapado. Los quales (dizque) dieron muchas gracias a Dios y hizieron muchas alegrías por las nuevas que sabían de aver el Almirante descubierto las yndias.[2] Dize el Almirante que aquella su navegaçión avía sido muy çierta y que avía carteado bien, que fuesen dadas muchas gracias a Nuestro Señor, aunque se hazía algo delantero, pero tenía por çierto que estava en la comarca de las yslas de los Açores y que aquella era una dellas. Y dizque fingió aver andado más camino por desatinar a los pilotos y marineros que carteavan, por quedar él señor de aquella derrota de las yndias como de hecho queda, porque ninguno de todos ellos traya su camino çierto por lo qual ninguno puede estar seguro de su derrota para las yndias.

Martes 19 de hebrero

Después del sol puesto vinieron a la ribera tres hombres de la ysla y llamaron. Enbióles la barca en la qual vinieron y truxeron gallinas y pan fresco y era día de Carnestolendas y truxeron otras cosas que le enbiava el capitán de la ysla, que se llamava Juan de Castañeda, diziéndole que lo cognosçía muy bien y que por ser noche no venía a vello, pero que en amaneçiendo vernía y traería más refresco, y traería consigo tres hombres que allá quedavan de la caravela, y que no los enbiava por el gran plazer que con ellos tenía oyendo las cosas de su viaje. El Almirante mandó hazer mucha honrra a los mensajeros y mandóles dar camas en que durmiesen aquella noche porque era tarde y estava la población lexos. Y porque el jueves passado quando se vido en la angustia de la tormenta hizieron el voto y votos susodichos, y el de que en la primera tierra donde oviese casa de Nuestra Señora saliesen en camisa, etc., acordó que la mitad de la gente fue[se] a complillo a una casita que estava junto con la mar como hermita, y él yría después con la otra mitad. Viendo que era tierra segura y confiando en las ofertas del capitán y en la paz que tenía Portogal con Castilla, rogó a los tres hombres que se fuesen a la población y hiziesen venir un clérigo para que les dixese una missa. Los quales ydos en camisa en complimiento de su romería y

---

[1] Margin: Tomó la ysla de Sancta María y así açertó en su navegación y todos los otros erraron.

[2] Margin: Pareçen fingidas estas alegrías que hizieron los Portogueses.

N, anchored where it seemed best and sent the boat ashore. They spoke to the islanders and learned that this was the island of Santa María, one of the Azores, and the islanders pointed out a harbour where they should take the caravel; they said that they had never seen such a storm as there had been during the last fifteen days, and they were amazed that they had escaped. He says that they gave many thanks to God and were very glad when they heard the news that the Admiral had discovered the Indies. The Admiral says that his course had been very accurate and that he had plotted it well, thanks be to God, although he was a little ahead of himself, but he was certain that he was in the neighbourhood of the islands of the Azores and that that island was one of them. And he says that he pretended to have sailed further to mislead the pilots and sailors who were plotting the course so that he would remain master of that route to the Indies, as he in fact remains, because none of the others was certain of the course and none can be sure of his route to the Indies.

Tuesday 19 February

After sunset three men from the island came to the shore and called. He sent the boat for them and they came out bringing chickens and fresh bread, and as it was Shrove Tuesday they brought other provisions sent by the Captain of the island, whose name was Juan de Castañeda,[250] together with a message saying that he knew the Admiral very well and had not come to see him because it was night, but that he would come at daybreak and bring more provisions, and would bring with him the three men from the caravel who had stayed there and whom he was not sending back yet because he enjoyed so much hearing them recount the events of the voyage. The Admiral ordered the messengers to be treated with great honour and given beds to sleep in that night as it was late and the village was a long way off. And because last Thursday when they were in the grip of the storm they made the vows described above, and because they had agreed that on the first land they reached they would process in their shirts, etc., he decided that half the men should go and fulfil the vow at a small house like a hermitage beside the sea, and that he would go later with the other half. Seeing that the land was safe and trusting in the Captain's offers and the peace which existed between Portugal and Castile, he asked the three men to go to the village and send a priest to say mass for them. When they had set off in their shirts in fulfilment of their vows, and while they were at their prayers, the whole village together with

estando en su oraçión, saltó contra ellos todo el pueblo a cavallo y a pie con el capitán, y prendiéronlos a todos.[1] Después estando el Almirante sin sospecha esperando la barca para salir él a complir su romería con la otra gente hasta las onze del día, viendo que no venían sospechó que los detenían, o que la barca se avía quebrado, por toda la ysla esta[r] çercada de peñas muy altas. Esto no podía ver el Almirante porque la hermita estava detrás de una punta. Levantó el ancla y dio la vela hasta en derecho de la hermita, y vido muchos de cavallo que se apearon y entraron en la barca con armas y vinieron a la caravela para prender al Almirante. Levantóse el capitán en la barca y pidió seguro al Almirante. Dixo que se lo dava, pero ¿qué ynnovaçión era aquella que no vía ninguno de su gente en la barca? Y añidió el Almirante que viniese y entrase en la caravela, que él haría todo lo que él quisiese. Y pretendía el Almirante con buenas palabras traello por prendello para recuperar su gente, no creyendo que violava la fe dándole seguro, pues él aviéndole ofreçido paz y seguridad lo avía quebrantado. El capitán como dizque traya mal propósito no se fió a entrar. Visto que no se llegava a la caravela, rogóle que le dixese la causa porqué detenía su gente y que dello pesaría al Rey de Portogal y que en tierra de los Reyes de Castilla reçebían los portogueses mucha honrra y entravan y estavan seguros como en Lisboa, y que los Reyes avían dado cartas de recomendaçión para todos los prínçipes y señores y hombres del mundo, las quales le mostraría si se quisiese llegar, y que él era su Almirante del mar Oçéano y Visorey de las Yndias que agora eran de Sus Altezas, de lo qual mostraría las provisiones firmadas de sus firmas y selladas con sus sellos, las quales le enseñó de lexos; y que los Reyes estavan en mucho amor y amistad con el Rey de Portogal y le avían mandado que hiziese toda la honrra que pudiese a los navíos que topase de Portugal, y que dado que no le quisiese darle su gente, no por eso dexaría de yr a Castilla pues tenía harta gente para navegar hasta Sevilla y serían él y su gente bien castigados haziéndole aquel agravio. Entonçes respondió el capitán y los demás no cognoscer acá Rey e Reyna de Castilla ni sus cartas ni le avían miedo; antes les darían a saber qué era Portugal, quasi amenazando. Lo qual oydo, el Almirante ovo mucho sentimiento y dizque pensó si avía passado algún desconçierto entre un reyno y otro después de su partida, y no se pudo çufrir que no les respondiese lo que era razón. Después tornóse dizque a levantar aquel capitán desde lexos y dixo al Almirante que se fuese con la caravela al puerto, y que todo lo que él hazía y avía hecho el Rey su Señor se lo avía embiado a mandar. De lo qual el

---

[1] Margin: Prendió el Portogués y los suyos a la gente del Almirante.

the Captain attacked them on horseback and on foot and imprisoned them all. Later, the Admiral, unsuspecting, waited until eleven o'clock in the morning for the boat to take him and the other men to fulfil their vow, and when he saw that they did not return, he suspected that they had been detained, or that the boat had been wrecked, as the whole island was surrounded by very high cliffs. The Admiral could not see if this was so because the hermitage was beyond a promontory. He weighed anchor and sailed towards the hermitage, and he saw many horsemen who dismounted and got into the boat carrying arms and came out to the caravel to arrest the Admiral. The Captain stood up in the boat and asked the Admiral for safe conduct. He said that he granted it, but why had there been a change of plan, and why were none of his men in the boat? And the Admiral added that if he came aboard the caravel, he would do everything the Captain wanted. The Admiral was trying to cajole him to come aboard in an attempt to capture him in exchange for his own men, believing that he was not acting in bad faith by promising him safe conduct since the Captain had offered peace and safety and had broken his word. Being up to no good, he says, the Captain did not dare come aboard. When he saw that the Captain would not approach the caravel, the Admiral asked him to explain why he had detained his men, saying that the King of Portugal would be displeased, and that in the lands of the Monarchs of Castile the Portuguese were treated very well and could come and go as safely as in Lisbon, and that the Monarchs had given him letters of introduction for all the princes and lords and men in the world, letters which he would show him if he cared to come aboard; and that he was their Admiral of the Ocean Sea and Viceroy of the Indies which were now the possessions of Their Highnesses, as proof of which he would show him documents signed by the Monarchs and sealed with their seals, and which he showed him at a distance; and that the Monarchs were great friends and allies of the King of Portugal and had ordered him to treat any ships of Portugal he might come across with all due respect; and that even if the Captain would not return his men, he would still go to Castile because he had sufficient men to sail as far as Seville, and that the Captain and his men would be well punished for causing him this offence. Then the Captain replied that he and the rest did not recognise the King and Queen of Castile, nor their letters, nor were they afraid of them; rather, they would show them who Portugal was, almost threatening. When he heard this the Admiral was very angry and says that he wondered if there had been some dispute between the two kingdoms since his departure, and he was unable to refrain from replying in the appropriate manner. Then the Captain stood up again in the distance, he says, and told the Admiral to take the caravel to the harbour, and that everything he was doing and had done was at the express orders of the King,

Almirante tomó testigos los que en la caravela estavan; y tornó el Almirante a llamar al capitán y a todos ellos y les dio su fe y prometió como quien era de no descender ni salir de la caravela hasta que llevase un çiento de portogueses a Castilla y despoblar toda aquella ysla. Y así se bolvió a surgir en el puerto donde estava primero porque el tiempo y viento era muy malo para hazer otra cosa.

Miércoles 20 de hebrero

Mandó adereçar el navío y hinchir las pipas de agua de la mar por lastre porque estava en muy mal puerto y temió que se le cortasen las amarras y así fue; por lo qual dio la vela hazia la ysla de Sant Miguel aunque en ninguna de las de los Açores ay buen puerto para el tiempo que entonçes hazía, y no tenía otro remedio sino huyr a la mar.

Jueves 21 de hebrero

Partió ayer de aquella ysla de Sancta María para la ysla de Sant Miguel para ver si hallara puerto para poder çufrir tan mal tiempo como hazía, con mucho viento y mucha mar y anduvo hasta la noche sin poder ver tierra una ni otra por la gran çerrazón y escurana que el viento y la mar causavan. El Almirante dize que estava con poco plazer porque no tenía sino tres marineros solos que supiesen de la mar porque los que más allí estavan no sabían de la mar nada. Estuvo a la corda toda esta noche con muy mucha tormenta y grande peligro y trabajo.[1] Y en lo que Nuestro Señor le hizo merced fue, que la mar o las ondas della venían de sola una parte, porque si cruzaran como las passadas muy mayor mal padeçiera. Después del sol salido, visto que no vía la ysla de Sant Miguel, acordó tornarse a la Sancta María por ver si podía cobrar su gente y la barca y las amarras y anclas que allá dexava.

Dize que estava maravillado de tan mal tiempo como avía en aquellas yslas y partes, porque en las yndias navegó todo aquel invierno sin surgir e avía siempre buenos tiempos y que una sola ora no vido la mar que no se pudiese bien navegar, y en aquellas yslas avía padeçido tan grave tormenta y lo mismo le acaeçió a la yda hasta las yslas de Canaria; pero passado dellas siempre halló los ayres y la mar con gran templança. Concluyendo dize el Almirante que bien dixeron los sacros theólogos y los sabios philósophos que el Parayso Terrenal

---

[1] Margin: Passó esta noche gran tormenta y peligro.

his lord. The Admiral called upon all those on board the caravel to witness what had happened and he called again to the Captain and all his men and promised them on his honour that he would not go ashore or leave the caravel until he had taken a hundred Portuguese to Castile and had depopulated that entire island. And so he returned to the first anchorage because the weather and the wind were too bad to do anything else.

Wednesday 20 February

He ordered the ship to be got ready and the casks to be filled with seawater as ballast, because he was in a very bad harbour and feared that they would cut his cables, which was what happened; so he set sail towards the island of San Miguel although there is no harbour in any of the Azores suitable for the weather as it was then, and he had no option but to flee into the open sea.

Thursday 21 February

Yesterday he left that island of Santa María for the island of San Miguel to see if he could find a harbour in which to shelter from the terrible weather, with strong winds and heavy seas, and he sailed until dark without seeing any land because of the thick clouds caused by the wind and the sea. The Admiral says he was not at all pleased because he had only three sailors who knew the sea, and the rest of the men with him had no knowledge of it. He beat about all this night in a severe storm and in great danger and difficulty. And Our Lord showed him mercy in that the sea, or the waves, came from one direction only, for if they had crossed each other as before he would have suffered much worse. After sunrise, finding that he could not see the island of San Miguel, he decided to return to Santa María to see if he could recover his men and the boat and the cables and anchors which he had left there.

He says that he was amazed at such bad weather as he experienced in those islands and in that area, because in the Indies he sailed the whole of that winter without dropping anchor and the weather was always good and that even for an hour he never saw a sea on which he could not sail with ease,[251] and among those islands he had suffered such a terrible storm and the same thing had happened on the outward voyage until he reached the Canary islands; but once past them he found the wind and the sea always very calm. In conclusion the Admiral says that the sacred theologians and wise philosophers were right in saying that the Terrestrial Paradise is at the far end of the Orient[252] because it is a very

está en el fin de Oriente porque es lugar temperadíssimo. Así que aquellas tierras que agora él avía descubierto es (dize él) el fin del Oriente.

Viernes 22 de hebrero

Ayer surgió en la ysla de Santa María en el lugar o puerto donde primero avía surgido, y luego vino un hombre a capear desde unas peñas que allí estavan fronteras diziendo que no se fuesen de allí. Luego vino la barca con çinco marineros y dos clérigos y un escrivano. Pidieron seguro, y dado por el Almirante, subieron a la caravela y porque era noche durmieron allí y el Almirante les hizo la honrra que pudo. A la mañana le requirieron que les mostrasse poder de los Reyes de Castilla para que a ellos les constase cómo con poder dellos avía hecho aquel viaje. Sintió el Almirante que aquello hazían por mostrar color que no avían en lo hecho errado sino que tuvieron razón, porque no avían podido aver la persona del Almirante la qual devieran de pretender coger a las manos, pues vinieron con la barca armada, sino que no vieron que el juego les saliera bien. Y con temor de lo que el Almirante les avía dicho y amenazado, lo qual tenía propósito de hazer, y creya que saliera con ello. Finalmente por aver la gente que lo tenían ovo de mostralles la carta general de los Reyes para todos los prínçipes y señores de encomienda y otras provisiones y dioles de lo que tenía y fuéronse a tierra contentos, y luego dexaron toda la gente con la barca, de los quales supo que si tomaran al Almirante nunca lo dexaran libre, porque dixo el capitán que el Rey su señor se lo avía así mandado.

Sábado 23 de hebrero

Ayer començó a querer abonançar el tiempo; levantó las anclas y fue a rodear la ysla para buscar algún buen surgidero para tomar leña y piedra para lastre, y no pudo tomar surgidero hasta oras de completas.

Domingo 24 de hebrero

Surgió ayer en la tarde para tomar leña y piedra y porque la mar era muy alta no pudo la barca llegar en tierra y al rendir de la primera guardia de noche començó a ventar güeste y sudueste. Mandó levantar las velas por el gran peligro que en aquellas yslas ay en esperar el viento sur sobre el ancla y en ventando

temperate place. So those lands which he had now discovered are (he says) the furthest Orient.

Friday 22 February

Yesterday he anchored off the island of Santa María in the place or harbour where he had first anchored, and a man came and signalled from some rocks facing them that they should not leave. Then the boat came with five sailors and two priests and a notary. They asked for safe conduct, and when the Admiral had granted it, they came aboard the caravel, and because it was night slept there, and the Admiral gave them the best reception he could. In the morning they asked him to show them the authorisation from the Monarchs of Castile so that they could confirm that he had undertaken that voyage on their behalf. The Admiral felt that they were doing this to give the impression that they had not acted wrongly but had been correct, because they had not been able to capture the Admiral in person, as they must have intended to do since they came armed in the boats; but it had not turned out as they intended, and they were afraid of what the Admiral had said and had threatened, which he fully intended to do and believed he could carry out. In the end, to regain the men they had captured he had to show them the general letter of authorisation from the Monarchs to all princes and lords, and other provisions, and he gave them what he had and they went ashore content, and then set free all the men with the boat, and from them he learned that if they had captured the Admiral they would never have let him go free, because the Captain said that those were the orders from the King, his lord.

Saturday 23 February

Yesterday the weather began to improve; he weighed anchors and went around the island to search for a good anchorage in order to take on firewood and stone for ballast, and could not find a place to anchor until the hour of compline.[253]

Sunday 24 February

He anchored yesterday afternoon to take on firewood and stones and because the sea was very heavy the boat could not reach the shore and at the end of the first night watch the wind began to blow from the W and SW. He ordered the sails to be hoisted because of the great danger in those islands of waiting at anchor with a S wind, and because if it is blowing from SW it will soon blow

sudueste luego vienta sur. Y visto que era buen tiempo para yr a Castilla,[1] dexó de tomar leña y piedra y hizo que governasen al leste y andaría hasta el sol salido, que avría seys oras y media, 7 millas por ora que son 45 millas y media. Después del sol salido hasta el ponerse, anduvo 6 millas por ora que en onze oras fueron 66 millas y quarenta y çinco y media de la noche fueron 111 y media y por consiguiente 28 leguas.

Lunes 25 de hebrero

Ayer después del sol puesto navegó al leste su camino çinco millas por ora; en treze oras desta noche andaría 65 millas que son 16 leguas y quarta. Después del sol salido hasta ponerse anduvo otras diez y seys leguas y media, con la mar llana gracias a Dios. Vino a la caravela un ave muy grande que pareçía águila.

Martes 26 de hebrero

Ayer después del sol puesto navegó a su camino al leste, la mar llana, a Dios gracias; lo más de la noche andaría 8 millas por ora. Anduvo 100 millas que son 25 leguas. Después del sol salido con poco viento; después tuvo aguaçeros. Anduvo obra de ocho leguas al lesnordeste.

Miércoles 27 de hebrero

Esta noche y día anduvo fuera de camino por los vientos contrarios y grandes olas y mar. Y hallávase çiento y veynte y çinco leguas del Cabo de San Viceynte, y ochenta de la ysla de la Madera, y çiento y seys de la de Santa María. Estava muy penado con tanta tormenta agora que estava a la puerta de casa.

Jueves 28 de hebrero

Anduvo de la mesma manera esta noche con diversos vientos al sur y al sueste y a una parte y a otra y al nordeste y al lesnordeste y desta manera todo este día.

Viernes 1º de março

Anduvo esta noche al leste quarta al nordeste doze leguas; el día corrió al leste quarta del nordeste 23 leguas y media.

---

[1] Margin: Partió de la ysla de Sancta María para Castilla.

from S. And seeing that it was good weather for going to Castile, he stopped loading firewood and stones and ordered them to steer E, and until sunrise, a matter of some six and a half hours, he made about 7 miles an hour, which is 45 miles and a half. Between sunrise and sunset he made 6 miles an hour, which in eleven hours is 66 miles, and then forty-five and a half during the night, which is 111 and a half, and so 28 leagues.

Monday 25 February

Yesterday after sunset he steered his course E at five miles an hour; in thirteen hours tonight he sailed about 65 miles, which is 16 leagues and a quarter. Between sunrise and sunset he sailed another sixteen leagues and a half, with a calm sea, thanks be to God. A very large bird which looked like an eagle came to the caravel.

Tuesday 26 February

Yesterday after sunset he steered on his course to the E, with a calm sea, thanks be to God; for most of the night he made about 8 miles an hour. He sailed 100 miles, which is 25 leagues. After sunrise there was little wind; later there were showers. He sailed a matter of eight leagues ENE.

Wednesday 27 February

This night and day he sailed off course because of contrary winds and heavy waves and sea. He reckoned that he was a hundred and twenty-five leagues off Cape St. Vincent, and eighty from the island of Madeira, and a hundred and six from Santa María. He was very distressed to have such stormy weather when he was so close to home.

Thursday 28 February

Likewise, tonight he sailed S and SE with variable winds, this way and that, and NE and ENE, and similarly throughout the day.

Friday 1 March

This night he sailed twelve leagues E by N; during the day he ran 23 leagues and a half E by N.

Sábado 2 de março

Anduvo esta noche a su camino al leste quarta del nordeste 28 leguas. Y el día corrió 20 leguas.

Domingo 3 de março

Después del sol puesto navegó a su camino al leste; vínole una turbiada que le rompió todas las velas[1] y vídose en gra[n] peligro, mas Dios los quiso librar. Echó suertes para enbiar un peregrino dizque a Santa María de la Çinta en Güelva, que fuese en camisa y cayó la suerte al Almirante. Hizieron todos también voto de ayunar el primer sábado que llegasen a pan y agua. Andaría sesenta millas antes que se le rompiesen las velas; después anduvieron a árbol seco, por la gran tempestad del viento y la mar que de dos partes los comía. Vieron señales de estar çerca de tierra. Hallávanse todo çerca de Lisboa.

Lunes 4 de março

Anoche padecieron terrible tormenta que se pensaron perder de las mares de dos partes que venían y los vientos que parecía que levantavan la caravela en los ayres, y agua del çielo y relámpagos de muchas partes;[2] plugo a Nuestro Señor de lo sostener, y anduvo así hasta la primera guardia, que Nuestro Señor le mostró tierra viéndola los marineros. Y entonces por no llegar a ella hasta cognoscella, por ver si hallava algún puerto o lugar donde se salvar, dio el papahígo por no tener otro remedio y andar algo aunque con gran peligro haziéndose a la mar, y así los guardó Dios hasta el día que dizque fue con infinito trabajo y espanto. Venido el día cognosçió la tierra que era la Roca de Sintra que es junto con el río de Lisboa,[3] adonde determinó entrar porque no podía hazer otra cosa, tan terrible era la tormenta que hazía en la villa de Casca[es], que es a la entrada del río. Los del pueblo dizque estuvieron toda aquella mañana haziendo plegarias por ellos, y después que estuvo dentro, venía la gente a verlos por maravilla de cómo avían escapado. Y así a ora de terçia vino a passar a Rastelo dentro del río de Lisboa, donde supo de la gente de la mar que jamás hizo invierno de tantas tormentas y que se avían perdido 25 naos en Flandes, y otras

---

[1] Margin: Padeció gran tormenta.
[2] Margin: Gran tormenta y espantable.
[3] Margin: Cognosçió la tierra que era la roca junto a Lisboa.

Saturday 2 March

This night he sailed 28 leagues on his course E by N. By day he ran 20 leagues.

Sunday 3 March

After sunset he sailed his course E; a squall blew up which tore all the sails and he was in great danger, but God willed that they should escape. He drew lots to send a pilgrim, he says, in his shirt to Santa María de la Cinta in Huelva, and it fell to the Admiral's lot. They all also made a vow to fast on bread and water on the first Saturday they reached land. He made about sixty miles before his sails were torn; then they sailed with bare masts because of the great storm and the wind and the sea which devoured them from opposite directions. They saw signs that they were near land. They reckoned they were quite near to Lisbon.

Monday 4 March

Last night they suffered a terrible storm and they thought they would perish in the sea which came from opposing quarters and the winds which seemed to lift the caravel into the air, and the rain and the lightning from many directions; it pleased Our Lord to sustain him, and he sailed in this way until the first watch when Our Lord showed him land sighted by the sailors. And then so as not to reach the land before identifying it and seeing if there were a harbour or somewhere to take refuge, his only option was to hoist the mainsail and to make a little progress by keeping out to sea although it was very dangerous, and so God protected them until daylight which was, he says, a time of infinite difficulty and terror. At daybreak he identified the land as the rock of Sintra close to the river at Lisbon where he decided to enter because he could do nothing else, so terrible was the storm over the town of Cascais which is at the entrance to the river. He says that the villagers spent all that morning offering prayers for them, and once he had entered the mouth of the river the people came to see them and marvelled at the way they had escaped. And so at the hour of terce he arrived at Rastelo[254] on the river near Lisbon, where he learned from the seafarers that there had never been a winter with such storms and that 25 ships had been lost

estavan allí que avía quatro meses que no avían podido salir. Luego escrivió el Almirante al Rey de Portogal[1] que estava nueve leguas de allí de cómo los Reyes de Castilla le avía[n] mandado que no dexase de entrar en los puertos de Su Alteza a pedir lo que oviese menester por sus dineros, y que el Rey le mandase dar lugar para yr con la caravela a la çiudad de Lisboa, porque algunos ruynes, pensando que traya mucho oro, estando en puerto despoblado, se pusiesen a cometer alguna ruyndad, y también porque supiese que no venía de Guinea sino de las yndias.

Martes 5 de março

Oy después que el patrón de la nao grande del Rey de Portogal, la qual estava también surta en Rastelo y la más bien artillada de artillería y armas que dizque nunca nao se vido, vino el patrón della que se llamava Bartolomé Díaz de Lisboa con el batel armado a la caravela y dixo al Almirante que entrase en el batel para yr a dar cuenta a los hazedores del Rey e al Capitán de la dicha nao.[2] Respondió el Almirante que él era Almirante de los Reyes de Castilla y que no dava él tales cuentas a tales personas ni saldría de las naos ni navíos donde estuviese, si no fuesse por fuerça de no poder çufrir las armas. Respondió el patrón que enbiase al maestre de la caravela; dixo el Almirante que ni al maestre ni a otra persona si no fuesse por fuerça. Porque en tanto tenía el dar persona que fuese como yr él, y que esta era la costumbre de los Almirantes de los Reyes de Castilla de antes morir que se dar ni dar gente suya. El patró[n] se moderó y dixo que pues estava en aquella determinaçión que fuese como él quisiese, pero que le rogava que le mandase mostrar las cartas de los Reyes de Castilla si las tenía. Al Almirante plugo de mostrárselas, y luego se bolvió a la nao y hizo relación al capitán que se llamava Alvaro Damán. El qual con mucha orden, con atabales y trompetas y añafiles haziendo gran fiesta, vino a la caravela, y habló con el Almirante y le ofreçió de hazer todo lo que él mandase.

Miércoles 6 de março

Sabido cómo el Almirante venía de las yndias oy vino tanta gente a verlo[3] y a ver los yndios de la çiudad de Lisboa que era cosa de admiración, y las

---

[1] Margin: Escrivió al rey de Portogal el Almirante.
[2] Margin: Querían los portogueses que el Almirante fuese a dar cuenta a los oficiales del rey de Portugal.
[3] Margin: Vino gran gente a ver al Almirante.

in Flanders and there were others there which had not been able to leave for four months. Then the Admiral wrote to the King of Portugal who was nine leagues from there, telling him how the Monarchs of Castile had ordered him not to be afraid of entering His Highness's ports to buy whatever he needed. He asked the King's leave to take the caravel to Lisbon, in case some villains, thinking that he was carrying a lot of gold, and seeing him in a deserted harbour, should take it into their heads to commit some act of villainy, and also so that the King might know that he had not come from Guinea but from the Indies.

Tuesday 5 March

Today the master of the great flagship of the King of Portugal, which was also anchored in Rastelo and was the best equipped with artillery and arms he says he had ever seen - the master, who was called Bartolomé Díaz de Lisboa,[255] came in the armed boat to the caravel and told the Admiral to enter the boat to go and give an account of himself to the King's factors and the captain of the said flagship. The Admiral replied that he was an admiral of the Monarchs of Castile and that he did not give accounts of himself to such people, nor would he leave any ship or vessel unless compelled to do so by superior force of arms. The master replied that he should send the master of the caravel; the Admiral said he would send neither the master nor anyone else unless he were forced to do so, because he regarded it as the same thing to send someone else as to go himself, and that it was the custom of the admirals of the Monarchs of Castile to die rather than surrender themselves or any of their men. The master compromised and said that since that was his determination, so be it, but he asked for the letters from the Monarchs of Castile to be shown to him if he had them. The Admiral was pleased to show them to him, and he then returned to the flagship and reported to the captain whose name was Alvaro Damán. He came to the caravel with great ceremony, with drums and trumpets and pipes, making a fine display, and spoke to the Admiral and offered to do whatever he wished.

Wednesday 6 March

When it was known today that the Admiral had returned from the Indies so many people came from the city of Lisbon to see him and to see the Indians that

maravillas que todos hazían, dando gracias a Nuestro Señor y diziendo que por la gran fe que los Reyes de Castilla tenían y deseo de servir a Dios que Su Alta Magestad los dava todo esto.

Jueves 7 de março

Oy vino infinitíssima gente a la caravela y muchos cavalleros y entre ellos los hazedores del Rey, y todos davan infinitíssimas gracias a Nuestro Señor por tanto bien y acreçentamiento de la cristiandad que Nuestro Señor avía dado a los Reyes de Castilla, el qual dizque apropiavan porque Sus Altezas se trabajavan y ejerçitava[n] en el acreçentamiento de la religión de Cristo.

Viernes 8 de março

Oy resçibió el Almirante una carta del Rey de Portugal con don Martín de Noroña por la qual le rogava que se llegase adonde él estava pues el tiempo no era para partir con la caravela. Y así lo hizo por quitar sospecha puesto que no quisiera yr, y fue a dormir a Sacamben. Mandó el Rey a sus hazedores que todo lo que oviese el Almirante menester y su gente y la caravela se lo diese sin dineros y se hiziese todo como el Almirante quisiese.

Sábado 9 de março

Oy partió de Sacanben para yr adonde el Rey estava que era el Valle del Parayso, nueve leguas de Lisboa; porque llovió no pudo llegar hasta la noche. El Rey le mandó resçebir a los principales de su casa muy honrradamente, y el Rey tanbién le resçibió con mucha honrra, y le hizo mucho favor y mandó sentar y habló muy bien ofreciéndole que mandaría hazer todo lo que a los Reyes de Castilla y a su servicio compliese complidamente y más que por cosa suya y mostró aver mucho plazer del viaje aver avido buen término y se aver hecho, mas que entendía que en la capitulaçión que avía entre los Reyes y él que aquella conquista le pertenecía. A lo qual respondió el Almirante que no avía visto la capitulaçión ni sabía otra cosa, sino que los Reyes le avían mandado que no fuese a la Mina ni en toda Guinea y que así se avía mandado apregonar en todos los puertos del Andaluzía antes que para el viaje partiese. El Rey

it was a great sight and marvellous to see the way they gave thanks to Our Lord, saying that it was for the great faith of the Monarchs of Castile and their desire to serve God that the Divine Majesty had given them all this.

Thursday 7 March

Today a huge number of people came to the caravel, including many gentlemen, and among them the King's factors, and they all gave infinite thanks to Our Lord for the great good fortune and advantage to Christendom which Our Lord had given to the Monarchs of Castile, which he says they attributed to the fact that Their Highnesses had worked hard and spared no effort in the furtherance of the religion of Christ.

Friday 8 March

Today the Admiral received a letter from the King of Portugal brought by don Martín de Noroña, in which he asked him to go to where he was because the weather was not suitable for setting out in the caravel. And he did so to avoid suspicion although he did not wish to go, and spent the night at Sacavém. The King ordered his factors to provide free of charge anything the Admiral and his men and the caravel required and that everything should be done as the Admiral wished.

Saturday 9 March

Today he left Sacavém to go to where the King was, which was the Valle do Paraiso, nine leagues from Lisbon;[256] because it rained he could not get there until night-time. The King ordered the principal members of his household to receive him with great honour, and the King also received him with great honour, paid him many respects, asked him to sit down and spoke very kindly, offering to ensure that everything that was in the interests of the Monarchs of Castile and the Admiral was done in due manner, and more so than if it were for himself, and he seemed to be very pleased that the voyage had been undertaken and had ended successfully, but that he understood that according to the treaty between the Monarchs and himself, those conquests belonged to him.[257] To which the Admiral replied that he had not seen the treaty or knew anything about it except that the Monarchs had ordered him not to go to Mina[258] or to anywhere in Guinea and that this had been proclaimed in all the ports of Andalusia before he set out on the voyage. The King graciously replied that he

graçiosamente respondió que tenía él por çierto que no avría en esto menester terçeros. Diole por güésped al prior del Clato que era la más prinçipal persona que allí estava, del qual el Almirante resçibió muy muchas honrras y favores.

Domingo 10 de março

Oy después de missa le tornó a dezir el Rey si avía menester algo que luego se lo daría, y departió mucho con el Almirante sobre su viaje, y siempre le mandava estar sentado y hazer mucha honrra.

Lunes 11 de março

Oy se despidió del Rey e le dixo algunas cosas que dixese de su parte a los Reyes mostrándole siempre mucho amor. Partióse después de comer, y enbió con él a don Martín de Noroña, y todos aquellos cavalleros le vinieron a acompañar y hazer honrra buen rato. Después vino a un monasterio de Sant Antonio que es sobre un lugar que se llama Villafranca donde estava la Reyna[1] y fuele a hazer reverençia y besarle las manos, porque le avía enbiado a dezir que no se fuese hasta que la viese. Con la qual estava el Duque y el Marqués donde resçibió el Almirante mucha honrra. Partióse della el Almirante [de] noche y fue a dormir [a] Allandra.

Martes 12 de março

Oy estando para partir de Allandra para la caravela llegó un escudero del Rey que le ofreçió de su parte que si quisiese yr a Castilla por tierra, que aquél fuese con él para lo aposentar y mandar dar bestias y todo lo que oviese menester. Quando el Almirante de él se partió le mandó dar una mula y otra a su piloto que llevava consigo y dizque al piloto mandó hazer merçed de veynte espadines segúnd supo el Almirante. Todo dizque se dezía que lo hazía porque los Reyes lo supiesen. Llegó a la caravela en la noche.

Miércoles 13 de março

Oy a las ocho oras con la marea de yngente y el viento nornorueste levantó las anclas y dio la vela para yr a Sevilla.[2]

---

[1] Margin: Fue a ver a la Reyna de Portugal.
[2] Margin: Partióse de Lisboa para Sevilla.

was certain that there would be no need for third parties in this matter. He appointed as his host the Prior of Crato[259] who was the principal person there, from whom the Admiral received many honours and favours.

Sunday 10 March

Today after mass the King again told him that if he needed anything it would immediately be given to him, and he discussed the voyage with the Admiral at great length, always inviting him to be seated, and treated him with great respect.

Monday 11 March

Today he took his leave of the King who gave him some messages to be given on his behalf to the Monarchs, always showing him great affection. He left after eating, along with don Martín de Noroña who had been sent with him, and all those gentlemen came to accompany him and do him great honour for a good while. Later he came to a monastery of San Antonio which is near a place called Villafranca where the Queen was in residence and he went to pay his respects and kiss her hands, because she had sent a message to him that he should not leave until he had seen her. The Duke and the Marquis[260] were with her and there the Admiral received great honour. The Admiral left her at nightfall and spent the night in Alhandra.

Tuesday 12 March

Today just as he was about to leave Alhandra for the caravel, a footman arrived with a message from the King saying that if he wished to go to Castile by land, the footman would go with him and arrange lodging and animals and anything he might need. When the Admiral left him he ordered a mule to be given to the Admiral and another to his pilot whom he had with him, and he says that he ordered that a present of twenty espadims[261] was to be given to the pilot, as the Admiral learned later. He says that all this was said to have been designed to come to the attention of the Monarchs. He arrived at the caravel at night.

Wednesday 13 March

Today at eight o'clock with a rough sea and the wind NNW he weighed anchors and set sail for Seville.

Jueves 14 de março

Ayer después del sol puesto sig[u]ió su camino al sur y antes del sol salido se halló sobre el Cabo de San Viçeynte que es en Portugal. Después navegó al leste para yr a Saltés y anduvo todo el día con poco viento hasta agora que está sobre Faro.

Viernes 15 de março

Ayer después del sol puesto navegó a su camino hasta el día con poco viento y al salir del sol se halló sobre Saltés, y a ora de mediodía con la marea de montante entró por la barra de Saltés hasta dentro del puerto de donde avía partido a tres de agosto del año passado. Y así dize él que acabava agora esta escriptura, salvo que estava de propósito de yr a Barçilona por la mar, en la qual çiudad le davan nuevas que Sus Altezas estavan. Y esto para les hazer relaçión de todo su viaje que Nuestro Señor le avía dexado hazer y le quiso alumbrar en él. Porque çiertamente allende que él sabía y tenía firme y fuerte sin escrúpulo que Su Alta Magestad haze todas las cosas buenas, y que todo es bueno salvo el pecado, y que no se puede abalar ni pensar cosa que no sea con su consentimiento, *esto deste viaje cognosco* (dize el Almirante) *que milagrosamente lo a mostrado así como se puede comprehender por esta escriptura por muchos milagros señalados que a mostrado en el viaje y de mí que a tanto tiempo que estoy en la corte de Vuestras Altezas con oppósito y contra sentençia de tantas personas prinçipales de vuestra casa los quales todos eran contra mí poniendo este hecho que era burla. El qual espero en Nuestro Señor que será la mayor honrra de la cristiandad que así ligeramente aya jamás apareçido.* Estas son finales palabras del Almirante don Cristóval Colón, de su primer viaje a las yndias y al descubrimiento dellas.

Deo Graçias.

Thursday 14 March

Yesterday after sunset he followed his course S and before sunrise he found himself off Cape St Vincent which is in Portugal. Then he steered E to go to Saltés and sailed all day with a light wind until now when he is off Faro.

Friday 15 March

Yesterday after sunset he pursued his course until daylight with little wind, and at sunrise he found himself off Saltés, and at midday with a rising tide he crossed the bar at Saltés and into the harbour from which he had set out on the third of August last year. And so he says that he is now finishing this account, except that he plans to go to Barcelona by sea, having been told that Their Highnesses were in that city. And this was in order to give them an account of the whole voyage which Our Lord had permitted him to carry out and for which He had inspired him. Because, certainly, besides the fact that he knew and believed firmly and strongly and without a trace of doubt that the Divine Majesty makes all things good, and that all things are good except sin, and that nothing can be imagined or thought without His consent, *I know from this voyage* (says the Admiral) *that He has miraculously shown it to be so, as can be understood from this account and from the many remarkable miracles manifested during the voyage, and from my own example, who for so long was opposed in Your Highnesses' court by the opinions of so many principal persons of your household, all of whom were against me saying that this undertaking was a jest. I trust in Our Lord that it will be the greatest honour for Christendom to have been brought to light so easily.* These are the final words of the Admiral don Christopher Columbus, in his first voyage of discovery to the Indies.

Thanks be to God.

NOTES TO THE JOURNAL

[1] The capture of Granada by Ferdinand and Isabella on 2 January 1492 marked the end of the Reconquest which had taken Christian Spain nearly 800 years to bring to completion. The Moorish king Boabdil formally surrendered by presenting the keys of the Alhambra to Ferdinand on 6 January. The prince referred to is the Infante don Juan (1478-97), the only son of the Catholic Monarchs. On the relationship between the conquest of Granada and the discovery of America, J.H. Elliott has written that they are 'at once an end and a beginning. While the fall of Granada brought to an end the *Reconquista* of Spanish territory, it also opened a new phase in Castile's long crusade against the Moor ... The discovery of the New World also marked the opening of a new phase - the great epoch of overseas colonisation - but at the same time it was a natural culmination of a dynamic and expansionist period in Castilian history which had begun long before. Both reconquest and discovery ... were in reality a logical outcome of the traditions and aspirations of an earlier age, on which the seal of success was now firmly placed. This success helped to perpetuate at home, and project overseas, the ideals, the values and the institutions of medieval Castile.' (*Imperial Spain 1469-1716*, London: Edward Arnold, 1963, pp. 33-4.) Columbus links the two events in his prologue in a way which shows how conscious he was of his own place in history.

[2] The mention of the 'Gran Can' is probably intended to refer to Kublai Khan (1215-94), grandson of Genghis Khan and first Mongol emperor of China. Kublai Khan was known to medieval Europe from the writings of Marco Polo, who visited China from 1271-95. Columbus owned a Latin edition of Marco Polo's *Travels*, which survives in the Colombina Library in Seville. The Mongol emperors were widely regarded as potential allies against Islam in view of their openness to Western influence. Franciscan missions were sent to China by Pope Nicholas IV as early as the mid 13th century, and an archbishopric was established at the court of the Grand Khan in Ta-tu (Cambaluc). One of the most celebrated Franciscan missionaries to China was Odoric of Pordenone, who travelled there in the 1320s. Contacts between the papacy and China were maintained at least until the time of Benedict XII's mission to the Mongol court in 1342. Columbus's assertion that the papacy had failed to provide the Mongol emperors with instruction in the Christian faith is therefore largely unfounded. The last Mongol emperor, Togon-temür, was overthrown with the establishment of the Ming dynasty in 1368, and Columbus's intention to establish diplomatic contacts with the 'Gran Can' was unfortunately over 120 years too late.

[3] On 30 March 1492 Ferdinand and Isabella signed an edict requiring all professed Jews to convert to Christianity or leave the country by midnight on 2

August that year. The expulsion marked the culmination of a long series of anti-Jewish measures designed to underpin the political unification of Spain by imposing a single religious culture.

[4] The 'Capitulations' or memorandum of agreement between Columbus and the Catholic Monarchs were granted by Ferdinand and Isabella on 17 April in Granada. The terms of the agreement were extremely generous to Columbus and appear to have been drawn up without any serious expectation that he would be successful in his undertaking. Under the agreement, the Crown would take 90% of the net proceeds from any merchandise, but Columbus would retain, as he points out in this timely reminder, jurisdiction in perpetuity over any newly discovered territory (*Capitulaciones del Almirante don Cristóbal Colón*, Madrid: Dirección General de Archivos y Bibliotecas, 1970).

[5] The three ships were the flagship, the Santa María, described throughout the Journal as a 'nao' and owned by Juan de la Cosa, and two caravels: the Pinta, owned by Cristóbal Quintero and captained by Martín Alonso Pinzón; and the Niña, owned by Juan Niño and captained by Vicente Yáñez Pinzón. The combined crew of the three ships amounted to some 90 men, 45 on the Santa María, 25 on the Pinta and 20 on the Niña (A.B. Gould, 'Nueva lista documentada de los tripulantes de Colón en 1492', *Boletín de la Real Academia de la Historia*, 85-88, 90, 92, 110, 111, Madrid, 1922-38).

[6] The safe conduct which Columbus took with him referred to the purpose of the voyage as 'pro aliquibus causis et negociis, servicium Dei ac fidei ortodoxe augmentum, necnon benefficium et utilitatem nostram, concernentibus' (*Capitulaciones*, p. 23). Although this makes it clear that Columbus sailed as an agent of the Crown, it is doubtful whether Their Majesties can be said to have commanded Columbus to undertake his voyage as is claimed here and elsewhere in the prologue. Nevertheless, Columbus was to remind the Monarchs several times throughout his career that it was they who initiated the undertaking in the first place.

[7] Until 22 November 1575 the master of a Spanish ship was not required to keep a daily log book (Julio F. Guillén Tato, *El primer viaje de Cristóbal Colón*, Madrid: Instituto Histórico de Marina, 1943, p. 16, n. 7).

[8] Las Casas's transcription of this prologue in the *Historia de las Indias* (I.35) ends with the word 'etc.' and may indicate that the text as we have it is incomplete.

[9] Columbus is usually thought to have reckoned in Roman miles of around 4850 feet, as against a modern nautical mile of 6080 feet. This view has been convincingly challenged by James E. Kelley, Jr., 'In the wake of Columbus on a portolan chart' in Louis De Vorsey, Jr., and John Parker, *In the Wake of Columbus. Islands and Controversy*, Detroit: Wayne State University Press, 1985, pp. 77-111. Kelley argues (p. 103) that Columbus used a mile which was about

5/6 the length of the Roman mile. This discrepancy would help to explain Columbus's apparent over-estimates of distances sailed at sea.

[10] Quintero's reluctance was due to the fact that the Crown had fined the town of Palos the use of two ships for the voyage, of which his was one.

[11] Las Casas adds 'or to Tenerife' in the margin. Columbus seems to have proceeded to Gomera, to look for a replacement for the Pinta, on Sunday 12 August and not having found one returned, according to Hernando Colón, to Gran Canaria to attend to the Pinta on 23 August. In the *Historia* (I.35), Las Casas implies that the sighting of the volcano at Tenerife, recorded later in the entry for 9 August, was made on the return from Gomera to Gran Canaria, which could account for this reference to Tenerife in the margin of the Journal. Hernando gives the date of the eruption as 23 August.

[12] The Pico de Teide is 12,198 feet. See note 161.

[13] The text has 'Pinta', but Hernando Colón says that it was the Niña that was re-rigged to improve her speed.

[14] Probably Fernan Domingues do Arco, to whom John II of Portugal granted the governorship of an island he intended to discover in 1484.

[15] There were many alleged sightings of legendary islands in the Atlantic due to freak atmospheric conditions. The tradition goes back to the voyage of the 6th-century Irish monk Brendan, and Brendan's island, together with others such as Antilia and Brasil, is frequently represented on medieval maps.

[16] This is unlikely. The crew were more familiar with the ships than he was and could not have been duped about their rate of progress. The interpretation which Las Casas puts on Columbus's decision to keep two records of the distance sailed is almost certainly based on a misunderstanding; see note 23.

[17] Columbus was the first to record the daily variation in the bearing of magnetic north relative to the Pole star. The observation is repeated in the entry for 17 September, where the explanation given may be attributed to Las Casas rather than Columbus himself.

[18] They had reached the outer edge of the Sargasso Sea, a large, relatively still area of the Atlantic 20°-35°N and 30°-70°W. The Sargasso Sea lies at the centre of the ocean current system and appeared on maps as early as Andrea Bianco's chart of 1436, and its position was almost certainly known to Columbus.

[19] If this comment is attributable to Las Casas, he can only have been referring with hindsight to the proximity of Puerto Rico or the Leeward Islands to the S.

[20] Las Casas (*Historia*, I.37) says that the men threatened to throw him overboard.

[21] The reference may be to Exodus 15.24 but the parallel is somewhat far-fetched, as are many of Columbus's biblical references. His identification with Moses, who led the Jews from captivity, has attracted some comment from historians in view of the fact that Columbus left Spain on 3 August, the very day on

which the edict of expulsion took effect.

22 For an informed discussion of the cartography of the first voyage see G.R. Crone, *The Discovery of America*, London: Hamish Hamilton, 1969, especially pp. 207-10. The chart which Columbus and Pinzón consulted on the outward voyage may have been the one which was sent to Columbus by Paolo Toscanelli; in any event it must have been derived from a group of world maps developed by the German cartographer Henricus Martellus and Francesco Roselli, on which Martin Behaim's globe of 1492 was based. All these maps illustrate the basic assumptions about the Atlantic on which Columbus based his own plan, particularly in respect of the width of the Ocean and the configuration of the eastern shore of Asia.

23 Las Casas's contention that Columbus was attempting to dupe his men about the true distance travelled, consistent with his use of the word 'fingía' ('pretended'), seems to have been based on a misunderstanding of the evidence of the Journal. It is clear that Columbus did record two estimates of the distance sailed each day, and Las Casas may even have found the explanation given: that Columbus gave the men the lower figure to keep their spirits up. However, the two reckonings have recently been reinterpreted by the American scholar James E. Kelley, Jr., *In the Wake of Columbus*, pp. 104-7. Kelley argues that as Columbus was an Italian he would have been used to a shorter league than the crew, all but four of whom were Spaniards. Columbus therefore converted his own reckoning to one which the crew would have been used to. Thus by 1 October Columbus's own reckoning was 707 leagues, and that which he announced to the crew was 584; the distance was the same, but the latter number is smaller by a factor of 5/6. It clearly made psychological sense to convert the distance into a measure with which the crew were familiar and not to tell them that they had sailed 707 leagues when they knew that they had sailed 584. Kelley's reinterpretation is supported by the fact that Las Casas inserted the word 'fingía' as an afterthought into the text, writing it above the line after crossing out the word 'dezía' ('said' or 'told').

24 The Guards are the two outermost stars (Kochab and Pherad) of the constellation Ursa Minor and the arms referred to are the arms of an imaginary human figure centred on the Pole star which was used to tell the time at night. A circle drawn around the human figure was divided into eight sections, each of 45°. The stars took three hours to move through each section, or 'line'; crossing three lines therefore marked the passage of nine hours.

25 Columbus (or Las Casas) here confirms the observation that the Pole star rotates around true north.

26 In fact, berry-like bladders which keep the gulfweed afloat.

27 The Spanish text has two spellings for the word 'west', which Las Casas says are the same; presumably the spelling 'vueste' is that which appears in the text

which Las Casas is summarising, while his own preferred spelling is 'güeste'.
²⁸ i.e. Japan.
²⁹ The name Rodrigo de Triana does not appear in the crew lists and the evidence of the *pleitos* suggests that the correct name was Juan Rodríguez Bermejo, a native of the town of Molinos. 'Rodrigo de Triana' may have been a nickname. In the event, Columbus claimed the reward on the grounds that he had seen a light the previous evening.
³⁰ This light has variously been attributed to Columbus's imagination, wishful thinking, or refusal to accept that someone else had spotted land first. If the light did exist, it may have been caused by a fire lit by native fishermen to attract fish to the vicinity at night. It would have to be a large fire, since the ships were over 50 miles from landfall at 10 pm the previous evening.
³¹ The exact location of this island remains uncertain, and no contemporary source identifies Guanahaní, presumably because its identity was common knowledge; there is no hint of doubt in Las Casas's mind about which island is being referred to. Many islands in the Bahamas and the Turks and Caicos group fit the description given by Columbus - low, green and surrounded by a reef, and connected to another island by a narrow isthmus. In all, nine islands have been suggested as the site of the first landfall, but only two, Watlings and Samana Cay, have attracted convincing arguments from respected scholars. The arguments have been summarised by John Parker in De Vorsey and Parker, *In the Wake of Columbus*, pp. 1-34, and by Robert Fuson in *The Log of Christopher Columbus*, Southampton: Ashford Press Publishing, 1987, pp. 199-221. Columbus visited four main islands before reaching Cuba on 28 October. He called these islands San Salvador, Santa María de la Concepción, Fernandina, and Isabela. If the first island is assumed to be Watlings, the others are Rum Cay, Long Island and Crooked Island; if the first island is Samana Cay, the others are Crooked Island, Long Island and Fortune Island. The Watlings theory is supported by S.E. Morison, *Admiral of the Ocean Sea*, Boston: Little, Brown and Co., 1942. The Samana Cay theory was first proposed by Gustavus V. Fox in 1882 and was endorsed by Joseph Judge in the November 1986 issue of the *National Geographic*. All landfall theories involve stretching the evidence of the text to some degree.
³² The original inhabitants of the Bahamas were Tainos, members of the Arawak cultural and linguistic group.
³³ Columbus assumed, along with his contemporaries, a correlation between skin colour and latitude. The idea derived from Aristotle via Pierre D'Ailly's *Imago Mundi*. The Canary Islanders referred to are presumably the original guanche inhabitants, not the Spanish settlers.
³⁴ Columbus is quick to form the notion that these people are preyed upon by another, superior, culture in the vicinity (identified as cannibals on 23 Novem-

ber) and the theory governs much of his decision-making for the rest of the first voyage (see Peter Hulme, 'Columbus and the Cannibals' in *Colonial Encounters*, London and New York: Methuen, 1986, pp. 13-43).

[35] Both Watlings and Samana Cay are 3-4° south of the latitude of Ferro, but Columbus is speaking in general terms.

[36] The native word 'canoa' is not used in the text at this point, but appears only in Las Casas's marginal note. The word used by Columbus to describe the Indian dugout, 'almadía', is of Arabic origin and was normally used to refer to a raft rather than a boat made from a single piece of wood.

[37] Or possibly: 'that they did not know the way'.

[38] Some authorities prefer to translate 'laguna' as 'lagoon'; the Samana Cay theory requires this interpretation.

[39] It should not be forgotten that 'cielo' also means 'sky'.

[40] This reconnoitre of the island is carried out in the ship's boat ('batel'), as Columbus makes clear later when he returns to the flagship.

[41] Columbus evidently has in mind the possibility of improving the defences of this fortified site by cutting a channel across the isthmus connecting it to the rest of the island.

[42] On 12 October he had mentioned six captives.

[43] Either Rum Cay (if the previous was Watlings) or Crooked Island (if Samana Cay).

[44] The Spanish text is corrupt at this point, but the meaning may be deduced from the context.

[45] Since this is much greater than they could have seen, it may be that Las Casas mistranscribed the original which may have read '28 miles'.

[46] Long Island.

[47] Tobacco, although Columbus does not mention smoking until 6 November.

[48] 9.00 am.

[49] Again, the original may have read '20 miles', since 20 leagues was further than they could have seen.

[50] The name of this island appears in four different spellings: Samaot, Samoet, Saomete, and Saometo. See note 54.

[51] Maize, although Columbus does not use the word in the Journal.

[52] Since there is no such plant, Columbus must have been mistaken either in his observation or his interpretation, or both.

[53] This is the first occasion on which Columbus refers to the native inhabitants as 'Indians'.

[54] Columbus names this island Isabela on 19 October, making it either Crooked Island or Fortune Island (see note 31), or both.

[55] Hammocks, although Columbus does not use the native word 'hamaca' until 3 November.

[56] Las Casas's marginal note points out that these were straw crowns, not chimneys; the Indians left other openings in the roof to allow smoke to escape.

[57] Columbus is recording at second hand at this point, but what the men saw could not have been dogs, which were unknown in the New World at this time. Las Casas explains in the ü

[58] The MS reading 'Yslabela', may be a pun ('isla bella') or a slip of the pen.

[59] Most authorities agree that Columbus anchored on Friday night off the southern end of Fortune Island. For this to be so, the text has to be assumed to be corrupt at this point, with 'güeste' being an error for 'sueste', and 12 leagues being an error for 12 miles.

[60] It is not clear how this cape relates to the one he had previously named Cabo Hermoso, although some authorities take it to be the southern point of Crooked Island. There is no explanation for the change of position from 19 to 20 October.

[61] Probably the north-western cape of Crooked Island, the 'isleo' being that referred to at the beginning of the entry for 19 October.

[62] Las Casas's marginal note presumably means that the animal was an iguana, as he makes clear in the *Historia* (I.43). Las Casas says that the iguana was highly prized by the Indians as food, while the Spaniards preferred it to breast of chicken; he could never bring himself to eat it, however, no matter how hungry he was.

[63] Aloe is a shrubby succulent plant of the family *Liliaceae*, native to Africa, whose juice is used as a purgative. It is not a native of the Caribbean. Columbus almost certainly confused this with the agave (family *Agavaceae*), which is.

[64] i.e. Cuba.

[65] First mention of Haiti, the island Columbus will call 'Española'. 'Bofío' or 'bohío' was in fact the Arawak word for 'house', which Columbus evidently mistook to be the name of the island.

[66] The Chinese city of Hang-chou, or Hangchow (Che-kiang province), described by Marco Polo as 'without doubt the finest and most splendid city in the world ... capital of the whole province of Manzi, a great repository of his [the Great Khan's] treasure and the source of such immense revenue that one who hears of it can scarcely credit it'. Ronald Latham, ed. and trans., *The Travels of Marco Polo*, London: The Folio Society, 1968, pp. 179-87.

[67] i.e. Japan, known to Columbus, as he implies in the entry for 24 October, through the wonderful account given by Marco Polo. As with the other islands, Columbus gives Cuba a new name (Juana), but he uses the Indian name for Cuba much more readily than with the others. A possible reason for this is suggested in Las Casas's marginal note to the entry for 30 October (see note 82).

[68] The text reads literally 'with the boat on the poop deck', but Morison interprets this as 'a bonaventure mizzen on the poop, contrived out of the boat's mast and sail' (*Admiral of the Ocean Sea*, pp. 251-2).

69 A line of cays, sometimes called the Ragged Islands, which mark the edge of the Great Bahama Bank, and which Columbus named the Islas de Arena (27 October).
70 This explanation is probably attributable to Las Casas, as in the entry for 13 October (see note 36). The word 'canoa' does not appear without its gloss until the entry for 28 November.
71 Now known as Columbus Bank.
72 Bahía Bariay (province of Holguín).
73 See note 57.
74 This clear indication that Cuba is an island is interesting in the light of Columbus's conclusion (reached as early as 1 November) that Cuba was the Chinese mainland, with Española subsequently being taken to correspond to Japan. No doubt because of this, Cuba was not circumnavigated until 1508. It is perhaps worth noting for the purposes of comparison that Cuba is a very large island, extending some 11° E-W, making it longer than the British Isles, with slightly less than half the surface area.
75 Río Jururú. Columbus's scheme for naming rivers seems to follow either the planets or the days of the week: Luna/lunes = Moon/Monday; Mares/martes = Mars/Tuesday.
76 Puerto Gibara. For the Spanish name see previous note.
77 See note 57.
78 Las Casas comments in the margin that they must have been skulls of manatees. There was no livestock on the islands at this time.
79 The Teta de Bariay. The Peña de los Enamorados, or Lovers' Leap, is in Granada.
80 The Silla de Gibara.
81 Punta Uvero.
82 Las Casas's marginal notes suggests that the Indians were referring to a province called Cubanacan. The final syllable of this word may help to explain why Columbus used the Indian name Cuba rather than his own coining 'Juana': he thought the name made it clear that this was the territory of the Gran Can. In the *Historia* (I.44) Las Casas explains that the suffix 'nacan' meant 'in the middle of'.
83 Given that Columbus's actual position was around latitude 21°N, there have been many attempts to explain this faulty reckoning. Morison argues that Columbus picked the wrong star when estimating his position (*Admiral of the Ocean Sea*, p. 258), and Fuson thinks that he read the wrong scale on the quadrant (*Log*, p. 43); but the error would have been so self-evidently absurd to an experienced navigator that it can only really be the result of scribal error, as Las Casas implies here and in the *Historia*, I.44.
84 Zaiton, now Chang-chou (Fu-kien province), the Chinese port described by

Marco Polo as 'the port for all the ships that arrive from India laden with costly wares and precious stones of great price and big pearls of great quality ... the total amount of traffic in gems and other merchandise entering and leaving this port is a marvel to behold ... it is one of the two ports in the world with the biggest flow of merchandise ... the revenue accruing to the Great Khan from this city and port is something colossal' (*Travels*, pp. 200-201). Note that Columbus has already come to a conclusion about the equivalence of Cuba and China from which he never subsequently deviates.

[85] Las Casas (*Historia*, I.45) points out that cinnamon was never found in the Caribbean, although the wild pepper of the area ('ajf') could be confused with the oriental variety they were seeking.

[86] This word usually occurs in the Journal in the form 'niames', and is used by Columbus to describe several species of root vegetables until 16 December, when he says that the Taino word is 'ajes'. 'Niame' or 'ñame' (English 'yam') is the Spanish form of a term brought back by the Portuguese from the Guinea coast where it refers to species of the genus *Dioscorea*; 'aje' refers to the species *Manihot esculenta*, known in English variously as manioc, yuca or cassava, though Columbus uses it also to refer to the sweet potato (*Ipomoea batatas*), as does Las Casas in his marginal note. See Fuson, *Log*, pp. 233-235.

[87] The mastic with which Columbus was familiar was from the small evergreen tree *Pistacia lentiscus*. What they found on Cuba was probably the gumbo-limbo (*Bursera simaruba*). See Fuson, *Log*, p. 103.

[88] In the *Historia* (I.46) Las Casas expands on the effects of tobacco, saying that it drugs the body and stops the Indians from feeling tired. He says that the Spaniards also caught the habit, and that when told it was a sin, they replied that they were powerless to stop. For his part, he says, he could not see what pleasure or profit they found in it.

[89] The nightingale (*Erithacus*) is not a native of the New World, but the term is often applied loosely to the mockingbird (*Mimus*).

[90] The background and significance of Columbus's decision to sail SE rather than NW, which had been his plan up to this point, is discussed by Peter Hulme, *Colonial Encounters*, pp. 28 ff.

[91] Great Inagua Island.

[92] Puerto Naranjo.

[93] Puerto Samá.

[94] See note 75.

[95] Pliny the Elder discusses the characteristics and properties of gum-producing trees in his *Historia naturalis*, Book XIII, but does not give the information in quite the way Columbus supposes.

[96] Cabo Lucrecia.

[97] At this point, Columbus seems to be assuming that Cabo Lucrecia is the east-

ernmost cape of Cuba, and that the area to the south across Nipe Bay is a separate island, i.e. Bohío or Española.

[98] Bahía de Tánamo.

[99] Columbus does not appear to have returned to Puerto del Príncipe on Saturday 24th. See note 109.

[100] Mandioc or yuca, which the Indians grated and pressed to remove the poisonous element, cyanogenetic glycoside, before making it into bread which they called 'caçabi' (26 December), hence 'cassava'.

[101] Probably an 'hutía', a large rodent of the Capromyidae family, native to the Antilles and Northern Venezuela. Badgers are not found in the Caribbean.

[102] Probably a trunkfish or coffer fish.

[103] The coconut palm was not a native of the Caribbean at this stage, but there was a local nut, *juglans insularis*.

[104] Morison, *Admiral of the Ocean Sea*, p. 265, comments on the accuracy of this observation, that low tide in Tánamo corresponds almost exactly to high tide in Huelva.

[105] This is Las Casas's comment, as Florida was not discovered until 1513.

[106] First mention of cannibals; for a full discussion of the background to this topic, see Hulme, *Colonial Encounters*, ch. 1.

[107] The assumption that a belligerent character correlates with intelligence accounts for Columbus's increasing interest in what lies to the E, on the Island he will call Española.

[108] Cayo Moa Grande. On November 24, Columbus passed it sailing E-W when looking for a harbour that evening, but was unable to enter 'because the wind was strong and the sea very rough'.

[109] This sentence seems confusing, as he is now some way E of Tánamo Bay, the 'Mar de Nuestra Señora'; but if it is taken to refer to the events of 14 November, the sense is clear: '[The last time he passed the Isla Llana] he eventually reached the Mar de Nuestra Señora ...' This reading contradicts the statement made in the entry for 14 November to the effect that Columbus returned to Puerto del Príncipe on Saturday 24th, since it is clear that the harbour he investigates today is Puerto Moa Grande or Santa Catalina, as he calls it.

[110] Punta de Mangle.

[111] *Arbutus unedo*, a small tree with dark, scaly bark and greenish-yellow leaves, producing white flowers and edible berries.

[112] Puerto de Jaragua.

[113] Puerto Moa Grande, presumably named after St. Catherine during the previous day, November 25, which was her feast day, although the act of naming is not recorded.

[114] Punta Guarico.

[115] Almost certainly an error for '6 miles'; 60 miles was more than they could see

and is not consistent with the itinerary which follows.
[116] Punta Plata.
[117] Bahía Cañete and Bahía Yamanigue.
[118] El Yunque.
[119] Puerto Baracoa, later named Puerto Santo (1 December).
[120] Las Casas (*Historia*, I.48) comments that the wax probably came from Yucatán.
[121] Las Casas (*Historia*, I.48) agrees with this assessment and specifically excludes the possibility of cannibalism.
[122] This cape is referred to in the entry for 4 December as 'Cabo del Monte'.
[123] Río Miel.
[124] Punta del Fraile.
[125] This cape, not previously named, is referred to in the entry for 3 December.
[126] Cape Maisí.
[127] After two false alarms, Columbus finally sights the island which he understands the Indians call 'Bohío', and he will call 'Española' (now Haiti and the Dominican Republic), see note 65.
[128] First mention of the naming of Cuba, in honour of the Infante don Juan, only son of the Catholic Monarchs.
[129] Port St. Nicolas; Columbus changes the name to Puerto de San Nicolao (Nicolás?) later in the day, in honour of the Saint whose feast day it was.
[130] Cap à Foux.
[131] Grande Pointe, part of the Haut Piton.
[132] Pointe Jean-Rabel.
[133] Las Casas wrongly gives the impression that the inlet is the Tortuga channel separating Tortuga (Ile de la Tortue) from Española. The Spanish is suspect at this point. The sense is that he called the island Tortuga, and that this sentence is unrelated to the previous one.
[134] This is the harbour he had called Puerto María earlier in the day. In his marginal note Las Casas claims not to understand the change of name. Having seen its size and importance later in the day, Columbus possibly felt that it warranted naming after the saint whose feast day it was.
[135] Some scholars detect in this name a reference to the Golden Chersonese, Ptolemy's name for the Malay Peninsula, but it seems unlikely that cape Cheranero, which must be quite small for it to be only 2 leagues from Puerto de San Nicolás, could bear any close relationship to such an important peninsula, or that Columbus would use such an important name on a relatively insignificant feature.
[136] Port à l'Écu. The distance ought possibly to be six miles, not leagues.
[137] Baie des Moustiques.
[138] Generally contracted to 'Española' or, in English, 'Hispaniola', a convenient

Notes to the Journal

shorthand for Haiti and the Dominican Republic.

[139] Not so, as Great Inagua lies N by W of his present position.

[140] Española is 29,418 square miles in area, to Cuba's 42,827, making Cuba nearly one and a half times larger.

[141] In fact the province around Cap Haïtien.

[142] It is interesting to note, in the light of the supposed cannibal practices of the Caribs, that Columbus consistently denies that the Caniba eat their victims (cf. the entry for 23 November).

[143] Trois Rivières.

[144] See note 86.

[145] i.e. the Indians asked the Spaniards not to return to the coast that night but invited them to stay with them as their guests.

[146] Literally '... they are whiter than the others and that among the others they saw two young women ...', evidently a confusion of perspective, since the Admiral's men appear to praising the fair skin of the Indians they had just returned from visiting, not 'the others' on the other islands.

[147] In fact he was nearly 20°N.

[148] Fuson, *Log*, p. 135, identifies Punta Pierna as Pointe de la Vallée; Punta Lanzada as Pointe des Oiseaux; and Punta Aguda, the eastern end of Tortuga, as Pointe Est.

[149] Les Trois Rivières.

[150] Diego de Arana.

[151] First mention in the text of the Journal itself of these roots which Las Casas had earlier (4 November) identified as 'batatas'. However, the description given is the same as that for 'niames' (see the entry for 13 December and note 86).

[152] The first mention of this word, which is retained in the translation because of the lack of an exact English equivalent. The island was divided into five main provinces each with its own cacique. The name of the local cacique is given as Guacanagarí in the entry for 30 December.

[153] Las Casas's marginal note suggests that this island, 'which never appeared', was Jamaica, even though Columbus's Indian guides always seem to site the island to the north of Española, in the vicinity of Great Inagua.

[154] Pointe des Icaques.

[155] Pointe Baril du Boeuf.

[156] Ile de Marigot.

[157] Pointe du Limbé.

[158] Mont Haut du Cap, above Cap Haïtien.

[159] On 5 December he had said that the nights were 15 hours long. Both figures are an over-estimate; the maximum length, including twilight, is never more than 13 hours.

[160] Baie d'Acul.

[161] Not so. Pico de Teide in Tenerife is 12,198 ft; the highest mountain on Española is Pico Duarte (Dominican Republic, Central Highlands) at 10,417 ft, but the northern range, which is all that Columbus has so far seen, barely reaches 4,000 ft.

[162] Las Casas comments (*Historia*, I.56) that Columbus took the fires to be warning signals, but that it was the Indians' custom to fire the fields, which they called 'sabanas' (from which the English word 'savannah'), in order to control the growth of grass and wildlife.

[163] Columbus also implies that he had visited England in a marginal note to his copy of Pierre d'Ailly's *Imago Mundi*, Louvain, 1483, f. 42r. The visit was probably made in or around 1477.

[164] Error for 5 miles.

[165] See note 86.

[166] Possibly the roots of *Cyperus esculentus*, which are eaten dried as nuts, or nowadays liquidised in the form of 'horchata'; or the fruit of *Arachis hypogea*, commonly known as the peanut and which the Tainos called 'manf'.

[167] See note 161.

[168] The Vega Real.

[169] Las Casas's marginal note identifies the cacique as Guacanagarí, named in the Journal on 30 December, chief of the Marién province where Columbus will establish the settlement of Navidad.

[170] Columbus notes the crossing of a language boundary.

[171] Possibly Rodrigo de Escobedo, described in the entry for 11 October as 'escribano de toda el armada' - 'secretary of the expedition'.

[172] A possible reference to cocoa.

[173] Previously (19 December) spelled 'Caribata'.

[174] Pointe Fort Picolet, on Cap Haïtien.

[175] Las Casas (*Historia*, I.58) explains that 'cacique' meant 'king', and that 'nitaino' meant 'caballero y señor principal', i.e. a nobleman, but subordinate to the king.

[176] The Cibao was the main gold-bearing area of the island, in the central highlands, and Columbus evidently took the initial syllable 'Ci-' to be related to the Cipango of Marco Polo's *Travels*.

[177] Ile des Rats.

[178] The sense appears to be that the Santa María missed the reefs, against which the sea could be heard breaking a league away, and drifted silently onto a sandbank. Columbus's flagship went aground off the beach of what is now Limonade Bord-de-Mer.

[179] Juan de la Cosa, owner of the Santa María.

[180] Called Navidad (4 January), the site of the settlement is a matter of conjecture. Archaeologists from the University of Florida have been excavating in the

vicinity of En Bas Saline, but without achieving definitive results (see Fuson, *Log*, pp. 231-2).

[181] The Pinta had left the other two vessels on 21 November and had not been seen since. Las Casas records (*Historia*, I.61) that Columbus sent a conciliatory letter to Martín Alonso Pinzón, overlooking his anger at having been deserted, and giving him the good news about the establishment of a settlement. Although Columbus was quick to see Divine Providence at work in the grounding of the Santa María, he must also have seen the re-appearance of the Pinta as a life-line, and was not anxious to antagonize Martín Alonso from a position of weakness (see the entries for 31 December and 3 January).

[182] Names of provinces, not islands (Las Casas, *Historia*, I.62). Guarionex was the cacique of the province which included the Cibao; Macorix was another province. Some of the other names had been garbled, according to Las Casas.

[183] Cornelian, a fine translucent red silica.

[184] Mistaken for a medicinal plant (genus *Rheum*) of Chinese origin. Rhubarb was imported into the West during the middle ages and used as a purgative due to the anthracene glycosides which it contains (and which give it its yellow colour, mentioned by Columbus at the end of the entry for 30 December).

[185] In fact, 26 December, unless Las Casas omitted to include the remark in his summary of the entry for 30 December.

[186] Las Casas records in the *Historia* (I.64) that he remembers seeing about a dozen or so Indians in Seville on Columbus's return, though he did not count them exactly.

[187] i.e. Punta Santa (Cap Haïtien), first named on 23 December.

[188] Pointe Yaquezi.

[189] Error for 3 miles.

[190] Siete Hermanos.

[191] Bahía de Manzanillo.

[192] Río Tapión.

[193] Cabra Island.

[194] The coloured stones were an outcrop of coral. Las Casas (*Historia*, I.64) expresses surprise that he did not mention the fine saltings to be found on this island.

[195] Punta Rucía.

[196] Las Casas's marginal note makes it clear that the winds from these directions are very strong, but that Columbus had not experienced them.

[197] Jamaica. Columbus investigated Jamaica in May 1494, during his second voyage.

[198] This legend is associated with the classical female warriors, the Amazons. Columbus would have been interested in this 'information' because it recalled a passage from Marco Polo about two islands, Masculina and Feminea, allegedly

located in the Arabian Sea: 'The inhabitants [of Male Island] are baptized Christians, observing the rule and customs of the Old Testament. For when a man's wife is pregnant he does not touch her again till she has given birth. After this he continues to abstain for another forty days. Then he touches her again as he will. But I assure you that in this island the men do not live with their wives or with any other women; but all the women live on the other island, which is called Female Island. You must know that the men of Male Island go over to Female Island and stay there for three months, that is March, April, and May. For these three months the men stay in the other island with their wives and take their pleasure with them. After this they return to their own island and get on with their business ... Male Island is about thirty miles distant from Female Island. According to their account, their reason for not staying all the year round with their wives is that if they did so they could not live. The sons who are born are nursed by their mothers in Female Island until they are fourteen years old, when they are sent to join their fathers in Male Island ...' (*Travels*, p. 252). Morison (*Admiral of the Ocean Sea*, p. 315) records a similar Arawak myth, in which the women were left on an island called Matinino, where they lived an Amazon-like existence visited annually by the men. On 13 January Columbus equates the female island with Matinino, now Martinique.

[199] Río Yaque del Norte, or Río Santiago.

[200] Las Casas notes in the margin that, since the Río Yaque was still in its virgin state, it is quite possible that Columbus found as much gold as he says, although Las Casas suspects that much of it was fool's gold (iron pyrites) and that the Admiral thought that everything which glistened was gold.

[201] Error for 7 leagues.

[202] Las Casas comments (*Historia*, I.66) that in fact they were no more than four leagues from the gold-bearing area of the Cibao.

[203] Punta Cabo Isabela.

[204] The 'mermaids' were manatees (*Trichechus*), large aquatic mammals native to the Caribbean; what he had seen in Guinea was the dugong or sea cow (*Dugong dugon*), another aquatic mammal native to the area from the Red Sea and Africa to Northern Australia.

[205] Where Martín Alonso Pinzón had been based for sixteen days. Pinzón called the river after himself; Columbus renamed it Río de Gracia, but it was always known subsequently as Río de Martín Alonso. It is now known as Río Chuzona Chico, and the harbour, Puerto Blanco or Puerto de Gracia.

[206] Punta Patilla.

[207] Isabel de Torrcs.

[208] Cabo Macorís. There is some doubt about the identification of place-names for the entries for 11 and 12 January. I follow Fuson, *Log*, pp. 170, 173, and agree with his identification of Puerto Rincón with the Golfo de las Flechas.

[209] Puerto de Plata.
[210] Punta Cabarete.
[211] Cabo de la Roca.
[212] Cabo Tutinfierno.
[213] Cabo Francés Viejo.
[214] Cabo Tres Amarras.
[215] Punta La Botella.
[216] Punta Pescadores.
[217] Cabo Cabrón.
[218] Puerto Escondido.
[219] Named on 16 January Cabo de San Theramo, now known as Cabo Samaná.
[220] Puerto Rincón. Fuson is more convincing in this identification than Morison, who locates the Golfo de las Flechas on the south side of Cape Samaná (*Admiral of the Ocean Sea*, pp. 311-12). Such a detour would have been against Columbus's expressed intentions to make headway to the east.
[221] As Las Casas's note says, the passage has been garbled. The conjunction was between Mars and Mercury, which Morison (*Admiral of the Ocean Sea*, p. 313) says was predicted in Regiomontanus's *Ephemerides*, an astronomical table Columbus carried with him.
[222] In fact a daub made from the seeds of the 'bija' tree (*Bixia orellana*).
[223] Las Casas identifies this man as a 'ciguayo'; Columbus had again correctly noted the crossing of a cultural boundary.
[224] See note 198.
[225] In fact an alloy of gold and copper ('guanin') smelted on the mainland.
[226] Note that Columbus no longer expresses doubts about their alleged cannbalism (see note 121).
[227] The sense appears to be that if his plans had been approved seven years before, in 1485, the Crown would have had seven more years' worth of revenue from his discoveries.
[228] Columbus was probably mistaking the eastern end of Española, across the Bahía de Samaná, as a separate island.
[229] See note 198.
[230] Puerto Rincón. See note 220.
[231] See note 198.
[232] Identified by Las Casas as Cabo del Engaño, the easternmost cape of Española. San Theramo is a variant of San Erasmo or St. Elmo, one of the patron saints of sailors.
[233] The Duke of Medina Sidonia, who held the tunny-fishing concession from the Crown.
[234] In fact, thirty-two miles or eight leagues.
[235] In fact, eight and a quarter leagues.

[236] Cape St. Vincent in Portugal is 37°07'N; Santa Maria in the Azores, where he will fetch up on 18 February, is 37°09'N. His observation, without the aid of astrolabe or quadrant, was therefore extremely accurate.
[237] Vicente Yáñez Pinzón, brother of Martín Alonso, and captain of the Niña.
[238] Bartolomé Roldán, who served as assistant pilot.
[239] Pero Alonso Niño, brother of Juan Niño who owned the Niña. He had been pilot of the Santa María.
[240] Sancho Ruiz de Gama, pilot of the Niña.
[241] Casablanca.
[242] Lacuna in text; if he had been on the latitude of Casablanca, he would have been north of Madeira. Morison reckons that the Admiral's estimate was 175 miles SE three quarters S of his true position (*Admiral of the Ocean Sea*, p. 322). In general terms, they were further north and much further west than any of their estimates.
[243] The Pinta landed at Bayona (Galicia) and Martín Alonso proceeded at Their Majesties' request to meet Columbus at Palos where he died a few days later (March 1493).
[244] Church in Guadalupe (Cáceres, Extremadura) which became a centre of pilgrimage from the twelfth century onwards when a shepherd found an image of the Virgin which had been hidden during the Moorish domination.
[245] Place of pilgrimage in the province of Ancona, on the Adriatic coast of Italy. The Madonna di Loreto is a small image of the Virgin and Child housed in the Santa Casa, said to have been carried from Nazareth in 1291 and to have been the site of an appearance of the Virgin which attested to its sanctity.
[246] In Columbus's day Ancona was a semi-independent republic under papal control; in 1532 it came under direct papal rule and in 1860 became part of Italy.
[247] Moguer (Huelva) was the home town of the Pinzón family and many of the crew.
[248] Diego's mother, Felipa Perestrello e Moniz, had died before 1485; Hernando's mother, Beatriz Enríquez de Harana, did not die until c. 1521. Columbus may have discounted Beatriz because she was his mistress, or he may simply have been feeling particularly maudlin in the difficult circumstances of the return voyage.
[249] Columbus wrote at least two other accounts of the first voyage at about this point on the return journey; the so-called *Carta a Luis de Santángel* is dated 'on board ship, off the Canaries, 15 February, 1493'. This letter was published in the same year in Spanish (Barcelona: Pedro Posa), Latin (9 editions throughout Europe), and Italian (3 editions). A German edition and a second Spanish edition appeared in 1497. Columbus also wrote to Their Majesties 'on the sea off Spain, 4 March, 1493' in a letter recently discovered and published by Antonio Rumeu de Armas, *Libro copiador de Cristóbal Colón*, Madrid: Testimonio, 1989,

Notes to the Journal

vol. II, pp. 435-443.

[250] i.e. Joao da Castanheira, acting governor of the island.

[251] This assertion contradicts the evidence of many entries written in the Caribbean, e.g. 6-12 November.

[252] The Earthly Paradise was frequently held to be located in the Far East and is often represented on medieval maps. Columbus's interest in the location of paradise comes to the fore during the third voyage (1498) when he reports that he has located paradise in the vicinity of the Gulf of Paria.

[253] 9 o'clock in the evening.

[254] Outport of Lisbon, now a suburb of the city.

[255] i.e. Bartolomeu Dias, who discovered the Cape of Good Hope in 1488 and later sailed with the fleet which discovered Brazil in 1497.

[256] The king, Joao II, who had rejected Columbus's project before it was offered to the Spanish, had left Lisbon to avoid an outbreak of plague and was in residence at the Franciscan monastery of Santa Maria das Virtudes, north of Lisbon.

[257] The treaty of Alcáçovas (1479), confirmed by the Papal Bull *Aeterni regis* (1481), had drawn a line of demarcation across the Atlantic giving the Portuguese jurisdiction over any islands which might be discovered south of the Canaries and in the area of Guinea. The king suspected that Columbus's discoveries were made in contravention of the terms of this treaty, and Las Casas speculates (*Historia*, I.74) that he must have been very angry at having lost the opportunity of the newly-discovered lands, and was looking for an excuse to deprive the Spanish crown of them. The dispute gave rise to further agreements between the two states based on lines of longitude (the Bull *Inter caetera*, 1493, and the Treaty of Tordesillas of 1494 which fixed the boundary at 46° 30' W).

[258] Sao Jorge da Mina was a trading post established by the Portuguese in what is now Ghana in 1481.

[259] A priory belonging to the order of St. John of Jerusalem near the Portuguese-Spanish border mid-way between Lisbon and Cáceres.

[260] The Duke was Manuel, duke of Béjar, the Queen's brother and King of Portugal from 1495. The Marquis was probably don Pedro de Noroña, marquis of Villareal and father of don Martín de Noroña.

[261] A gold coin.